I0046999

Charles Dadant

Langstroth on the Hive and Honey Bee

Charles Dadant

Langstroth on the Hive and Honey Bee

ISBN/EAN: 9783743324503

Manufactured in Europe, USA, Canada, Australia, Japa

Cover: Foto ©berggeist007 / pixelio.de

Manufactured and distributed by brebook publishing software
(www.brebook.com)

Charles Dadant

Langstroth on the Hive and Honey Bee

LANGSTROTH

ON THE

HIVE AND HONEY BEE

REVISED, ENLARGED, AND
COMPLETED BY

CHAS. DADANT AND SON.

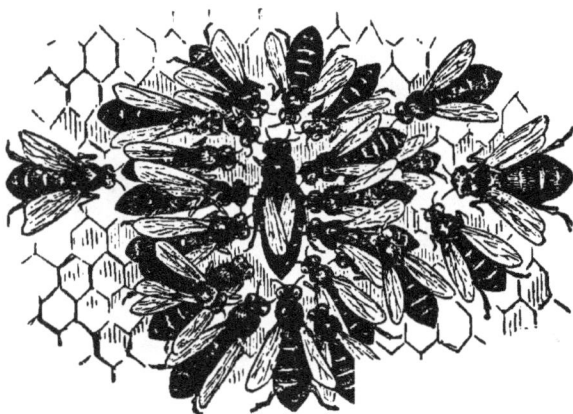

PUBLISHED BY CHAS. DADANT & SON,
HAMILTON, HANCOCK COUNTY, ILLINOIS, U. S. A.

1889.

COPYRIGHTED, 1888, BY

CHAS. DADANT & SON.

ALL RIGHTS RESERVED.

PREFACE.

By his invention of the most practical movable-frame hive, and by his book, "'The Hive and Honey-Bee", — a book as attractive as a novel, — Mr. Langstroth has laid the foundation of American Apiculture, whose methods and implements have become popular throughout the world.

The re-writing of the "Hive and Honey-Bee" was entrusted to us, in 1885, by Mr. Langstroth, as his feeble health rendered him unable to attend to it since its last revision in 1859.

In this difficult work, which demanded a review of the progress accomplished in the past thirty years, we have had to introduce more new matter than we had anticipated. This will probably please the Apiarists who have already read former editions, and who have been waiting for this long-promised revision. Yet, we have retained as much as possible of Mr. Langstroth's writings, and all who are conversant with his style will readily recognize his masterly pen.

Our thanks are due to Mr. C. F. Muth, of Cincinnati, for the enthusiastic interest which he has taken in this book,· and to the able teacher and writer, Miss Favard, of Keokuk, for her criticism of the literary part of the work.

As bee-keeping, like all other sciences, is but an accumulation of former discoveries, ·we have borrowed much from all sides, but we have tried to give due credit to all. Some of the engravings given are not original with the works

I

from which we take them. Those of Girard, for instance, are reduced copies of the beautiful chromos of Clerici, after the microscopic studies of Count G. Barbo, of Milan. Text-books are never entirely free from compilations of this kind.

Having spared neither time nor expense to produce a book worthy of the father of American Apiculture, we hope that our work will be favorably received and will prove of some use in helping progress.

THE REVISERS.

DECEMBER, 1888

BIOGRAPHY OF L. L. LANGSTROTH.

LORENZO LORAIN LANGSTROTH, the "father of American Apiculture," was born in the city of Philadelphia, December 25, 1810. He early showed unusual interest in insect life. His parents were intelligent and in comfortable circumstances, but they were not pleased to see him " waste so much time " in digging holes in the gravel walks, filling them with crumbs of bread and dead flies, to watch the curious habits of the ants. No books of any kind on natural history were put into his hands, but, on the contrary, much was said to discourage his " strange notions." Still he persisted in his observations, and gave to them much of the time that his playmates spent in sport.

In 1827, he entered Yale College, graduating in 1831. His father's means having failed, he supported himself by teaching, while pursuing his theological studies. After serving as mathematical tutor in Yale College for nearly two years, he was ordained Pastor of a Congregational church in Andover, Massachusetts, in May, 1836, and was married in August of that yearto Miss A. M. Tucker of New Haven.

Strange to say, notwithstanding his passion in early life for studying the habits of insects, he took no interest in such pursuits during his college life. In 1837, the sight of a glass vessel filled with beautiful comb honey, on the table of a friend, led him to visit the attic where the bees were kept. This revived all his enthusiasm, and before he went home he purchased two colonies of bees in old box hives.

III

The only literary knowledge which he then had of bee-culture was gleaned from the Latin writings of Virgil, and from a modern writer, " who was somewhat skeptical as to the existence of a *queen-bee.*"

In 1839, Mr. Langstroth removed to Greenfield, Massachusetts. His health was much impaired, and he had resigned his pastorate. Increasing very gradually the number of his colonies, he sought information on all sides. The " Letters of Huber " and the work of Dr. Bevan on the honey bee (London, 1838), fell into his hands and gave him an introduction to the vast literature of bee-keeping.

In 1848, having removed to Philadelphia, Mr. Langstroth, with the help of his wife, began to experiment with hives of different forms, but made no special improvements in them until 1851, when he devised the movable frame hive, used at the present day in preference to all others. This is recorded in his journal, under the date of October 30, 1851, with the following remarks: " The use of these frames will, I am persuaded, give a new impetus to the easy and profitable management of bees."

This invention, which gave him perfect control over all the combs of the hive, enabled him afterwards to make many remarks and incidental discoveries, the most of which he recorded in his book, on the habits and the natural history of the honey-bee. The first edition of the work was published in 1852, and in its preparation he was greatly assisted by his accomplished wife. A revised edition was published in 1857, another in 1859, and large editions, without further revisions, have since been published.

In January, 1852, Mr. Langstroth applied for a patent on his invention. This was granted him; but he was deprived of all the profits of this valuable discovery, by infringements and subsequent law-suits, which impoverished him and gave him trouble for years; though no doubt remains *now* in the mind of any one, as to the originality and priority of his discoveries.

From the very beginning, his hive was adopted by such men as Quinby, Grimm and others, while the inventions of Munn and Debeauvoys are now buried in oblivion.

Removing to Oxford, Ohio, in 1858, Mr. Langstroth, with the help of his son, engaged in the propagation of the Italian bee. From his large apiary he sold in one season $2,000 worth of Italian queens. This amount looks small at the present stage of bee-keeping, but it was enormous at a time when so few people were interested in it.

The death of his only son, and repeated attacks of a serious head trouble, together with physical infirmities caused by a railroad accident, compelled Mr. Langstroth to abandon extensive bee-culture in 1874; but he has always, since then, kept a few colonies on which to experiment.

Mr. Langstroth is now " venerated" by American bee-keepers, who are aware of the great debt due him by the fraternity. He is to them what Dzierzon* is to German Apiarists. When his health permits him to attend one of the meetings of the North American Bee-Keepers' Society, the leaders of Apiculture feel proud and happy to see and to hear him.

Mr. Langstroth is an eminent scholar. His bee library is one of the most extensive in the world. He learned French without a teacher, simply through his knowledge of Latin, for the sole purpose of reading the many valuable works on bees, in the French language. He is a pleasant and eloquent speaker. His writings are praised by all, and we can not close his biography better than by quoting an able writer, who called him the " Huber of America."

---

* Pronounce *Tseertsone.*

TABLE OF CONTENTS

BY PARAGRAPHS.

VII

Chapter II.—Buildings of Bees.

Chapter III.—Food of Bees.

Chapter IV.—The Bee-Hives.

Chapter V.—Handling Bees.

CHAPTER VI.—Natural Swarming.

CHAPTER VII—Artificial Swarming.

CHAPTER VIII.—Queen Rearing.

CHAPTER IX.—Races of Bees.

CHAPTER X.—The Apiary.

CHAPTER XI.—Shipping and Transporting Bees.

CHAPTER XVI—Pasturage and Overstocking.

CHAPTER XVII—Production.

CHAPTER XVIII.—Diseases of Bees.

CHAPTER XIX.—Enemies of Bees.

CHAPTER XX.—Honey Handling, and Marketing. Uses of Honey.

CHAPTER XXI.—Beeswax and its Uses.

CHAPTER XXII.—Bees and Fruits and Flowers.

CHAPTER XXIII.—Bee Keeper's Calendar. Mistakes and Axioms.

THE HIVE AND HONEY-BEE.

CHAPTER I.

1. All the leading facts in the natural history, and the breeding of bees, ought to be as familiar to the Apiarist, as the same class of facts in the rearing of his domestic animals. A few crude and half-digested notions, however satisfactory to the old-fashioned bee-keeper, will no longer meet the wants of those who desire to conduct bee-culture on an extended and profitable system. Hence we have found it advisable to give a short description of the principal organs of this interesting insect, and abridged passages taken from various scientific writers, whose works have thrown an entirely new light on many points in the physiology of the bee. If the reader will bear with us in this arduous task, he will find that we have tried to make the descriptions plain and simple, avoiding, as much as possible, scientific words unintelligible to many of us.

2. Honey-bees are insects belonging to the order *Hymenoptera;* thus named from their four membranous, gauzy wings. They can flourish only when associated in large numbers, as in a colony. Alone, a single bee is almost as helpless as a new-born child, being paralyzed by the chill of a cool summer night.

1

3. The habitation provided for bees is called a hive. The inside of a bee-hive shows a number of combs about half-an-inch apart and suspended from its upper side. These combs are formed of hexagonal cells of various sizes, in which the bees raise their young and deposit their stores.

4. In a family, or colony of bees, are found (Plate II) —

1st, One bee of peculiar shape, commonly called the *Queen*, or *mother-bee*. She is the only *perfect female* in the hive, and all the eggs are laid by her ;

2d, Many thousands of *worker-bees*, or incomplete females, whose office is, while young, to take care of the brood and do the inside work of the hive ; and when older, to go to the fields and gather honey, pollen, water, and propolis or bee-glue, for the needs of the colony ; and

3d, At certain seasons of the year, some hundreds and even thousands of large bees, called *Drones*, or male-bees, whose sole function is to fertilize the young queens, or virgin females.

Before describing the differences that characterize each of these three kinds, we will study the organs which, to a greater or less extent, they possess in common, and which are most prominently found in the main type, the worker-bee.

General Characteristics.

5. In bees, as in all insects, the frame-work or skeleton that supports the body is not internal, as in mammals, but mostly external. It is formed of a horny substance, scientifically called *chitine*, and well described in the following quotation :

6. " Chitine is capable of being moulded into almost every conceivable shape and appearance. It forms the hard back of the repulsive cockroach, the beautiful scale-like feathers of the gaudy butterfly, the delicate membrane which supports the lace-

PLATE 2.

QUEEN, DRONE, AND WORKER—Magnified and Natural Size.

wing in mid air, the transparent cornea covering the eyes of all insects, the almost impalpable films cast by the moulting larvæ, and the black and yellow rings of our native and imported bees, besides internal braces, tendons, membranes, and ducts innumerable. The external skeleton, hard for the most part, and varied in thickness in beautiful adaptation to the strain to which it may be exposed, gives persistency of form to the little wearer; but it needs, wherever movement is necessary, to have delicate extensions joining the edges of its unyielding plates. This we may understand by examining the legs of a lobster or crab, furnished like those of the bee, with a shelly case, but so large that no magnifying glass is required. Here we see that the thick coat is reduced to a thin and easily creased membrane, where, by flexion, one part is made to pass over the other."..........

"Again, almost every part of the body is covered by hairs, the form, structure, direction, and position of which, to the very smallest, have a meaning." (Cheshire, "Bees and Bee-keeping," p. 30. London, 1887.)

7. Mr. Cheshire explains that, as the skeleton or framework of the bee is not sensitive, these hairs act as organs of touch, each one containing a nerve. They also act as clothing and aid in retaining heat—

"and give protection, as the stiff, straight hairs of the eyes, whilst some act as brushes for cleaning, others are thin and webbed for holding pollen grains; whilst by varied modifications, others again act as graspers, sieves, piercers, or mechanical stops to limit excessive movement."

8. The three sections of the body of the honey-bee are perfectly distinct: the head; the thorax, or centre of locomotion, bearing the wings and the legs; and the abdomen, containing the honey-sack, stomach, bowels, and the main breathing organs.

The principal exterior organs of the head are the antennæ, the eyes, and the parts composing the mouth.

9. The eyes are five in number, two composite eyes, one on each side of the head, which are but clusters of small eyes or facets, and three *convex* eyes, or *ocelli*, arranged in a triangle at the top of the head.

10. The facets of the composite eyes, thousands in number, are six-sided, like the cells of the honey-comb, and being directed towards nearly every point, they permit the insect to see in a great number of directions at the same time.

A.   B.   C.

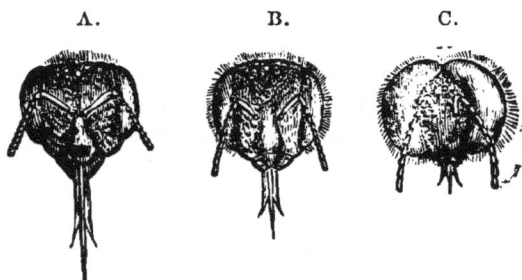

Fig. 1.

A, Head of worker.   B, Head of queen .   C, Head of drone.   (Magnified.)
(From ''Les Abeilles'' of Maurice Girard.)

11. In comparing the eyes of worker, queen and drone, Mr. Cheshire says :

'' The worker spends much of her time in the open air. Accurate and powerful vision are essentials to the proper prosecution of her labours, and here I found the compound eye possessing about 6,300 facets. In the mother of this worker I expected to find a less number, for queens know little of daylight. After wedding they are out of doors but once, or at most twice, in a year.* This example verified my forecast, by showing 4,920 facets on each side of the head. A son of this mother, much a stay-at-home also, was next taken. His facets were irregular in size, those at the lower part of the eye being much less than those near the top; but they reached the immense number of 13,090 on each side of the head. Why should the visual apparatus of the drone be so extraordinarily developed beyond that of the worker, whose need of the eye seems at first to be much more pressing than his ? ''

* When going out with a swarm.

This question Mr. Cheshire answers, as will be seen farther, in considering the antennæ. (26)*

12. The three small eyes, *ocelli*, are thought by Maurice Girard ("Les Abeilles," Paris, 1878), and others, to have a microscopic function, for sight at short distances. In the hive, the work is performed in the dark, and possibly (?) these eyes are fitted for this purpose.

13. Their return from long distances, either to their hive or to the place where they have found food, proves that bees can see very far. Yet, when the entrance to their hive has been changed, even only a few inches, they cannot readily find it.

Their many eyes looking in different directions, enable them to guide themselves by the relative position of objects, hence they always return to the identical spot they left.

14. If we place a colony in a forest where the rays of the sun can scarcely penetrate, the bees, at their exit from the hive, will fly several times around their new abode, then, selecting a small aperture through the dense foliage, they will rise above the forest, in quest of the flowers scattered in the fields. And like children in a nutting party, they will gather their crop here and there, a mile or more away, without fear of being lost or unable to return.

As soon as their honey-sack is full, or, if a threatening cloud passes before the sun, they start for home, without any hesitation, and, among so many trees, even while the wind mingles the leafy twigs, they find their way; so perfect is the organization of their composite eyes.

15. Bees can notice and remember colors. While experimenting on this faculty, we placed some honey on small pieces of differently colored paper. A bee alighted on a yellow paper, sucked her load and returned to her hive.

---

* The reader will readily understand that the numbers between parentheses refer to the paragraphs bearing those numbers. This is for the convenience of the student.

While she was absent, we moved the paper. Returning, she came directly to the spot, but, noticing that the yellow paper was not there, she made several inquiring circles in the air, and then alighted upon it. A similar experiment was made by Lubbock. (A. J. Cook, "Bee-keepers' Guide," Lansing, 1884.)

16. We usually give our bees flour, in shallow boxes, at the opening of Spring, before the pollen appears in the flowers. These boxes are brought in at night. Every morning they are put out again, after the bees have commenced flying and hover around the spot. If by chance, some bits of white paper are scattered about the place, the bees visit those papers, mistaking them for flour, on account of the color.

17. But " the celebrated Darwin was mistaken in saying that the colorless blossoms, which he names obscure blossoms, are scarcely visited by insects, while the most highly colored blossoms are very fondly visited by bees." (Gaston Bonnier, " Les Nectaires," Paris, 1879.)

18. For, although color attracts bees, it is only one of the means used by nature to bring them in contact with the flowers. The smell of honey is, certainly, the main attraction, and this attraction is so powerful, that frequently, at daybreak in the summer, the bees will be found in full flight, gathering the honey which has been secreted in the night, when nothing, on the preceding evening, could have predicted such a crop. This happens especially when there is a production of honey-dew, after a storm. We have even known bees to gather honey from the tulip trees, (*Liriodendron tulipifera*) on very clear moonlight nights.

19. The antennæ (fig. 2, A, B), two flexible horns which adorn the head of the bee, are black, and composed of twelve joints, in the queen and the worker, and thirteen in the drone. The first of these joints, the scape, next to the head, is longer than the others, and can move in every direction. The antenna is covered with hairs.

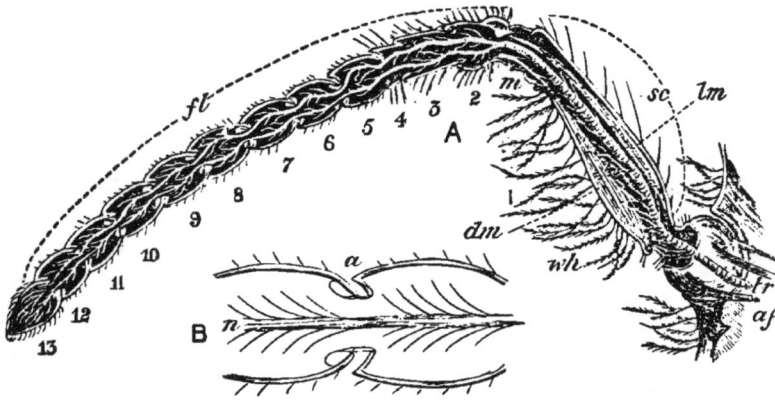

FIG. 2.

LONGITUDINAL SECTION OF DRONE ANTENNA, NERVE STRUCTURES REMOVED.

(Magnified 20 times. From Cheshire.)

A. *sc*, scape; *fl*, flagellum; 1, 2, &c., number of joints: *af*, antennary fossa, or hollow; *tr*, trachea; *m*, soft membrane; *wh*, webbed hairs; *lm*, levator muscle; *dm*, depressor muscle.

B, small portion of flagellum (magnified 60 times); *n*, nerve; *a*, articulation of joint.

"These hairs, standing above the general surface, constitute the antennæ marvelous touch organs; and as they are distributed all round each joint, the worker-bee in a blossom cup, or with its head thrust into a cell in the darkness of the hive, is, by their means, as able accurately to determine as though she saw; while the queen, whose antenna is made after the same model, can perfectly distinguish the condition of every part of the cell into which her head may be thrust. The last joint, which is flattened on one side, near the end, is more thickly studded, and here the hairs are uniformly bent towards the axis of the whole organ. No one could have watched bees without discovering that, by the antennæ, intercommunication is accomplished; but for this purpose front and side hairs alone are required; and the drone, unlike the queen and worker, very suggestively, has no others, since the condition of the cells is no part of his care, if only the larder be well furnished." (Cheshire.)

20. The celebrated François Huber, of Geneva, made a

number of experiments on the antennæ, and ascertained that they are the organs of smell and feeling.

Before citing his discoveries, we must pay our tribute of admiration to this wonderful man. (Plate III.)

Huber, in early manhood, lost the use of his eyes. His opponents imagined that to state this fact would materially discredit his observations. And to make their case still stronger, they asserted that his servant, Francis Burnens, by whose aid he conducted his experiments, was only an ignorant peasant. Now this so-called "ignorant peasant" was a man of strong native intellect, possessing the indefatigable energy and enthusiasm indispensable to a good observer. He was a noble specimen of a self-made man, and rose to be the chief magistrate in the village where he resided. Huber has paid a worthy tribute to his intelligence, fidelity, patience, energy and skill.*

Huber's work on bees is such an admirable specimen of the inductive system of reasoning, that it might well be studied as a model of the only way of investigating nature, so as to arrive at reliable results.

21. Huber was assisted in his researches, not only by Burnens, but by his own wife, to whom he was betrothed before the loss of his sight, and who nobly persisted in marrying him, notwithstanding his misfortune and the strenuous dissuasions of her friends. They lived longer than the ordinary term of human life in the enjoyment of great domestic happiness, and the amiable naturalist, through her assiduous attentions, scarcely felt the loss of his sight.

22. Milton is believed by many to have been a better poet in consequence of his blindness; and it is highly probable that Huber was a better Apiarist from the same cause.

---

* A single fact will show the character of the man. It became necessary, in a certain experiment, to examine separately all the bees in two hives. "Burnens spent *eleven* days in performing this work, and during the whole time he scarcely allowed himself any relaxation, but what the relief of his eyes required."

PLATE 3.

FRANÇOIS HUBER,

Author of the " *Nouvelles Observations sur les Abeilles,*" published in
Geneva, Switzerland, 1792-1814.

This writer is mentioned pages 7. 8, 9, 13, 14, 44, 48, 49, 50, 51, 53, 54, 55, 66,
67, 74, 76, 81, 94, 99, 100, 106, 119, 120, 139, 177, 201, 204,
239, 274, 290, 291, 376, 460.

His active, yet reflective mind, demanded constant employment; and he found, in the study of the habits of the honeybee, full scope for his powers. All the observations and experiments of his faithful assistants being daily reported, many inquiries and suggestions were made by him, which might not have suggested themselves, had he possessed the use of his eyes.

Few, like him, have such command of both time and money, as to be able to prosecute on so grand a scale, for a series of years, the most costly experiments. Having repeatedly verified his most important observations, we take great delight in holding him up to our countrymen as the PRINCE OF APIARISTS.

23. Huber, having imprisoned a queen in a wire cage, saw the bees pass their antennæ through the meshes of the cage, and turn them in every direction. The queen answered these tokens of love by clinging to the cage and crossing her antennæ with theirs. Some bees were trying to draw the queen out, and several extended their tongues to feed her through the meshes.* Huber adds:

" How can we doubt now that the communication between the workers and the queen was maintained by the touch of the antennæ."

24. That bees can hear, either by their antennæ or some other organ, few will now deny, even although the sound of a gun near the hive is entirely unnoticed by them.

" Should some alien being watch humanity during a thunderstorm, he might quite similarly decide that thunder was to us inaudible. Clap might follow clap without securing any external sign of recognition; yet let a little child with tiny voice but shriek for help, and all would at once be awakened to activity. So with the bee : sounds appealing to its instincts meet with immediate response, while others evoke no wasted emotion." (Cheshire.)

* Wonderful as the experiment seemed at that time, the fact is verified now by daily occurrences in queen-rearing.

" The sound that bees produce by the vibrating of their wings
is often the means of calling one another. If you place a bee-hive
in a very dark room, their humming will draw the scattered bees
together. In vain do you cover the hive, or change its place, the
bees will invariably go towards the spot whence the sound comes."
(Collin, " Guide du Propriétaire d'Abeilles," Paris, 1875.)

25. To prove that bees can hear is easy, but to determine
the location of the organ is more difficult. The small holes
which were discovered on the surface of the antennæ, have
been considered as organs of hearing by Lefébure (1838),
and by others later. Cheshire has noticed these small holes
in the six or seven last articulations of the antennæ : holes

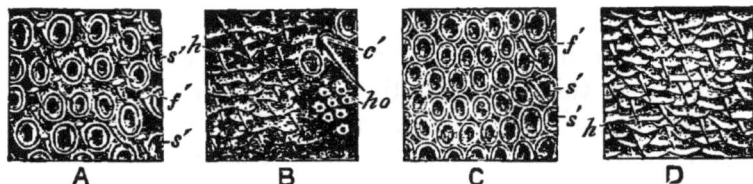

Fig. 3.

PARTS OF SURFACE OF ANTENNÆ.

(Magnified 360 times.   From Cheshire.)

A, portion of front surface of one of the lower members of the flagellum
(worker or queen). *s'*, smelling organ; *f'*, feeling hair.

B, portion of the side and back of same (worker). *h*, ordinary hair; *c'*,
conoid hair; *ho* (auditory ?) hollows.

C, portion of one of the lower members of flagellum (drone).

D, portion of lower member of flagellum (back, worker or queen).

which become more numerous towards the end of the antenna,
so that the last joint carries perhaps twenty. He, also, considers these as the organs of hearing, especially because they
are larger in the drones, who may need to distinguish the
sounds of the queen's wings.*   On this question, Prof. Cook,
in his '' Bee-keepers' Guide," says:

" No Apiarist has failed to notice the effect of various sounds
made by the bees upon their comrades of the hive, and how con-

---

* The queens and the drones, in flight, each have a peculiar and easily dis-
tinguishable sound.

tagious the sharp note of anger, the low hum of fear, and the pleasant tone of a swarm as they commence to enter their new home. Now, whether insects take note of these vibrations as we recognize pitch, or whether they just distinguish the tremor, I think no one knows."

26. It is well proven that bees can smell with their antennæ, and Cheshire carefully describes the "*smell hollows*," not to be mistaken for the "ear holes," which are smaller, but also located on the antenna.

"In the case of the worker, the eight active joints of the antenna have an average of fifteen rows, of twenty smell-hollows each, or 2,400 on each antenna. The queen has a less number, giving about 1,600 on each antenna. If these organs are olfactory, we see the reason. The worker's necessity to smell nectar explains all. We, perhaps, exclaim—Can it be that these little threads

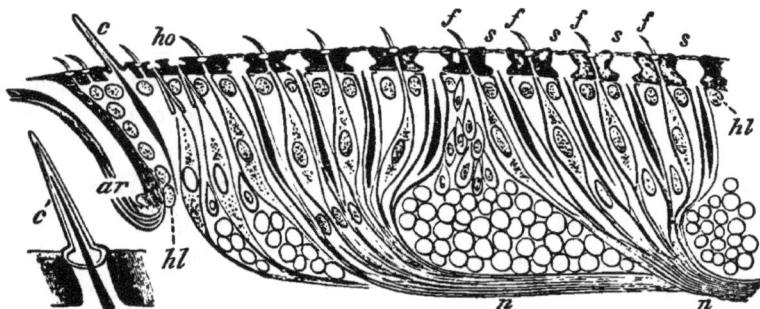

Fig. 4.

LONGITUDINAL SECTION THROUGH PORTION OF FLAGELLUM OF ANTENNA OF WORKER.

(Magnified 300 times. From Cheshire.)

f, feeling hair; *s*, smelling organ; *ho*, hollow; *c*, conoid or cone-shaped hair; *hl*. hypodermal or under-skin layer; *n*,*n*, nerves in bundles; *ar*, articulation; *c'*, conoid hair, magnified 800 times.

we call antennæ can thus carry thousands of organs each requiring its own nerve end? But greater surprises await us, and I must admit that the examinations astonished me greatly. In the drone antenna we have thirteen joints in all, of which nine are barrel-shaped and special, and these are covered completely by smell-hollows. An average of thirty rows of these, seventy in a row, on the nine joints of the two antennæ, give the astounding

number of 37,800 distinct organs. When I couple this development with the greater size of the eye of the drone, and ask what
is his function, why needs he such a magnificent equipment? and
remember that he has not to scent the nectar from afar, nor spy
out the coy blossoms as they peep between the leaves, I feel forced
to the conclusion that the pursuit of the queen renders them necessary." (Cheshire.)

27. While giving these short quotations and beautiful
engravings from Cheshire's anatomy of the bee, we earnestly
advise the scientific bee-student to procure and read his
work. Mr. Cheshire shows us those minute organs so beautifully and extensively magnified, that in reading his book we
feel as though we were transported by some Genius inside
of the body of a giant insect, every detail of whose organization was laid open before us. However wonderful the
statement made above, of the existence of nearly 20,000
organs in such a small thing as the antenna of a bee, this
fact will not be disputed. Those of our bee-friends, who have
had the good luck to meet the sympathetic editor of the
British Bee-Journal, Mr. Cowan, during his trip to America,
in 1887, will long remember the wonderful microscopical
studies, and the microscope which he brought with him.
This instrument, the most powerful by far that we ever had
seen, gave us a practical peep into the domain of the infini-
tesimal.

28. Better than any other description of the smallness of
atoms is that given by Flammarion, in his "Astronomie
Populaire ":

"It is proven," he says, "that an atom cannot be larger than
one ten-millionth of a millimeter. It results from this, that the
number of atoms contained in the head of a pin, of an ordinary
diameter, would not be less than

8,000,000,000,000.000,000,000,000.

And if it was possible to count these atoms, and to separate them
at the rate of one billion per second, it would take 250,000 years
to number them."

29. Girard reports, as follows, an experiment on the olfactory organs of our little insects:

" While a bee was intently occupied sucking honey, we brought near her head a pin dipped in ether. She at once showed symptoms of a great anxiety; but an inodorous pin remained entirely unnoticed."

30. Whatever be the location of their olfactory organs, they are unquestionably endowed with a marvelous power of detecting the odor of honey in flowers or elsewhere.

One day we discovered that some bees had entered our honey-room, through the key-hole. We turned them out, and stopped it up. Some time after, more bees had entered, and we vainly searched for the crevice that admitted them. Finally a feeble hum caused us to notice that they were coming down the chimney to the fire-place, which was closed by a screen. The wedge which held this screen having become somewhat loose, the motion of the screen in windy weather opened a hole just large enough for a bee to crawl through. A few bees were waiting behind the screen, and as soon as its motion allowed one to pass, she manifested her joy by the humming which led to the discovery. These bees, escaping with a load, when the door was opened, had become customary and interested visitors.

31. Every bee-keeper has noticed that their flight is guided by the scent of flowers, though they be a mile or more away. In the city of Keokuk, situated on a hill in a curve of the Mississippi, the bees cross the river, a mile wide, to find the flowers on the opposite bank.

32. " Not only do bees have a very acute sense of smell, but they add to this faculty the remembrance of sensations. Here is an example: We had placed some honey on a window. Bees soon crowded upon it. Then the honey was taken away, and the outside shutters were closed and remained so the whole winter. When, in Spring, the shutters were opened again, the bees came back, although there was no honey on the window. No doubt, they remembered that they got honey there before. So, an inter-

val of several months was not sufficient to efface the impression
they had received.— (Huber, "Nouvelles Observations sur les
Abeilles," Genève, 1814.)

33. It is well known, also, that bees wintered in cellars
(**646**) remember their previous location when taken out in
the Spring.

If food is given to a colony, at the same hour, and in the
same spot, for two days in succession, they will expect it
the third day, at the same time and place.

34. When one of her antennæ is cut off, no change takes
place in the behavior of the queen. If you cut both antennæ
near the head, this mother, formerly held in such high consideration by her people, loses all her influence, and even the maternal
instinct disappears. Instead of laying her eggs in the cells, she
drops them here and there.—(Huber.)

The experiments made by Huber on workers and drones,
in regard to the loss of the antennæ, are equally conclusive.
The workers, deprived of their antennæ, returned to the
hive, where they remained inactive and soon deserted it forever, light being the only thing which seemed to have any
attraction for them.

In the same way, drones, deprived of their antennæ, deserted the observatory hive, as soon as the light was excluded
from it, although it was late in the afternoon, and no drones
were flying out. Their exit was attributed to the loss of
this organ, which helps to direct them in darkness.

35. The inference is obvious, that a bee deprived of her
antennæ loses her intellect.

" If you deprive a bird, a pigeon, for instance, of its cerebral
lobe, it will be deprived of its instinct, yet it will live if you stuff
it with food. Furthermore, its brain will eventually be renewed,
thus bringing back all the uses of its senses."—(Claude Bernard,
" Science Expérimentale.")

Bees, however, cannot live without their antennæ, and
these organs would not grow again, like the brains of birds,
the legs of crawfishes, or the tails of lizards.

36. Let us notice, in reference to the sensorial organs, that the brain of workers is very much larger than that of either the queen or the drone, who need but a very common instinct to perform their functions; while the various occupations of the workers, who act as nurses, purveyors, sweepers, watchful wardens, and directors of the economy of the bee-hive, necessitate an enlargement of faculties very extraordinary in so small an insect.

37. We cannot better close this chapter than by quoting the celebrated Hollander, Swammerdam, as Cheshire does:

"I cannot refrain from confessing, to the glory of the immense, incomprehensible Architect, that I have but imperfectly described and represented this small organ; for to represent it to the life in its full perfection, far exceeds the utmost efforts of human knowledge."

38. We have now come to the most difficult organ to describe — the mouth of the bee. But we will first visit the interior of the head and of the thorax, to find the nursing and salivary* glands, and explain their uses.

39. The workers have three pairs of glands: two pairs, different in form, placed in the head (a, a, fig. 5), and one larger pair, located in the thorax or corselet. The upper

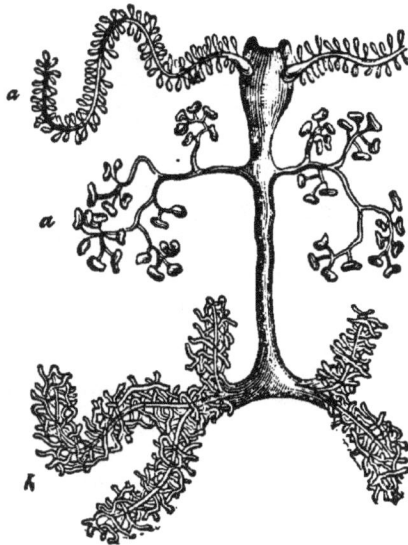

Fig. 5.

SALIVARY GLANDS OF THE WORKER-BEE.

(Magnified. From Maurice Girard.)

a, a, glands of the head; b, glands of the thorax.

pair, which resembles a string of onions, is absent in the

_____

* In plainer words, spittle-producing tubes.

drones and queens. According to Girard, these upper glands were discovered by Meckel in 1846. They are very large and dilated in the young worker bees, while they act as nurses, but are slim in the bees of a broodless colony. In the old bees, that no longer nurse the brood, they wither more and more, till they become shrunken and seemingly dried. Hence Maurice Girard, and others before him, have concluded very rationally that these upper glands produce the milky food given to the larvæ, during the first days of their development. Mr. Cheshire has advanced the very reasonable theory that the queen, during the time of egg-laying, is fed by the workers from the secretions of this gland.

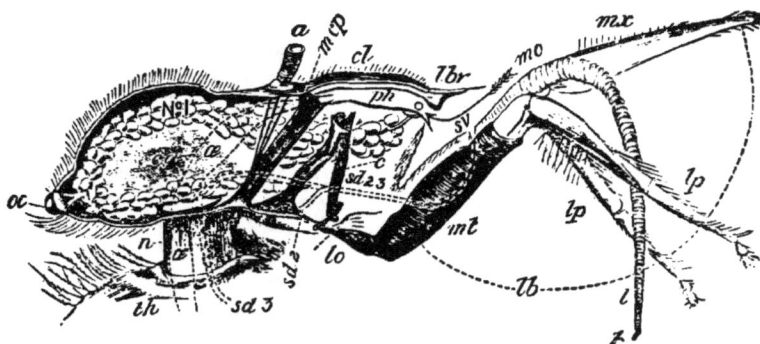

Fig. 6.

LONGITUDINAL SECTION THROUGH HEAD OF WORKER.

(Magnified 14 times. From Cheshire.)

a, antenna, with three muscles attached to *mcp*, meso-cephalic pillar; *cl*, clypeus; *lbr*, labrum or upper lip; No. 1, upper salivary or chyle gland (this gland really runs in front of the meso-cephalic pillars, but here the latter are kept in view); *o*, opening of same in the mouth; *oc*, ocellus or simple eye; *cg*, cephalic ganglion, or brain system; *n*, neck; *th*, thorax; *œ*, œsophagus or gullet: *sd* 2, 3, salivary ducts of glands two and three; *sv*, salivary valve; *ph*, pharynx; *lb*, labium or lower lip, with its parts separated for display: *mt*, mentum or chin; *mo*, mouth; *mx*, maxilla; *lp*, labial palpi; *l*, ligula or tongue; *b*, bouton.

40. "The queen at certain periods has the power of producing between 2,000 and 3,000 eggs daily (**98**). A careful calculation shows that 90,000 of these would occupy a cubic inch and

weigh 270 grains. So that a good queen, for days or even weeks*
in succession, would deposit, every twenty-four hours, between
six and nine grains of highly-developed and extremely rich tissue-
forming matter. Taking the lowest estimate, she then yields the
incredible quantity of twice her own weight daily, or more accurately four times, since at this period, more than half her weight
consists of eggs. Is not the reader ready to exclaim : What
enormous powers of digestion she must possess! and since pollen is the only tissue-forming food of bees, what pellets of this
must she constantly keep swallowing, and how large must be the
amount of her dejections! But what are the facts ? Dissection
reveals that her chyle stomach is smaller than that of the worker,
and that at the time of her highest efforts, often scarcely a pollen
grain is discoverable within it, its contents consisting of a transparent mass, microscopically indistinguishable from the so-called
"royal jelly"; while the most practiced bee-men say they
never saw the queen pass any dejections at all. These contradic-
tions are utterly inexplicable, except upon the theory I propound
and advocate. She does pass dejections, for I have witnessed
the fact; but these are very watery.".........—(Cheshire.)

Thus according to Cheshire, the food eaten by the queen,
during egg-laying, is already digested and assimilated by
the bees, for her use. Her dejections which are scanty
and liquid, are licked up by the workers, as are also the dejections of the drones, if not too abundant.

41. The other two pairs of glands, which are common to
workers, queens, and drones, evidently produce the saliva.
The functions of both must be the same, for they unite in
the same canal (*sd*, *2*, *3*, fig. 6), terminated by a valvule,
which, passing though the mentum or chin (*mt*), opens at
the base of the tongue. The saliva produced by them is
used for different purposes. It helps the digestion; it
changes the chemical condition of the nectar (**246**) harvested from the flowers ; it helps to knead the scales of wax
(**201**) of which the combs are built, and perhaps the propolis (**236**) with which the hives are varnished. It is used

---

* These facts have been demonstrated so repeatedly, that they are as well
established as the most common laws in the breeding of our domestic animals.

also to dilute the honey when too thick, to moisten the
(**263**) pollen grains, to wash the hairs when daubed with
honey, etc.

These glands yield their saliva while the tongue of the
bees is stretched out; but the upper glands (No. 1, fig. 6),
which open on both sides of the pharynx or mouth (*ph*), can
yield their chyle only when the tongue is bent backwards,
to help feed the larva (**64**) lying at the bottom of the cell.

42. The mouth of the bee has mandibles or outer jaws,
which move sidewise, like those of ants and other insects,
instead of up and down as in higher animals. These jaws
are short, thick, without teeth, and beveled inside so as to
form a hollow when joined together, as two spoons would do.
With them, they manipulate the wax to build their comb,
open the anthers of flowers to get the honey, and seize and
hold, to drag them out, robbers or intruders, or débris of
any kind.

Fig. 7.
Head of honey-
hornet.
(Magnified.)

Fig 8.
Head of honey-
bee.
(Magnified.)

Fig. 9.
Mandible of honey-
hornet.
(Magnified.)

Fig. 10.
Mandible of honey-
bee.
(Magnified.)

43. Fig. 9 shows the jaws of the Mexican hornet highly
magnified. Fig. 10 shows the jaws of the honey-bee, highly
magnified. Notice the difference in the shape of the two,
the saw-like appearance of the one, and the spatula shape
of the other. A glance at these figures is enough to convince any intelligent horticulturist of the truth of Aristotle's
remark — made more than two thousand years ago — that
" bees hurt no kinds of sound fruit, but wasps and hornets
are very destructive to them."

We shall give further evidence concerning the correctness of this statement. (**871**)

44. Below the antennæ, the clypeus or shield (*cl*, fig. 6) projects, which is prolongated by an elastic rim called labrum or upper lip (*lbr*). The pharynx is the mouth (*ph*), and the œsophagus (*œ*) the gullet, through which the food goes into the stomach.

As we have already seen, the canals of the upper glands open on each side of the mouth, and discharge their chyle into it at will.

45. The chin or mentum (*mt*) is not literally a part of the mouth. It can move forward and backward, and supports several pieces, among which is the tongue, or proboscis, or ligula (*l*). The tongue is not an extension of the chin, but has its root in it, and can only be partly drawn back into it, its extremity, when at rest, being folded back under the chin.

46. There are, on each side of the tongue, the labial palpi or feelers* (*b*, fig. 11, and *lp*, fig. 6), which are fastened to the chin by hinged joints. They are composed of four pieces each, the first two of which are broad, and the other two small and thin, and provided with sensitive hairs of a very fine fabric. Outside of the palpi are the maxillæ (*c*, fig. 11, and *mx*, fig. 6) which in some insects have the function of jaws, but which, in the bee, only serve, with the palpi, to enfold the tongue in a sort of tube, formed and opened at the will of the insect, and which, by a certain muscular motion, as also by the ability of the tongue to move up and down in this tube, force the food up into the mouth.

47. The tongue is covered with hairs, which are of graded sizes, so that those nearest the tip or bouton are thin and flexible. It — the tongue — is grooved like a trough, the edges of which can also unite to form a tube, with perfect

* Organs of taste according to Leydig and Jobert.

joints.   It is easily understood that if this tongue was a tube, the pollen grains when conveyed through it would obstruct it, especially when daubed with very thick honey.

48. "A most beautiful adaptation here becomes evident. Nectar gathered from blossoms needs conversion into honey. Its cane sugar must be changed into grape sugar, and this is accomplished by the admixture of the salivary secretions of Systems Nos. 2 and 3 (*sd*, *2*, *3*, fig. 6), either one or both.   The tongue is drawn into the mentum by the shortening of the retractor linguæ muscle, which, as it contracts, diminishes the space above the salivary valve, and so pumps out the saliva, which mixes with the nectar as it rises, by methods we now understand.   Bees, it has often been observed, feed on thick syrup slowly ; the reason is simple. The thick syrup will not pass readily through minute passages without thinning by a fluid. This fluid is saliva, which is demanded in larger quantities than the poor bees can supply. They are able, however, to yield it in surprising volume, which also explains how it is that these little

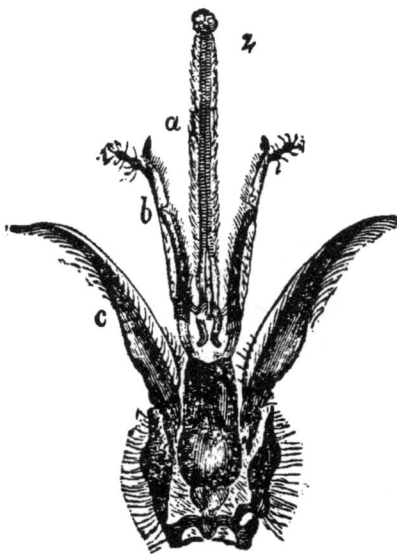

Fig. 1'.

TONGUE AND APPENDAGES.

(Magnified.   From Maurice Girard.)

a, tongue; *b*, labial palpi; *c*, maxilla.

marvels can so well clean themselves from the sticky body honey. The saliva is to them both soap and water, and the tongue and surrounding parts, after any amount of daubing, will soon shine with the lustre of a mirror."—(Cheshire.)

49. The length of the tongue of the honey-bee is of great importance to bee-keepers.   Some flowers, such as red clover, have a corolla so deep, that few bees are able to gather the honey produced in them.   Therefore, one of the chief aims of progressive bee-keepers, should be to raise bees with longer tongues. This can undoubtedly be done sooner or later, by careful selection, in the same way that all our domestic plants and animals have been improved in the past. For this, *patience* and *time* are required.

50. The thorax is the intermediate part of the body. It is also called " corselet." It is formed of three rings soldered into one. Each of the three rings bears one pair of legs, on its under side; and each of the last two rings bears a pair of wings, on its upper side; making four wings and six legs, all fastened on the thorax.

51. Each leg is composed of nine joints (B, Plate IV), the two nearest the body (*c*, *tr*) being short. The next three are the femur (*f*), tibia (*ti*), and planta (*p*) also called metatarsus. The last four joints form the tarsus (*t*) or foot.

52. The last joint of the tarsus, or tip of the foot, is provided with two claws (*an*, fig. 12), that cling to objects or to the surfaces on which the bee climbs. These claws can be folded, somewhat like those of a cat (A, fig. 12), or can be turned upwards (B, fig. 12) when the bees are hanging in clusters. When they walk on a polished surface, like the pane of a window, which the claws cannot grasp, the latter are folded down; but there is between them a small rubber-like pocket, pulvillus (*pv*, A, B,) which secretes a sticky, " clammy " substance, that enables the bee to cling to the smoothest surfaces. House-flies and other insects cling to walls and windows by the same process. It was formerly asserted that insects cling to the smooth surfaces by air suction, but the above explanation is correct, and you can actually see " the footprints of a fly " on a pane of glass, with the help of a microscope, remnants of the " clammy " substance being quite discernible. By this ingenious arrangement, bees can walk indifferently upon almost anything, since wherever the claws fail, the pulvilli take their place.

53. "But another contrivance, equally beautiful, remains to be noticed. The pulvillus is carried folded in the middle (as at C, fig. 12), but opens out when applied to a surface, for it has at its upper part an elastic and curved rod (*cr*) which straightens as the pulvillus is pressed down, C and D, fig. 12, making this clear. The flattened-out pulvillus thus holds strongly while pulled, by the weight of the bee, along the surface, to which it adheres, but comes up at once if lifted and rolled off from its opposite sides, just as we should peel a wet postage stamp from its envelope. The bee, then, is held securely till it attempts to lift the leg, when it is freed at once; and, by this exquisite yet simple plan, it can fix and release each foot at least twenty times per second."—(Cheshire.)

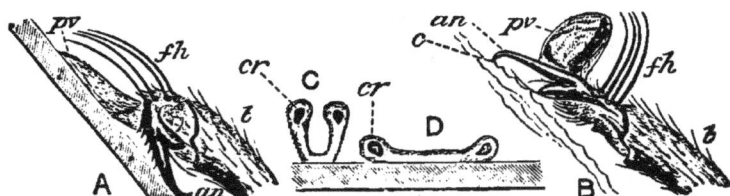

Fig. 12.

BEE'S FOOT IN CLIMBING, SHOWING ACTION OF PULVILLUS.
(Magnified 30 times. From Cheshire.)

A, position of the foot in climbing slippery surface or glass; *pv*, pulvillus; *fh*, feeling hairs; *an*, anguiculus, or claw; *t*, tarsal joint.
B, position of the foot in climbing rough surface.
C, section of pulvillus just touching flat surface; *cr*, curved rod.
D, pulvillus applied to surface.

54. The legs of bees, like all other parts of their body, are covered with hairs of varied shapes and sizes, the description of which is beyond the limits of this work. We will confine ourselves to a short explanation of the uses, which have a direct bearing upon the work of the bee.

The hairs of the front, or first, pair of legs (C, Plate IV) are especially useful in cleaning the eyes and the tongue, and gathering the pollen grains.

55. On the metatarsus, the lower of the two largest joints of these front legs, is a rounded notch (E, *a*, Plate IV), closed when the leg is folded, by a sort of spur or velum, (*v*, C, E, H) fastened to the tibia, or upper large joint. The learned Dr. Dubini, of Milan (L'Ape, Milan, 1881), speaks of it as being used to cleanse the antennæ and the tongue of the pollen that sticks to them. Mr. Cheshire thinks it is used only to cleanse the antennæ, from the fact that this notch, which has teeth like a comb (F, Plate IV), is found as well in the queen and the drone as in the worker, and that its aperture corresponds exactly to the different sizes of the antenna of each sex. (H, Plate IV.)

56. The second pair of legs have no notch, but the lower

A B   C   'D

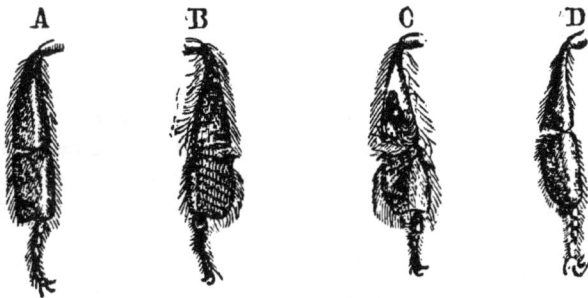

Fig. 13.

POSTERIOR LEGS.

(Magnified. From Maurice Girard.)

A, of the queen; B, of the worker (under side); C. of the worker (upper side); D, of the drone.

extremity of the tibia bears a spur (D *s*, Plate IV) or spine, which is used in loosening the pellets of pollen, brought to the hive on the tibias of the posterior legs (Plate IV). This spur also helps in cleaning the wings.

57. The posterior or hind legs are very remarkable, in several respects. Between the tibia and the metatarsus (B, *wp*, Plate IV) they have an articulation, whose parts close like pincers, and which serve to loosen from the abdomen the scales of wax to be mentioned farther on (**201**). As neither the queen nor the drone produces wax, they are destitute of this implement.

58. " But the chief interest centers on the two joints last mentioned (*ti*, *p*, A. B.. Plate IV), as a device for carrying the pollen of the blossom home to the hive. The metatarsus is enlarged into a sub-quadrangular form, constituting a flattish plate, slightly convex on both surfaces. The outer face (*p*. A, Plate IV) is not remarkable, but the one next the body (*p*, B) is furnished with stiff combs, the teeth of which are horny, straight spines, set closely, and arranged in transverse rows across the joint, a little projecting above its plane, and the tips of one comb slightly overlapping the basis of the next. Their colour is reddish-brown ; and entangled in the combs, we almost invariably discover pollen granules, which have been at first picked up by the thoracic hairs, but combed out by the constant play of the legs over the breast— in which work, the second pair, bearing a strong resemblance to the third, performs an important part."

59. " So soon as the bees have loaded these combs, they do not return to the hive, but transfer the pollen to the hollow sides of the tibia, seen at *ti*, A. This concavity, corbicula, or pollen basket. is smooth and hairless, except at the edges, whence spring long, slender, curved spines, two sets following the line of the bottom and sides of the basket, while a third bends over its front. The concavity fits it to contain pollen, while the marginal hairs greatly increase its possible load, like the sloping stakes which the farmer places round the sides of his waggon when he desires to carry loose hay. the set bent over (see G, Plate IV) accomplishing the purpose of the cords by which he saves his property from being lost on the road. But a difficulty arises : How can the pollen be transferred from the metatarsal comb to the basket above ? Easily ; for it is the left metatarsus that charges the right basket, and *vice versa*. The legs are crossed, and the metatarsus naturally scrapes its comb-face on the upper edge of the opposite tibia, in the direction from the base of the combs towards their tips. These upper hairs standing over *wp*, B, or close to *ti*, A (which are opposite sides of the same joint), are nearly straight, and pass between the comb teeth. The pollen, as removed, is caught by the bent-over hairs, and secured. Each scrap adds to the mass until the face of the joint is more than covered, and the hairs just embrace the pellet as we see it in the cross-section at G. The worker now hies homewards, and the spine, as a crow-bar, does its work."—(Cheshire.)

60. The four wings, in two pairs, are supported by hol-

PLATE 4.

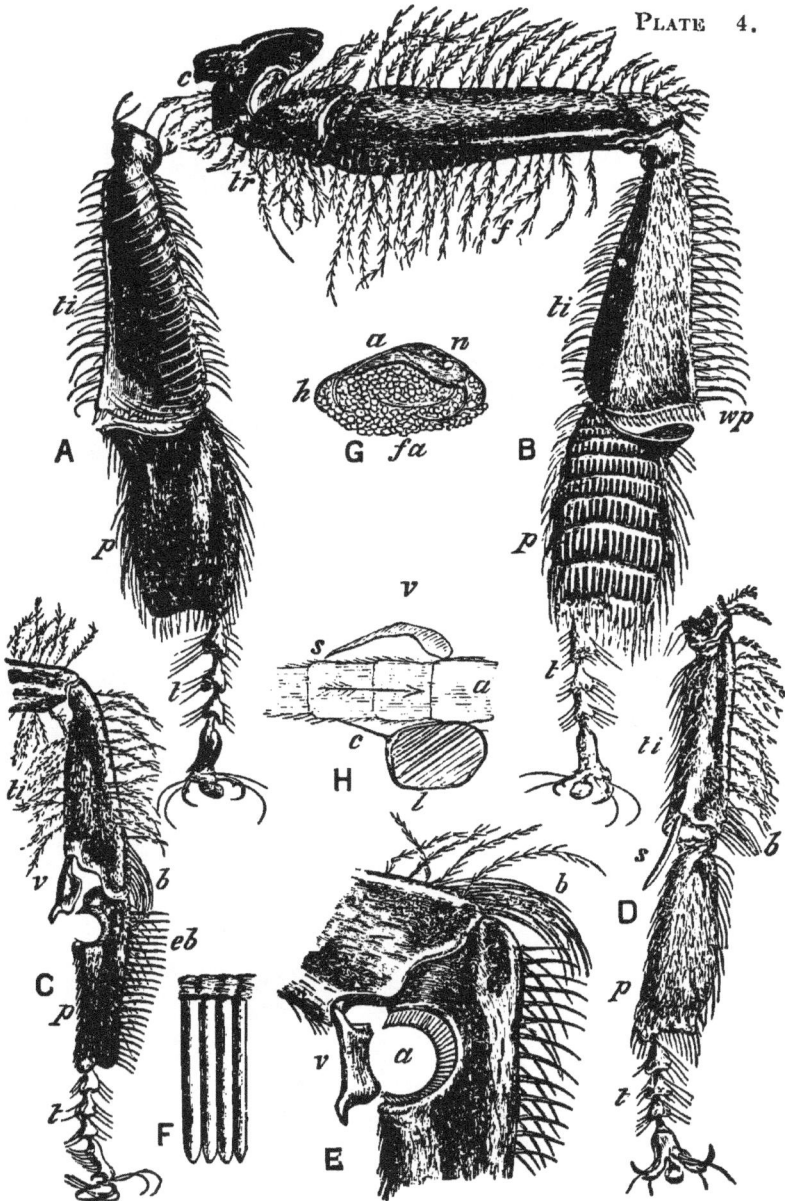

LEGS OF WORKER-BEE.

(Magnified 10 times. From Cheshire.)

A, third right leg, side from the body. ti, tibia, showing pollen basket; p, planta or metatarsus; t, tarsus. B, third right leg, side next the body. c, coxa; tr, trochanter; wp, pincers. C, front right leg. v, velum; b, brush; eb, eye-brush. D, second right leg. b, brush; E, joint of first leg, more enlarged. v, velum; a, antenna comb; b, brush. F, teeth of antenna comb, magnified 200 times. G, cross-section of tibia through pollen-basket. n; nerve; h, holding hairs; fa, farina or pollen. H, antenna in process of cleaning. v, velum; s, scraping edge; a, antenna; l, section of leg; c, antenna comb.

low *nervures* or ribs, and have a great power of resistance. In flight, the small wings are fastened to the large ones by small hooks (fig. 14), located on the edge of their outer nervure, that catch in a fold of the inner edge of the large wings. Thus united, they present to the air a stronger surface and give the bees a greater power of flight. No doubt, a single pair of wings of the same surface would have better attained the desired aim, but their width would have annoyed the bees in going inside of the cells, either to feed

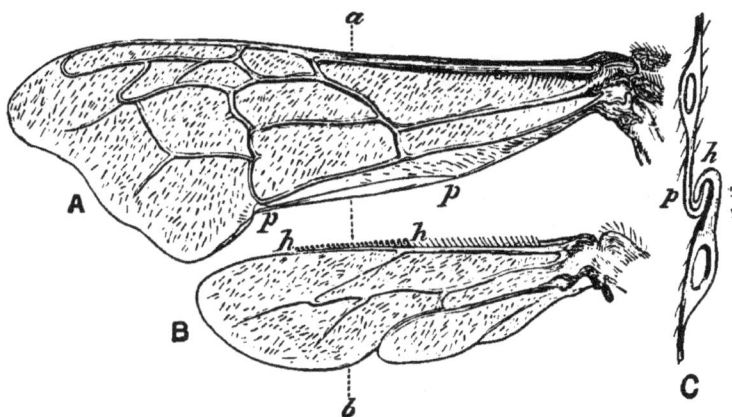

Fig. 14.

WINGS OF THE HONEY-BEE.

(Magnified.   From Cheshire.)

A, anterior wing, under side; *p,p*, plait.

B, posterior wing, under side; *h,h*, hooklets.

C, cross-section of wings through line *a,b*, showing hooklets in plait.

the larvæ or to deposit supplies.   Imagine a blue fly trying, with its wide wings, to go inside of a cell!

61. "Mr. Gaurichon has noticed that when the bees fan, or ventilate the entrance of the hive, their wings are not hooked together as they are in flight, but act independently of one another." (Dubini, 1881.) A German entomologist, Landois, states that, according to the pitch of their hum, the bees' flight must at times be equal to 440 vibrations in a second, but he noticed that this speed could not be kept up without fatigue.   It is well known that the more rapid the vibrations, the higher the pitch.

62. DIGESTING APPARATUS.—The honey obtained from the blossoms, after mixing with the saliva (**41**), and passing through the mouth and the œsophagus, is conveyed into the honey-sack.

63. This organ, located in the abdomen, is not larger than a very small pea, and so perfectly transparent as to appear, when filled, of the same color as its contents; it is properly the first stomach, and is surrounded by muscles which enable the bee to compress it, and empty its contents through her proboscis into the cells. She can also, at will, keep a supply, to be digested, at leisure, when leaving with a swarm, (**418**), or while in the cluster during the cold of winter (**620**), and use it only as fast as necessary. For this purpose, the honey-sack is supplied at its lower extremity, inside, with a round ball, which Burmeister has called the *stomach-mouth*, and which has been beautifully described by Schiemenz (**1883**). It opens by a complex valve and connects the honey-sack

Fig. 15.

DIGESTING APPARATUS.

(Magnified. From Maurice Girard.)

a, tongue; *b*, œsophagus; *c*, honey-sack; *d*, stomach; *e*, malpighian tubes; *f*, small intestine; *g*, large intestine.

with the digesting-stomach, through a tube or canal, projecting inside the latter.   This canal is lined with hairs pointing downward, which prevent the solid food, such as pollen grains, from returning to the honey-sack. Cheshire affirms that this stomach-mouth, which protrudes into the honey-sack, acts as a sort of sieve, and strains the honey from the grains of pollen floating in it, appropriating them for digestion, and allowing the honey to flow back into the sack. The bee could thus, at will, "eat or drink from the mixed diet she carries."

64. According to Schonfeld, (*Illustrierte Bienenzeitung*) the chyle, or milky food which is used to feed the young larvæ,—and which we have shown to be, most probably, the product of the upper pair of glands (**39–40**),—would be produced from the digesting-stomach, which he and others call chyle-stomach. Although we are not competent in the matter, we would remark that the so-called chyle-stomach produces *chyme*, or digested food, from which the *chyle*, or nourishing constituent, is absorbed by the cell-lining of the stomach and of the intestines, and finally converted into blood. We do not see how this chyle could be regurgitated, by the stomach, to be returned to the mouth.

65. In mammals, the chyliferous vessels do not exist in the stomach, but in the intestine, the function of the stomach being only to digest the food by changing it into chyme, from which the chyle is afterwards separated, for the use of the body.

66. Again, in the mammals, the glands which produce milk are composed of small clusters of *acini*, which take their secretions from the blood and empty them into vessels terminating at the surface of the breast. The action of the upper or chyle gland (**40**), in the bee, is exactly similar to the action of those lacteal glands, and the fact that this gland is absent in the queen and in the drone is, to us, positive evidence that the chylous or lacteal food (given the larvæ) is produced by these glands alone, and not by the direct action of the digesting-stomach.

67. The food arriving in the stomach is mixed with the

gastric juice, which helps its transformation, and the undu-
lating motion of the stomach sends it to its lower extremity,
toward the intestines. But, before entering into them, the
chyme receives the product of several glands which have
been named Malpighian tubes (e, fig. 15) from the scientist
Malpighi, who was the first to notice them. A grinding
motion of the muscles placed at the junction of the stomach

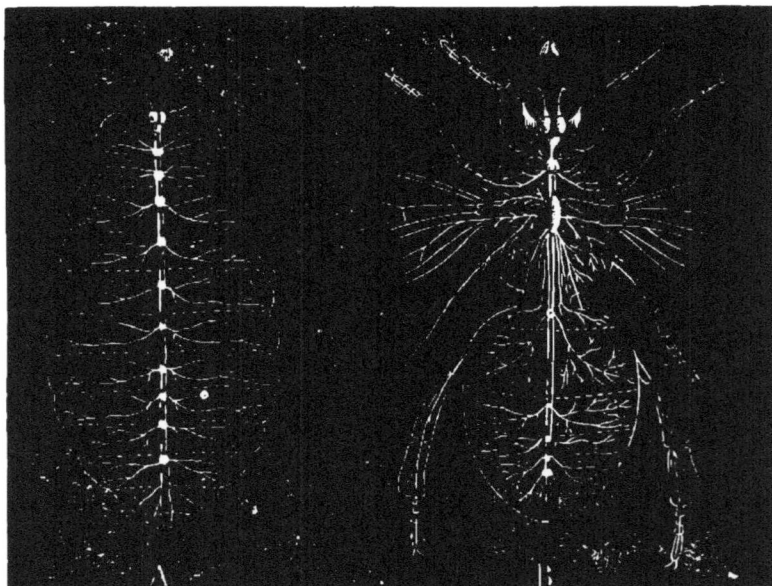

Fig. 16. (From Girard.)
NERVOUS SYSTEM OF THE HONEY-BEE. (Magnified.)
A, in the larva; B, in the bee.

with the intestines, acting on the grains of pollen not
yet sufficiently dissolved, prepares them to yield their
assimilable particles to the absorbing cells in the walls of
the small intestine. Thence they go into the large intestine, from which the refuse matter is discharged by the
worker-bees, *while on the wing*. We italicize the words,

because this fact has considerable bearing on the health of
the bees, when confined by cold or other causes, as will be
seen farther on. (**639.**)

68. "The nervous system (fig. 16) of the honey-bee, the seat
of sensation and of the understanding, is very interesting, on account of the profound difference which it presents when compared
with the nervous system of the larva. The honey-bee, more perfect in organization than the butterfly, begins as a larva deficient
in legs, very much inferior to the caterpillar from which the butterfly proceeds. It is very interesting to notice, that the drones,
although larger than the workers, especially in the head, have a
smaller brain. This state of things coincides with the fact that
the drones are not intelligent, while no one can refuse gleams
of intelligence to the worker-bees, as nurses and builders."
—(Girard.)

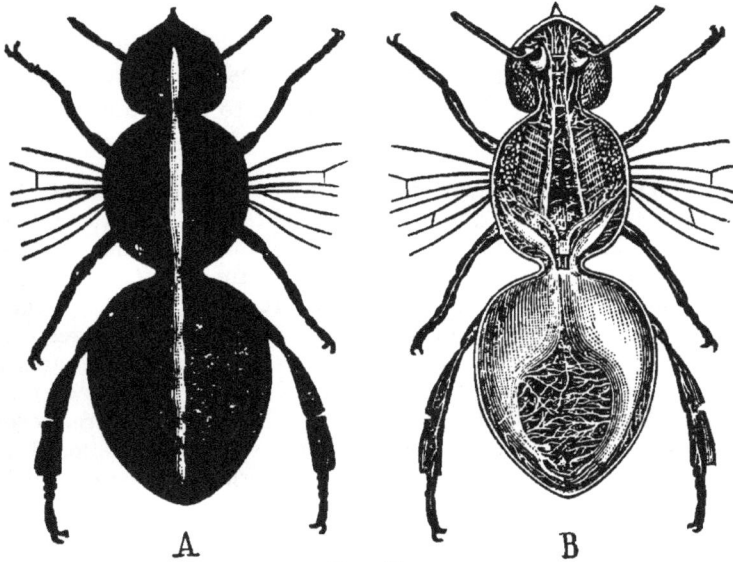

Fig. 17.

A, HEART OF THE HONEY-BEE. B, RESPIRATORY SYSTEM.
(Magnified. From Girard.)

69. The heart, or organ of the circulation of the blood,
formed of five elongated rooms, in the abdomen, is terminated in the thorax, and in the head, by the aorta, which is

not contractible.   Each room of the heart presents, on
either side, an opening for the returning blood.   The blood,
"soaking through the body" (Cheshire), comes in contact
with the air contained in the tracheal ramifications, where it
is arterialized, or in plainer words, renovated, before coming back to the heart.

The bee is not provided with any discernible blood or
lymphatic vessels save the aorta, and its blood is colorless.

70. The breathing organ of the bee is spread through its
whole body.   It is formed of membranous vessels, or tracheæ, whose ramifications spread and penetrate into the
organs, as the rootlets of a plant sink down into the soil.
Connected with these, there is, on each side of the abdominal cavity, a large tracheal bag, variable in form and dimensions, according to the quantity of air that it contains.
Bees breathe through holes, or spiracles, which are placed
on each side of the body, and open into the tracheal bags
and tracheæ.

71. "The act of respiration consists in the alternate dilatation
and contraction of the abdominal segments. By filling, or emptying the air-bags, the bee can change her specific gravity.
When a bee is preparing herself for flight, the act of respiration
resembles that of birds, under similar circumstances. At the moment of expanding her wings, which is indeed an act of respiration, the spiracles or breathing holes are expanded, and the air,
rushing into them, is extended into the whole body, which by
the expansion of the air-bags, is enlarged in bulk, and rendered
of less specific gravity ; so that when the spiracles are closed, at
the instant the insect endeavors to make the first stroke with,
and raise itself upon, its wings, it is enabled to rise in the air,
and sustain a long and powerful flight, with but little muscular
exertion." * * * "Newport has shown that the development of heat in insects, just as in vertebrates, depends on the
quantity and activity of respiration and the volume of circulation."—(Packard, Salem, 1869.)

72. Mr. Cheshire notices that bees, even in full, vigorous youth and strength, are not at all times able to take

flight. The reader may have noticed that if they are frightened, or even touched with the finger, they will occasionally move only by slight jumps. This temporary inability to fly, is due to the small quantity of air that their tracheal sacs contain. They were at rest, their blood circulated slowly, their body was comparatively heavy; but when their wings were expanded, the tracheal bags, that were as flat as ribbons, were soon filled with air, and they were ready to take wing.

Practical Apiarists well know that they may be shaken off the comb, and gathered up, with a shovel, with a spoon, or even with the hands, to be weighed or measured in open vessels, like seeds. The foregoing remarks give the explanation of this fact.

73. When the tracheal bags are filled with air, bees, owing to their peculiar structure, can best discharge the residues contained in their intestines.

The queen is differently formed, her ovaries occupying part of the space belonging to the air-sacks in the worker, hence her discharges, like those of the drones (**190**), take place in the hive. (**40.**)

74. "The tracheous bags of the abdomen, which we would be tempted to name *abdominal lungs,* hold in reserve the air needed to arterialize the blood and to produce muscular strength and heat, in connection with the powerful flight of the insect. Heat is indispensable, to keep up the high temperature of the hive, for the building of comb and rearing of brood. The aerial vesicles increase, by their resonance, the intensity of the humming, and are used also like the valve of a balloon, to slacken or increase the speed of the flight, by the variation of density, according to the quantity or weight, of the air that they contain. This accumulated air is also the means of preventing asphyxy, which the insects resist a long time. Lastly, these air-bags help in the mating of the sexes, which takes place in the air; the swelling of the vesicles being indispensable to the bursting forth of the male organs."—(Girard.)

75. The hum that is produced by the vibration of the

wings is different in each of the three kinds of inhabitants of the hive, and easily recognizable to a practiced ear. The hum of the drone is the most sonorous. But worker-bees, when angry or frightened, or when they call each other, emit different and sharper sounds. On the production of these sounds, bee-keepers and entomologists are far from being agreed.

"Inside of every opening of the aerial tubes is a valvular muscle, which helps to control the mechanism of respiration. This can be opened or closed at will, by the bee, to prevent the ingress, or egress, of air. It is by this means that the air is kept in the large tracheous bags and decreases the specific gravity of the insect. The main resonant organ of the bee is placed in front of this stopping muscle, at the entrance of the trachea."

" The humming is not produced solely by the vibrating of the wings, as is generally admitted. Chabrier, Burmeister, Landois, have discovered in the humming, three different sounds: the first, caused by the vibrating of the wings; the second, sharper, by the vibration of the rings of the abdomen ; the third, the most intense and acute, produced by a true vocal mechanism, placed at the orifices of the aerial tubes."—(Girard.)

76. The bee-keeper who understands the language of bees, can turn it to his advantage. Here are some examples :

" When something seems to irritate the bees, who are in front of a hive, on the alighting-board, they emit a short sound, z-z-z-, jumping at the same time towards the hive. This is a warning. Then they fly and examine the object of their fears, remaining sustained by their wings, near the suspected object, and emitting at the same time, a distinct and prolonged sound. This is a sign of great suspicion. If the object moves quickly, or otherwise shows hostile intent, the song is changed into a piercing cry for help, in a voice whistling with anger. They dash forward violently and blindly, and try to sting.

" When they are quiet and satisfied, their voice is the humming of a grave tune ; or, if they do not move their wings, an allegro murmur. If they are suddenly caught or compressed, the sound is one of distress. If a hive is jarred at a time when all the bees are quiet, the mass speedily raise a hum, which ceases as suddenly. In a queenless hive, the sound is doleful, lasts longer and at times increases in force. When bees swarm, the tune is clear and gay, showing manifest happiness."—(Œttl-Klauss, 1836.)

77. The German pastor Stahala has published a very complete study on the language of bees, which has appeared in some of the bee-papers of Italy, France and America. We do not consider it as altogether accurate; but there are some sounds described that all bee-keepers ought to study, especially the doleful wail of colonies which have lost their queen, and have no means of rearing another.

78. THE STING.—The sting of the bee, a terror to so many, is indispensable to her preservation. Without it, the attraction, which honey presents to man and animals, must have caused the complete destruction of this precious insect, years ago.

79. This organ is composed, *1st*, of a whitish vesicle, or poison sack, about the size of a small mustard seed, located in the abdomen, in which the venomous liquid is stored. This liquid is elaborated in two long canals, similar in appearance to the Malpighian tubes, each of which is terminated at its upper extremity, by a small round bag or enlargement. It is similar to formic acid, although perhaps more poisonous.

80. *2d*, In the last ring of the abdomen, and connected with the poison sack, is a firm and sharp sheath, open in its whole length, which supports the sting proper, and acts independently of it. The bee can force this sheath out of the abdomen, or draw it in, at will.

81. *3d*, The sting is composed of two spears of a polished, chestnut-colored, horny substance, which, supported by the sheath, make a very sharp weapon. In the act of stinging, the spears emerge from the sheath, about two-thirds of their length. Between them and on each of them, is a small groove, through which the liquid, coming from the poison sack, is ejected into the wound.

3

82. Each spear of the sting has about nine barbs, which are turned back like those of a fish hook, and prevent the sting from being easily withdrawn. When the insect is prepared to sting, one of these spears, having its point a little longer than the other, first darts into the flesh, and

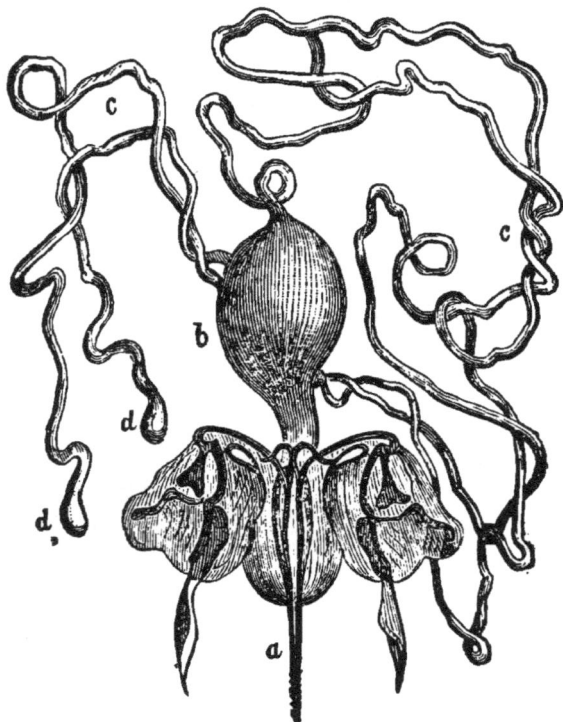

Fig. 18.
THE STING OF THE WORKER-BEE, AND ITS APPENDAGES.
(Magnified. From Girard.)
a, sting; b, poison-sack; c,c, poison glands; d,d, secreting bags.

being fixed by its foremost barb, the other strikes in also, and they alternately penetrate deeper and deeper, till they acquire a firm hold of the flesh with their barbed hooks.

"Meanwhile, the poison is forced to the end of the spears, by much the same process which carries the venom from the tooth of a viper when it bites."—(Girard.)

83. The muscles, though invisible to the eye, are yet strong enough to force the sting, to the depth of one-twelfth of an inch, through the thick skin of a man's hand.

"The action of the sting," says Paley, "affords an example of the union of *chemistry* and mechanism; of chemistry, in respect to the *venom* which can produce such powerful effects; of mechanism, as the sting is a compound instrument. The machinery would have been comparatively useless, had it not been for the chemical process by which, in the insect's body, *honey* is converted into *poison;* and on the other hand, the poison would have been ineffectual, without an instrument to wound, and a syringe to inject it."

"Upon examining the edge of a very keen razor by the microscope, it appears as broad as the back of a pretty thick knife, rough, uneven, and full of notches and furrows, and so far from anything like sharpness, that an instrument as blunt as this seemed to be, would not serve even to cleave wood. An exceedingly small needle being also examined, it resembled a rough iron bar out of a smith's forge. The sting of a bee viewed through the same instrument, showed everywhere a polish amazingly beautiful, without the least flaw, blemish, or inequality, and ended in a point too fine to be discerned."

84. As the extremity of the sting is barbed like an arrow, the bee can seldom withdraw it, if the substance into which she darts it is at all tenacious. A strange peculiarity of the sting and the muscles pertaining to it, is their spasmodic action, which continues quite a while, even after the bee has torn herself away, and has left them attached to the wound. In losing her sting, she often parts with a portion of her intestines, and of necessity soon perishes. Wasps and hornets are different from bees in this respect, for they can sting repeatedly without endangering their lives.

Although bees pay so dearly for the exercise of their patriotic instincts, still, in defense of home and its sacred treasures, they

"Deem life itself to vengeance well resign'd,
"Die on the wound and leave their sting behind."

85. The sting is not, however, always lost. When a bee prepares to sting, she usually curves her abdomen so that she can drive in her sting perpendicularly. To withdraw it, she turns around the wound. This probably rolls up its barbs, so that it comes out more readily. If it had been driven obliquely instead of perpendicularly, as sometimes happens, she could never have extracted it by turning around the wound.

86. Sometimes, only the poison-bag and sting are torn off, then she may live quite a while without them, and strange to say, seems to be more angry than ever, and persists in making useless attempts to sting.

87. If a hive is opened during a Winter day, when the weather does not permit the bees to fly, a great number of them raise their abdomens, and thrust out their stings, in a threatening manner. A minute drop of poison can be seen on their points, some of which is occasionally flirted into the eyes of the Apiarist, and causes severe irritation. The odor of this poison is so strong and peculiar, that it is easily recognized. In warm weather it excites the bees, and so provokes their anger, that when one has used its sting in one spot on skin or clothes, others are inclined to thrust theirs in the same place.

88. The sting, when accompanied by the poison-sack, may inflict wounds hours, and even days, after it has been removed, or torn, from the body of the bee. But when buried in honey, its poison is best preserved, for it is very volatile, and when exposed to the air, evaporates in a moment. The stings of bees, which, perchance, may be found in broken combs of honey, often retain their power, and we have known of a person's being stung in the mouth, by carelessly eating honey in which bees had been buried by the fall of the combs.

Mr. J. R. Bledsoe, in the *American Bee Journal*, for 1870, writes:

89. "It may often happen that one or both of the chief parts of the sting are left in the wound, when the sheath is withdrawn, but are rarely perceived, on account of their minuteness; the person stung congratulating himself, at the same time, that the sting has been extracted. I have had occasion to prove this fact repeatedly in my own person and in others. * * * The substance of the sting, on account of its nature is readily dissolved by the fluids of the body, consequently giving irritation as a foreign body for only a short time comparatively. The sting when boiled in water becomes tender and easily crushed."

For further particulars concerning the sting, we will refer our readers to the chapter entitled "Handling Bees."— (**378.**)

90. Before terminating this comparatively short, but perhaps, to many of our readers, tedious study of the organs of the bee, we desire to commend Messrs. Girard, Packard, Cook, Schiemenz, Dubini, and especially Mr. F. Cheshire, who, by their writings, have helped us in this part of our undertaking. We must add also that the more we study bees, the more persuaded we are that Mr. Packard was right when he wrote:

91. "Besides these structural characters as animals, endowed with instinct, and a kind of reason, differing, perhaps, only in degree, from that of man, these insects outrank all the articulates. In the unusual differentiation of the individual into males, females, and sterile workers, and a consequent subdivision of labor between them; in dwelling in large colonies; in their habits and in their relation to man as domestic animals, subservient to his wants, the bees possess a combination of characters which are not found in any other sub-order of insects, and which rank them first and highest in the insect series."—("Guide to the Study of Insects.")

92. One of their peculiarities, is, especially, the care given by most of the hymenopters to their progeny. We will show how bees nurse their young. Other insects of the same sub-order construct their nests of clay or paper, or burrow in the wood, or in the earth. All prepare for

their young a sufficient supply of food ; some of pollen and
honey, others of animal substance.   Several kinds of wasps
provide their nests with living insects, spiders, caterpillars,
etc., that they have previously paralyzed, but without killing them, by piercing them with their stings.

Ants seem to possess even a greater solicitude.   When
their nests are overthrown, they carry their larvæ to some
hidden place out of danger.

We have exhibited the use of the organs of bees as a
race.   We will now examine the character of each of the
three kinds of inhabitants of the bee-hive.

The Queen.

93. Although honey-bees have attracted the attention

Fig. 19.

of naturalists for ages, the sex
of the inmates of the bee-hive
was, for a long time, a mystery.
The Ancient authors, having noticed in the hive, a bee, larger
than the others, and differently
shaped, had called it the " King
Bee."

94. To our knowledge, it was an English bee-keeper,
Butler, who, first among bee-writers, affirmed in 1609, that
the King Bee was really a queen, and that he had seen her
deposit eggs. (" Feminine Monarchy.")

95. This discovery seems to have passed unnoticed, for
Swammerdam, who ascertained the sex of bees by dissection, is held as having been the first to proclaim the sex of
the Queen bee. (Leyde, 1737.) A brief extract from the
celebrated Dr. Boerhaave's Memoir of Swammerdam, showing the ardor of this naturalist, in his study of bees, should
put to blush the arrogance of those superficial observers,

who are too wise to avail themselves of the knowledge of others:

"This treatise on Bees proved so fatiguing a performance, that Swammerdam never afterwards recovered even the appearance of his former health and vigor. He was most continually engaged by day in making observations, and as constantly by night in recording them by drawings and suitable explanations.

"His daily labor began at six in the morning, when the sun afforded him light enough to survey such minute objects; and from that hour till twelve, he continued without interruption, all the while exposed in the open air to the scorching heat of the sun, bareheaded, for fear of intercepting his sight, and his head in a manner dissolving into sweat under the irresistible ardors of that powerful luminary. And if he desisted at noon, it was only because the strength of his eyes was too much weakened by the extraordinary afflux of light, and the use of microscopes, to continue any longer upon such small objects.

"He often wished, the better to accomplish his vast, unlimited views, for a year of perpetual heat and light to perfect his inquiries; with a polar night, to reap all the advantages of them by proper drawings and descriptions."

96. The name of queen was then given to the mother bee, although she in no way governs, but seems to reign like a beloved mother in her family.

97. She is the only perfect female in the hive, the laying of eggs being her sole function; and she acquits herself so well of this duty, that it is not uncommon to find queens, who lay more than 3,500 eggs per day, for several weeks in succession during the height of the breeding season. In our *observing hives* we have seen them lay at the rate of six eggs in a minute. The fecundity of the female of the white ant is, however, much greater than this, being at the rate of sixty eggs a minute; but her eggs are simply extruded from her body, and carried by the workers into suitable nurseries, while the queen-bee herself deposits her eggs in their appropriate cells.

98. This number of 3,500, that a good queen can lay

per day, will seem exaggerated to many bee-keepers, owners of small hives. They will perhaps ask how such laying can be ascertained. Nothing is easier. Let us suppose
that we have found a hive, with 1,200 square inches of
comb occupied by brood. As there are about 55 worker-
cells to the square inch of comb (**217**), 27 to 28 on each
side, we multiply 1,200 by 55, and we have 66,000 as the
total number of cells occupied at one time. Now, it takes
about 21 days for the brood to develop from the egg to the
perfect insect, and we have 3,145 as the average number of
eggs laid daily by that queen, in 21 days. Of course, this
amount is not absolutely accurate, as the combs are not
always entirely filled, but it will suffice to show, within
perhaps a few hundred, the actual fecundity of the queen.

Such numbers can be found every year, in most of the
good colonies, provided that the limited capacity of the
hive will not prevent the queen from laying to the utmost
of her ability.

99. The laying of the queen is not equal at all seasons.
She lays most during the spring and summer months, previous to the honey crop and during its flow. In late autumn
and winter months, she lays but little.

100. Her shape is widely different from that of the
other bees. While she is not near so bulky as a drone, her
body is longer; and as it is considerably more tapering, or
sugar-loaf in form, than that of a worker, she has a somewhat wasp-like appearance. Her wings are much shorter
in proportion than those of the drone, or worker; * the
under part of her body is of a golden color, and the upper
part usually darker than that of the other bees.† Her motions are generally slow and matronly, although she can,
when she pleases, move with astonishing quickness. No
colony can long exist without the presence of this all-impor-

---

* The wings of the queen are *in reality* longer than those of the worker.
† This applies only to queens of the black or common race.

tant insect; but must as surely perish, as the body without the spirit must hasten to inevitable decay.

101. The queen is treated with the greatest respect and affection by the bees. A circle of her loving offspring often surround her, testifying in various ways their dutiful regard; some gently embracing her with their antennæ, others offering her food from time to time, and all of them politely backing out of her way, to give her a clear path when she moves over the combs. If she is taken from them, the whole colony is thrown into a state of the most intense agitation as soon as they ascertain their loss; all the labors of the hive are abandoned; the bees run wildly over the combs, and frequently rush from the hive in anxious search for their beloved mother. If they cannot find her, they return to their desolate home, and by their sorrowful tones reveal their deep sense of so deplorable a calamity. Their note at such times, more especially when they first realize their loss, is of a peculiarly mournful character; it sounds somewhat like a succession of wailings on the minor key, and can no more be mistaken by an experienced bee-keeper, for their ordinary happy hum (**76**), than the piteous moanings of a sick child could be confounded by the anxious mother with its joyous crowings when overflowing with health and happiness. We shall give, in this connection, a description of an interesting experiment.

102. A populous stock was removed, in the morning, to a new place, and an empty hive put upon its stand. Thousands of workers which were ranging the fields, or which left the old hive after its removal, returned to the familiar spot. It was truly affecting to witness their grief and despair; they flew in restless circles about the place where once stood their happy home, entering the empty hive continually, and expressing in various ways, their lamentations over so cruel a bereavement. Towards evening, ceasing to take wing, they roamed in restless platoons, in and out of the hive, and

over its surface, as if in search of some lost treasure. A
small piece of brood-comb was then given to them, containing worker-eggs and worms. The effect produced by its
introduction took place much quicker than can be described.
Those which first touched it raised a peculiar note, and in a
moment, the comb was covered with a dense mass of bees;
as they recognized, in this small piece of comb, the means
of deliverance, despair gave place to hope, their restless
motions and mournful voices ceased, and a cheerful hum
proclaimed their delight. If some one should enter a building filled with thousands of persons tearing their hair, beating their breasts, and by piteous cries, as well as frantic
gestures, giving vent to their despair, and could by a single
word cause all these demonstrations of agony to give place
to smiles and congratulations, the change would not be more
instantaneous than that produced when the bees received
the brood-comb!

The Orientals called the honey-bee "*Deborah;* She that
speaketh." Would that this little insect might speak, in
words more eloquent than those of man's device, to those
who reject any of the doctrines of revealed religion, with the
assertion that they are so improbable, as to labor under a
fatal *a priori* objection. Do not all the steps in the development of a queen from the worker-egg, labor under the
very same objection? and have they not, for this reason been
formerly regarded, by many bee-keepers, as unworthy of
belief? If the favorite argument of infidels will not stand
the test, when applied to the wonders of the bee-hive, is it
entitled to serious weight, when, by objecting to religious
truths, they arrogantly take to task the Infinite Jehovah for
what He has been pleased to do or to teach? With no
more latitude than is claimed by such objectors, it were
easy to prove that a man is under no obligation to believe
any of the wonders of the bee-hive, even although he is him-

self an intelligent eye-witness to their substantial truth.*

103. The process of rearing Queen-bees will now be particularly described. Early in the season, if a hive becomes very populous, and if the bees make preparations for swarming, a number of royal cells are begun, being commonly constructed upon those edges of the combs which are not attached to the sides of the hive. These cells somewhat resemble a small pea-nut, and are about an inch deep, and one-third of an inch in diameter: being very thick, they require much wax for their construction. They are seldom seen in a perfect state after the hatching of the queen, as the bees cut them down to the shape of a small acorn-cup (fig. 20). These queen-cells, while in progress, receive a very unusual amount of attention from the workers. There is scarcely a second in which a bee is not peeping into them; and as fast as one is satisfied, another pops in her head to report progress, or increase the supply of food. Their importance to the community might easily be inferred from their being the center of so much attraction.

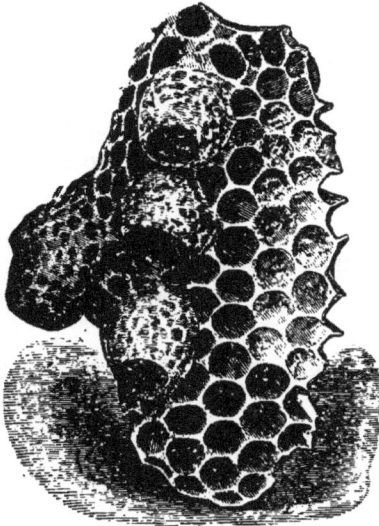

Fig. 20.

QUEEN-CELLS IN PROGRESS.

104. While the other cells open sideways, the queen-cells always hang with their mouth *downwards*. Some Apiarists

* The passages referring to religious subjects have been nearly all retained in this revision, *at Mr. Langstroth's request*, even when not in accordance with our views. As intelligent men are always tolerant, we know our readers will not object to them.

think that this peculiar position affects, in some way, the development of the royal larvæ; while others, having ascertained that they are uninjured if placed in any other position, consider this deviation as among the inscrutable mysteries of the bee-hive. So it seemed to us until convinced, by a more careful observation, that they open downwards simply *to save room.* The distance between the parallel ranges of comb in the hive is usually too small for the royal cells to open sideways, without interfering with the opposite cells. To economize space, the bees put them on the unoccupied edges of the comb, where there is plenty of room for such very large cells.

105. The number of royal cells in a hive varies greatly; sometimes there are only two or three, ordinarily not less than five; and occasionally, more than a dozen.

Some races of bees have a disposition to raise a greater number of queen-cells than others. At the Toronto meeting of the North American Bee-keepers' Association, in September, 1883, Mr. D. A. Jones, the noted Canadian importer of Syrian and Cyprian bees, and publisher of the *Canadian Bee Journal,* exhibited a comb containing about eighty queen-cells, built by a colony of Syrian bees (**560**). Such cases are rare in the hive of any other race.

106. As it is not intended that the young queens should all be of the same age, the royal-cells are not all begun at the same time. It is not fully settled how the eggs are deposited in these cells. In some few instances, we have known the bees to transfer the eggs from common to queen-cells; and this may be their general method of procedure. Mr. Wagner put some queenless bees, brought from a distance, into empty combs that had lain for two years in his garret. When supplied with brood, they raised their queen in this old comb! Mr. Richard Colvin, of Baltimore, and other Apiarian friends, have communicated to us instances almost as striking. Yet, Huber has proved that bees do not ordinarily transport the eggs of the queen from one cell to another. We shall hazard the conjecture, that, in a crowded state of the hive, the queen deposits her eggs in cells on the edges of the comb, some of which are afterwards changed by the workers into royal cells. Such is a queen's instinctive hatred of her own kind, that it seems improbable that she should be intrusted with even the initiatory steps for securing a race of successors.

(For further particulars concerning the raising of large numbers of queen-cells, see **515.**)

107. The egg which is destined to produce a queen-bee does not differ from the egg intended to become a worker; but the young queen-larvæ are much more largely supplied with food than the other larvæ; so that they seem to lie in a thick bed of jelly, a portion of which may usually be found at the base of their cells, soon after they have hatched, while the food given to the worker-larvæ after three days, and for the last days of their development, is coarser and more sparingly given, as will be seen farther on.

108. The effects produced on the royal larvæ by their peculiar treatment are so wonderful, that they were at first rejected as idle whims, by those who had neither been eye-witnesses to them, nor acquainted with the opportunities enjoyed by others for accurate observation. They are not only contrary to all common analogies, but seem marvelously strange and improbable. The most important of these effects we shall briefly enumerate.

1st. The peculiar mode in which the worm designed for a queen is treated causes it to arrive at maturity almost one-third earlier than if it had been reared a worker. And yet, as it is to be much more fully developed, according to ordinary analogy, it should have had a slower growth.

2d. Its organs of reproduction are completely developed, so that it can fulfill the office of a mother.

3d. Its size, shape, and color are greatly changed; its

lower jaws are shorter, its head rounder, and its abdomen without the receptacles for secreting wax ; its hind legs have neither brushes nor baskets, and its sting is curved (fig. 21), and one-third longer than that of a worker.

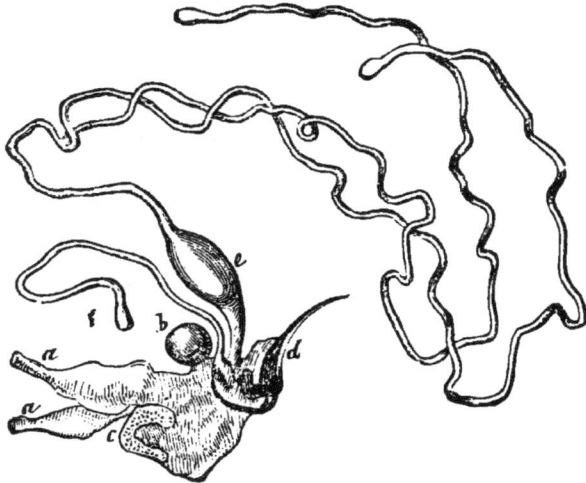

Fig. 21.
THE STING OF THE QUEEN.
(Magnified. From Girard.)
a,a, branches of the oviduct; *c*, oviduct; *b*, spermatheca; *d*, sting;
e, poison-sack; *f*, gland.

4th. Its instincts are entirely changed. Reared as a worker, it would have thrust out its sting at the least provocation ; whereas now, it may be pulled limb from limb without attempting to sting. As a worker, it would have treated a queen with the greatest consideration ; but now, if brought in contact with another queen, it seeks to destroy her as a rival. As a worker, it would frequently have left the hive, either for labor or exercise ; as a queen, it never leaves it after impregnation, except to accompany a new swarm.

5th. The term of its life is remarkably lengthened. As a worker, it would not have lived more than six or seven months ; as a queen, it may live seven or eight times as long. All these wonders rest on the impregnable basis of demonstration, and instead of being witnessed only by a select few, are now, by the use of the movable-comb hive, familiar sights to any bee-keeper who prefers an acquaintance with facts, to caviling and sneering at the labors of others.

109. The process of rearing queens, to meet some special emergency, is even more wonderful than the one already described. If the bees have worker-eggs, or worms not more than three days old, they make one large cell out of three, by nibbling away the partitions of two cells adjoining a third. Destroying the eggs or worms in two of these cells, they place before the occupant of the other, the usual food of the young queens; and by enlarging its cell, give it ample space for development.* As a security against failure, they usually start a number of queen-cells, for several days in succession.

110. DURATION OF DEVELOPMENT.—The eggs hatch in three days after they are laid. The small worm which is intended to produce a queen, is six days in its larval state, and seven in its transformation into a chrysalis and winged insect. These periods are not absolutely fixed; being of shorter or longer duration, according to the warmth of the hive and the care given by the bees. In from ten to sixteen days† they are in possession of a new queen, in all respects resembling one reared in the natural way; while the eggs in the adjoining cells, which have been developed as workers, are nearly a week longer in coming to maturity.

111. THE VIRGIN QUEEN.—Feeble and pale, in the first moments after her birth, the young queen, as soon as she

---

* It was a German bee-keeper, Schirach, who discovered that a queen can be raised from a worker-egg. (''The New Natural and Artificial Multiplication of Bees,'' Bautzen, 1761.)

† In ten days, if the larva selected is about six days old; in sixteen, if they have selected newly-laid eggs.

has acquired some strength, travels over the combs, looking for a rival, either hatched or unhatched.

112. "Hardly ten minutes had elapsed since the young queen had emerged from her cell, when she began to look for sealed queen-cells. She rushed furiously upon the first that she met, and, by dint of hard work, made a small opening in the end. We saw her drawing, with her mandibles, the silk of the cocoon, which covered the inside. But, probably, she did not succeed according to her wishes, for she left the lower end of the cell, and went to work on the upper end, where she finally made a wider opening. As soon as this was sufficiently large, she turned about, to push her abdomen into it. She made several motions, in different directions, till she succeeded in striking her rival with the deadly sting. Then she left the cell; and the bees, which had remained, so far, perfectly passive, began to enlarge the gap which she had made, and drew out the corpse of a queen just out of her nymphal shell. During this time, the victorious young queen rushed to another queen-cell, and again made a large opening, but she did not introduce her abdomen into it; this second cell containing only a royal-pupa not yet formed. There is some probability that, at this stage of development, the nymphs of queens inspire less anger to their rivals; but they do not escape their doom; for, whenever a queen-cell has been prematurely opened, the bees throw out its occupant, whether worm, nymph, or queen. Therefore, as soon as the victorious queen had left this second cell, the workers enlarged the opening and drew out the nymph that it contained. The young queen rushed to a third cell; but she was unable to open it. She worked languidly and seemed tired of her first efforts."—(Huber.)

113. Huber did not allow this experiment to go on any further, as he wished to use the remainder of the queen-cells. Had he left these cells untouched, the bees would have finished the work of destruction.

114. We have noticed repeatedly, that the queen-cells are always destroyed a few hours after the birth of the queen, unless the colony has determined to swarm. In the latter case, the workers prevent the newly-hatched queen from approaching the queen-cells, till she is old enough and strong enough to leave with the swarm. **(443.)**

115. Like some human beings who cannot have their own way, she is highly offended when thus repulsed, and utters, in a quick succession of notes, a shrill, angry sound, not unlike the rapid utterance of the words, "peep, peep." If held in the closed hand, she will make a similar noise. To this angry note, one or more of the unhatched queens, imprisoned and nursed in their cells by the bees, answer by the sound "kooa, kooa"; the difference in their voices, being due to the confinement of the latter in the cell.

These sounds, so entirely unlike the usual steady hum of the bees, are almost infallible indications that a swarm will soon issue. They are occasionally so loud as to be heard at some distance from the hive.

The reader will understand that all these facts relate to a hive of bees, from which the old queen has been previously and suddenly removed, either by the Apiarist for some purpose, or by swarming, or accident.

116. Sometimes two queens hatch at the same time. We translate the narrative of Huber in such an emergency:

"On the 15th of May, 1790, two queens emerged from their cells, at about the same time, in one of our observing hives. They rushed quickly upon one another, apparently in great anger, and grasped one another's antennæ, so that the head, corselet and abdomen of the one, were touching the head, corselet and abdomen of the other. Had they curved the posterior extremity of their bodies, they could have stung each other, and both would have perished. But it seems that Nature has not wished that their duels should result in the death of both combatants, and that it is prescribed to queens. while in this position, to flee instantly with the greatest haste. As soon as both rivals understood that they were in danger from one another, they disentangled themselves and fled apart. . . . . . A few minutes after, their fears ceased and they attacked one another again, with the same result. The worker bees were much disturbed, all this time, and more so while the combatants were separated. Each time, the bees stopped the queens in their flight, keeping them prisoners for a minute." "At last, in a third attack, the stronger, or more savage, of the queens, ran to her unsuspecting

4

rival, seized her across the wings, and, climbing upon her, pierced her with her sting. The vanquished queen, crawled languidly about, and soon after died."—("Nouvelles Observations.")

117. Although it is generally admitted that two queens cannot inhabit the same hive, it happens, sometimes, that mother and daughter, are found living peaceably together, and even laying eggs at the same time. This is when the bees, having noticed the decrease in fecundity of the old queen, have raised a young queen to replace her. But this abnormal state lasts only a few weeks, or a few months at most.

118. Our junior partner was, one day, hunting for a queen with his sister. "What a large and bright-colored queen!" exclaimed he, on finding her. "Why, no! she is dark and small," said his sister. Both were right, for there were two queens, mother and daughter, on the same comb, and not six inches apart. At another time we were looking for an old queen, whose prolificness had decreased, intending to supersede her. To our wonder, the hive was full of brood. We found the old queen. Evidently a queen so small, so ragged and worn, could not be the mother of such a quantity of brood. We continued our search and found another queen, daughter of the first, large and plump. Had we introduced a strange queen into this hive, after having destroyed the old one, thinking that we had made the colony queenless, she would have been killed.

119. We could relate a number of such instances. The most interesting case was the simultaneous laying of two queens of different breeds in the same hive, one black, the other Italian. The colony had two queens, when we introduced our Italian queen. We found the younger one and killed her, and the old one was so little considered by her bees, that they accepted our imported queen and allowed both to remain together. To our astonishment there were some black bees hatching among the pure Italians, and it was not till we accidentally discovered the old black queen that we understood the matter.

There are more such cases than most bee-keepers would imagine, and when these happen to buyers of improved races of bees, if they are not very close observers, they are apt to accuse the venders of having cheated them. Such instances make the business of queen selling quite disagreeable.

120. IMPREGNATION.—The fecundation of the queen bee has occupied the minds of Apiarists and savants for ages. A number of theories were advanced. If a number of drones are confined in a small box, they give forth a strong odor: Swammerdam supposed that the queen was impregnated by this scent (*aura seminalis*) of the drones. Réaumur, a renowned entomologist, in 1744, thought that the mating of the queen was effected inside of the hive. Others advanced that the eggs were impregnated by the drones in the cells.

After making a number of experiments to verify these theories, and finding all false, Huber finally ascertained that, like many other insects, the queen was fecundated in the open air and on the wing; and that the influence of this connection lasts for several years, and probably for life.

121. Five days or more after her birth, the virgin queen goes out to have intercourse with a drone. Several bee-keepers of note, such as Neighbour of England ("Cook's Manual," 1884, page 72) and Dzierzon of Germany, wrote that a queen may go out on her marriage-flight when only three days old. The shortest time we have ever noticed between the birth of a queen and her first bridal-flight was five days, and on this we are in accordance with Mr. Alley of Massachusetts, one of the most extensive queen breeders in the world. The average time is six or seven days. Earlier bridal-trips are probably due to

the disturbing of the colony by the Apiarist, for we have noticed that this disturbing hastens the maturity of the workers. The bridal-flight takes place about noon, at which
time, the drones are flying most numerously.

122. On leaving her hive, the queen flies with her head
turned towards it, often entering and departing several
times before she finally soars into the air. Such precautions
on the part of a young queen are highly necessary, that she
may not, on her return, lose her life, by attempting, through
mistake, to enter a strange hive. Many queens are lost in
this way.

123. As the mating of the queen and the drone takes
place in the air, very few persons have witnessed it. The
following narration will please our readers:

" A short time ago, during one of those pleasant days of May, I
was roaming in the fields, not far from Courbevoie. Suddenly I
heard a loud humming and the wind of a rapid flight brushed my
cheek. Fearing the attack of a hornet, I made an instinctive motion with my hand to drive it away. There were two insects,
one of which pursued the other with eagerness. coming from high
in the air. Frightened no doubt, by my movements, they arose
again. flying vertically to a great height, still in pursuit of each
other. I imagined that it was a battle, and desiring to know the
result, I followed, at my best, their motions in the air, and got
ready to lay hold of them, as soon as they would be within reach.

" I did not wait long. The pursuing insect rose above the other,
and suddenly fell on it. The shock was certainly violent, for both
united, dropped with the swiftness of an arrow and passed by me,
so near that I struck them down, with my handkerchief. I then
discovered that this bitter battle was but a love-suit. The two
insects, stunned and motionless. were coupled. The copulation
had taken place in the air, at the instant when I had seen one of
them falling on the other, twenty or twenty-five feet above the
ground.

" It was a queen-bee and a drone. Persuaded that I had killed
them, I made no scruple of piercing them both with the same pin.
But the pain recalled them to life again, and they promptly separated. This separation was violent, and resulted in the tearing
off of the drone's organ (**188**) which remained attached to the

queen. The queen was yet alive on the following morning. For some time after her separation from the drone, she brushed the last ring of her abdomen, as though trying to extract the organ of the drone. She endeavored to bend herself, probably in order to bring this part within reach of her jaws, which were constantly moving, but the pin prevented her from attaining her aim. Her activity soon decreased and she ceased to move."—(Alex. Levi, *Journal Des Fermes*, Paris, 1869.)

Messrs. Cary and Otis had witnessed a similar occurrence in July 1861. (*American Bee Journal*, Vol. I, page 66.)

124. It is now well demonstrated that in a single mating, a queen is fertilized for life, although in a few rare instances they have been said to mate two days in succession, perhaps because the first mating was insufficient.

125. After the queen has re-entered the hive, she gets rid of the organ of the drone by drawing it with her claws, and she is sometimes helped in this work by the workerbees. The drone dies in the act of fertilization. (**188.**)

126. Although fertilization of the queen in confinement has been tried by many, it has never been successful. Those who, from time to time, claimed to have succeeded were evidently deceiving themselves through ill-made experiments. (**187.**)

127. Having ascertained that the queen-bee is fecundated in the open air and on the wing, Huber still could not form any satisfactory conjecture how eggs were fertilized which were not yet developed in her ovaries. Years ago, the celebrated Dr. John Hunter (1792), and others, supposed that there must be a permanent receptacle for the male sperm, opening into the oviduct. Dzierzon, who must be regarded as one of the ablest contributors of modern times to Apiarian science, maintained this opinion, and stated that he had found such a receptacle filled with a fluid resembling the semen of the drones. He does not seem to have then demonstrated his discoveries by any microscopic examinations.

128. In the Winter of 1851–2, the writer submitted for scientific examination several queen-bees to Dr. Joseph Leidy, of Philadelphia, who had the highest reputation both at home and abroad, as a naturalist and microscopic anatomist. He found, in making his dissections, a small globular sac, about $\frac{1}{38}$ of an inch in diameter, communicating with the oviduct, and filled with a whitish fluid; this fluid, when examined under the microscope, abounded in the spermatozoids* which characterize the seminal fluid. A comparison of this substance, later in the season, with the semen of a drone, proved them to be exactly alike.†

129. These examinations have settled, on the impregnable basis of demonstration, the mode in which the eggs of the queen are fecundated. In descending the oviduct to be deposited in the cells, they pass by the mouth of this seminal sac, or "*spermatheca*," and receive a portion of its fertilizing contents. Small as it is, it contains sufficient to impregnate millions of eggs. In precisely the same way, the mother-wasps and hornets are fecundated. The females only of these insects survive the Winter, and often a single one begins the construction of a nest, in which at first only a few eggs are deposited. How could these eggs hatch, if the females had not been impregnated the previous season? Dissection proves that they have a spermatheca similar to that of the queen-bee. It never seems to have occurred to the opponents of Huber, that the existence of a permanently-impregnated mother-wasp is quite as difficult to be accounted for, as the existence of a similarly impregnated queen-bee.

130. The celebrated Swammerdam, in his observations

---

* *Spermatozoids* are the living germs of the seminal fluid.

† Prof. Siebold, in 1843, examined the spermatheca of the queen-bee, and found it after copulation, filled with the seminal fluid of the drone. At that time, Apiarists paid no attention to his views, but considered them, as he says, to be only "*theoretical stuff*." It seems, then, that Prof. Leidy's dissection was not, as we had hitherto supposed, the first, of an impregnated spermatheca.

upon insects, made in the latter part of the seventeenth century, has given a highly magnified drawing of the ovaries of the queen-bee, a reduced copy of which we present (Plate V) to our readers. The small globular sac (D), communicating with the oviduct (E), which he thought secreted a fluid for sticking the eggs to the base of the cells, is the seminal reservoir, or spermatheca. Any one who will carefully dissect a queen-bee, may see this sac, even with the naked eye.

It will be seen that the ovaries (G and II) are double, each consisting of an amazing number of ducts filled with eggs, which gradually increase in size.*

131. Huber, while experimenting to ascertain how the queen was fecundated, confined some young ones to their hives by contracting the entrances, so that they were more than three weeks old before they could go in search of the drones. To his amazement, the queens whose impregnation was thus retarded never laid any eggs but such as produced drones!

He tried this experiment repeatedly, but always with the same result. Bee-keepers, even from the time of Aristotle, had observed that all the brood in a hive were occasionally drones.

132. Dzierzon appears to have been the first to ascertain the truth on this subject; and his discovery must certainly be ranked among the most astonishing facts in all the range of animated nature.

Dzierzon asserted that *all impregnated eggs produce females, either workers or queens; and all unimpregnated ones,*

* Since the first edition of this work was issued, we have ascertained that Posel (page 54) describes the oviduct of the queen, the spermatheca and its contents, and the use of the latter in impregnating the passing egg. His work was published at Munich, in 1784. It seems also from his work ("A Complete Treatise of Forest and Horticultural Bee-Culture," page 36), that before the investigations of Huber, Jausha, the bee-keeper royal of Maria Theresa, had discovered the fact that the young queens leave their hive in search of the drones.

males, or drones! He stated that in several of his hives he
found drone-laying queens, whose wings were so imperfect
that they could not fly, and which, on examination, proved
to be unfecundated. Hence, he concluded that the eggs
laid by an unimpregnated queen-bee, had sufficient vitality to produce drones.

133. *Parthenogenesis*, meaning "generation of a virgin,"
is the name given to this faculty of a female, to produce
offspring without having been fecundated, and is not at all
rare among insects.

134. In the Autumn of 1852, our assistant found a young
queen whose progeny consisted entirely of drones. The
colony had been formed by removing a few combs containing bees, brood, and eggs, from another hive, and had
raised a new queen. Some eggs were found in one of the
combs, and young bees were already emerging from the
cells, all of which were drones. As there were none but
worker-cells in the hive, they were reared in them, and not
having space for full development, they were dwarfed in
size, although the bees had pieced the cells to give more
room to their occupants.

We were not only surprised to find drones reared in
worker-cells, but equally so that a young queen, who at
first lays only the eggs of workers, should be laying
drone-eggs, and at once conjectured that this was a case of
an unimpregnated drone-laying queen, sufficient time not
having elapsed for her impregnation to be unnaturally retarded. All necessary precautions were taken to determine
this point. The queen was removed from the hive, and
although her wings appeared to be perfect, she could not
fly. It seemed probable, therefore, that she had never been
able to leave the hive for impregnation.

135. To settle the question beyond the possibility of
doubt, we submitted this queen to Professor Leidy for microscopic examination. The following is an extract from

(Plate 5.)

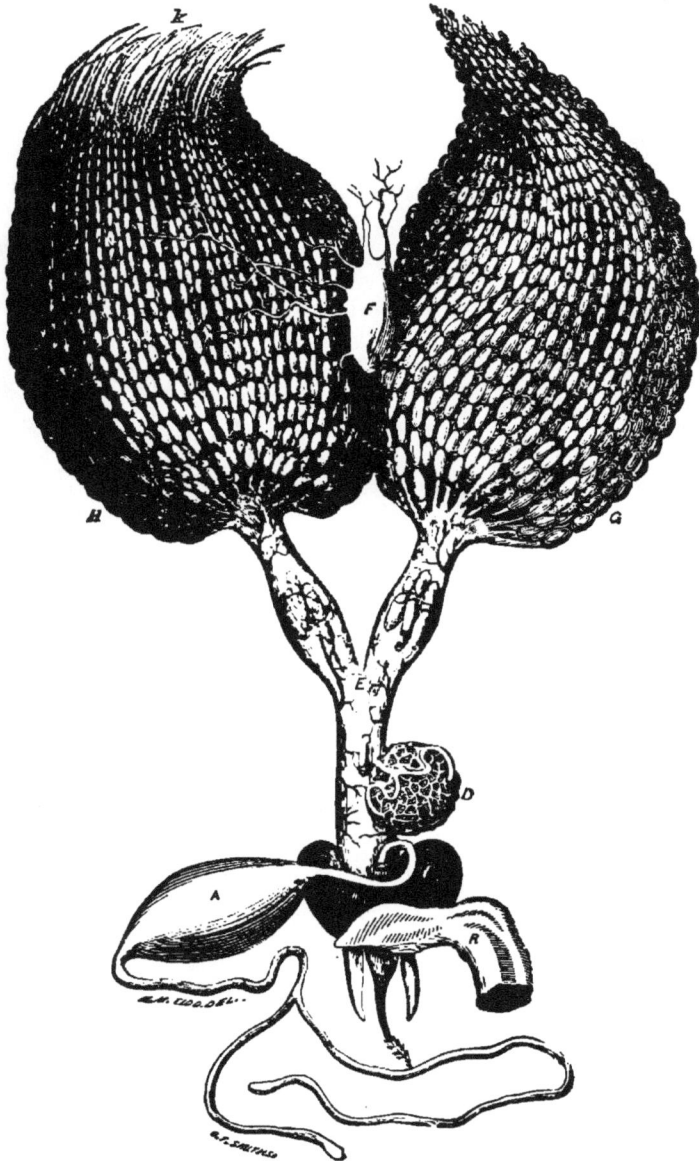

THE OVARIES OF THE QUEEN, IN COMBINATION WITH THE STING.
(Magnified.)

H and *G*, ovaries uniting in a common oviduct *E*; *D*, spermatheca; *A*, poison-sack, *R*, rectum; *C*, muscles.

his report: "The ovaries were filled with eggs, the poison-sac full of fluid; and the spermatheca distended with a per-fectly colorless, transparent, viscid liquid, *without a trace of spermatozoids.*"

136. On examining this same colony a few days later, we found satisfactory evidence that these drone-eggs were laid by the queen which had been removed. No fresh eggs had been deposited in the cells, and the bees on missing her had begun to build royal cells, to rear, if possible, another queen. Two of the royal cells were in a short time discon-tinued; while a third was sealed over in the usual way, to undergo its changes to a perfect queen. As the bees had only a drone-laying queen, whence came the female egg from which they were rearing a queen?

At first we imagined that they might have stolen it from another hive; but on opening this cell it contained only *a dead drone!* Huber had described a similar mistake made by some of his bees. At the base of this cell was an unus-ual quantity of the peculiar jelly fed to develop young queens. One might almost imagine that the bees had dosed the unfortunate drone to death; as though they had hoped by such liberal feeding to produce a change in his sexual organization.

137. In the Summer of 1854, we found another drone-laying queen in our Apiary, with wings so *shrivelled* that she could not fly. We gave her successively to several queen-less colonies, in all of which she deposited only drone-eggs.

138. In Italy there is a variety of the honey-bee differing in size and color from the common kind. If a queen of this variety is crossed with the common drones, her drone-prog-eny will be *Italian* (**551**), and her worker-brood a cross between the two; thus showing that the kind of drones she will produce has no dependence on the male by which she is fecundated.

" The following interesting experiment was made by Berlepsch, in order to confirm the drone-productiveness of a virgin queen. He contrived the exclusion of queens at the end of September, 1854, and, therefore, at a time when there was no longer any males ; he was lucky enough to keep one of them through the Winter, and this pro ·uced drone-offspring on the 2d of March, in the following year, furnishing fifteen hundred cells with brood. That this drone-bearing queen remained a virgin, was proved by the dissection which Leuckart undertook, at the request of Berlepsch. He found the state and contents of the seminal pouch of this queen to be exactly of the same nature as those found in virgin queens. The seminal receptacle in such females never contains semen-masses, with their characteristic spermatozoids, but only a limpid fluid, destitute of cells and granules which is produced from the two appendicular glands of the seminal capsule ; and, as I suppose, serves the purpose of keeping the semen transferred into the seminal capsule in a fresh state, and the spermatozoids active, and, consequently, capable of impregnation."—(Siebold, " Parthenogenesis.")

139. Again, to prove that Dzierzon was right, Professor Von Siebold, in 1855, dissected several eggs at the Apiary of Baron Von Berlepsch, and he found spermatozoids in every female egg, or egg laid in worker-cell, but although he examined thirty-two male eggs, or eggs laid in drone-cells, he could not discover a single spermatozoid either in or around them. In the act of copulation, the sperm of the drone is received into the spermatheca (Plate V, *D*), which is placed near and can empty itself into the oviduct. When an egg passes by the spermatheca, if the circumstances are such that a few spermatozoids empty out of the bag on the egg, the sex of it is changed from male to female.

It appears that there is in each egg a small opening (*micropyle*, *i* and *j*, fig. 24), through which the living spermatozoids enter, when the circumstances are such that a few of them can slip out the seminal bag and slide into the oviduct. Such is the process of impregnation.

140. Aristotle noticed, more than 2,000 years ago, that

the eggs which produce drones are like the worker-eggs.*
With the aid of powerful microscopes we are still unable to
detect any difference in the size or outside appearance of
the eggs of the queen.

141. These facts, taken in connection, constitute a perfect demonstration that unfecundated queens are not only
able to lay eggs, but that their eggs have sufficient vitality
to produce drones.

It seems to us probable, that after fecundation has been
delayed for about three weeks, the organs of the queen-bee
are in such a condition that it can no longer be effected;
just as the parts of a flower, after a certain time, wither
and shut up, and the plant becomes incapable of fructifica-
tion. Perhaps, after a certain time, the queen loses all desire to go in search of the male.

There is something analogous to these wonders in the
"*aphides*" or green lice, which infest plants. We have undoubted evidence that a fecundated female gives birth to
other females, and they in turn to others, all of which without impregnation are able to bring forth young; until, after
a number of generations, perfect males and females are produced, and the series starts anew!

However improbable it may appear that an unimpregnated
egg can give birth to a living being, or that sex can depend
on impregnation, we are not at liberty to reject facts because we cannot comprehend the reasons of them. He who
allows himself to be guilty of such folly, if he aims to be consistent, must eventually be plunged into the dreary gulf of
atheism. Common sense, philosophy, and religion alike
teach us to receive, with becoming reverence, all undoubted
facts, whether in the natural or spiritual world; assured

---

* Cheshire says that "worker-egg" is a misnomer, since all worker-eggs
are impregnated, and hence female-eggs. But the term is too intelligible and
popular, for us to change it; since Cheshire himself bows before custom, and
uses it.

that however mysterious they may appear to us, they are
beautifully consistent in the sight of Him whose " understanding is infinite."

142. It had long been known that the queen deposits
drone-eggs in the large or drone-cells, and worker-eggs in
the small or worker-cells, and that she usually makes no
mistakes. Dzierzon inferred, therefore, that there was some
way in which she was able to decide the sex of the egg before it was laid, and that she must have such a control over
the mouth of the seminal sac as to be able to extrude her
eggs, allowing them at will to receive or not a portion of its
fertilizing contents. In this way he thought she determined
their sex, according to the size of the cells in which she
laid them.

143. Our lamented friend, Mr. Samuel Wagner, had advanced a highly ingenious theory, which accounted for all
the facts, without admitting that the queen had any special
knowledge or will on the subject. He supposed that, when
she deposited her eggs in the worker-cells, her body was
slightly compressed by their size, thus causing the eggs as
they passed the spermatheca to receive its vivifying influence.

144. But this theory was overthrown by the fact that
the queen sometimes lays eggs in cells that are built only to
a third of their length, whether worker-cells or drone-cells,
and in which no compression can take place. Yet, it is
very difficult to admit that the queen is endowed with a
faculty that no other animal possesses, that of knowing and
deciding the sex of her progeny beforehand. It seems to
us that she must be guided by her instinct like all other
beings, for she always begins, in the Spring, by laying in
small cells, using large cells only when no others are in reach
in the warm part of the hive. Sometimes, however, when
she is very heavy with eggs, she lays in drone-cells as she
comes to them. Usually it is only when the hive is warm

throughout, and worker-cells all occupied, that she fills the unoccupied drone-cells. This has given rise to the popular theory that the bees raise drones whenever they intend to swarm. It is possible that the width of the cells and the position of her legs when laying in drone-cells (**224**) prevents the action of the muscles of her spermatheca.

145. The preference of the queen for worker-cells can not be disputed. If all the drone-combs are removed from a hive and replaced with worker-combs, she will not show any displeasure. She will live in that hive for years, without laying any drone-eggs, except, perhaps, here and there, in odd-shaped junction-cells. Mr. A. I. Root, of Medina, O., makes the same remark:

"By having a hive furnished entirely with worker-comb, we can so nearly prevent the production of drones, that it is safe enough to call it a complete remedy."—("A. B. C. of Bee Culture," page 134, Medina, 1883.)

146. If, on the other hand, we furnish a swarm with nothing but drone-comb, already built, they would soon leave the hive. But, if a few worker-cells are among the drone-cells, the queen will find them and will lay in them. On this question Mr. Root says:

147. "Bees sometimes rear worker-brood in drone-comb when compelled to from want of room, and they always do it by contracting the mouth of the cells, and leaving the young bee a rather large berth in which to grow and develop." ("A. B. C.," page 133.) "If you give a young laying queen a hive supplied only with drone-combs, she will rear worker-brood in these drone-cells. The mouth of the cells will be contracted with wax as mentioned before." (Page 188.)

148. An experiment, made in Bordeaux, under the supervision of Mr. Drory, editor of the "*Rucher*," has proven that the queen may lay worker-eggs in drone-cells. A piece of drone-comb containing worker-brood, was sent us by him. The eggs were laid irregularly and the mouth of the

cells had been contracted, as mentioned by Mr. Root. This
contraction of the cell mouth seems indispensable to enable
the queen to put in motion the muscles of her spermatheca.

149. We will add, with Mr. Root, that in the Spring, or
late in the Fall, when the crop is not abundant, the queen
will travel over drone-combs without depositing a single egg
in them.   Even by feeding the colony, when in these conditions, the queen cannot be readily induced to lay in
drone-cells.   Our conclusions on this point differ from those
of Mr. Root.   We think that the queen prefers worker-
cells to drone-cells, because the fecundation of the eggs by
the action of the muscles of the spermatheca probably gives
her a pleasant sensation, which she does not experience in
laying drone-eggs.

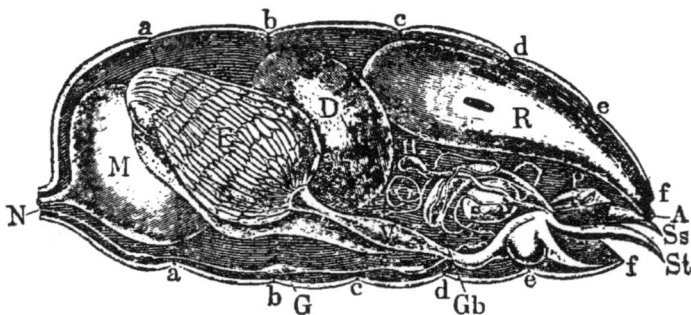

Fig. 22.

ABDOMEN OF THE QUEEN-BEE.

(Magnified.   From the *"Illustrierte Bienen-zeitung."*)

a, b, c, d, e, rings of the abdomen; *N*, nerve-chain; *M*, honey-sack; *E*,
ovaries; *D*, stomach; *R*, rectum; *G*, ganglions; *A*, anus; *Sx*, ovipositor;
St, sting; *P*, muscles; *H*, gland; *S*, poison-sack.

150. Some very prolific queens occasionally lay drone-
eggs in worker-cells.   It may be due to fatigue.   This will
readily be admitted when we consider the number of eggs
laid in one day. **(98.)**

151. Dzierzon found that a queen which had been *refrigerated* for a long time, after being brought to life by warmth,

laid only male eggs, whilst previously she had also laid female eggs. Berlepsch refrigerated three queens by placing them thirty-six hours in an ice-house, two of which never revived, and the third laid, as before, thousands of eggs, but *from all of them only males were evolved.* In two instances, Mr. Mahan has, at our suggestion, tried similar experiments, and with like results. A short exposure of a queen, to pounded ice and salt, answers every purpose. The spermatozoids are in some way rendered inoperative by severe cold.

152. The queen begins laying about two days after impregnation. She is seldom treated with much attention by the bees until after she has begun to replenish the cells with eggs; although if previously deprived of her, they show, by their despair, that they fully appreciated her importance to their welfare.

The extraordinary fertility of the queen-bee has already been noticed. The process of laying has been well described by the Rev. W. Dunbar, a Scotch Apiarist:

153. " When the queen is about to lay, she puts her head into a cell, and remains in that position for a second or two, to ascertain its fitness for the deposit she is about to make. She then withdraws her head, and curving her body downwards, inserts the lower part of it into the cell; in a few seconds she turns half round upon herself and withdraws, leaving an egg behind her."

In the Winter, or early Spring, she lays first in the middle of the cluster, and continues in a circle, around the first eggs laid, till she has filled the whole warm space. She then crosses over to the next comb and does the same thing; as the bees always cluster on different combs in groups exactly opposite, to produce the utmost possible concentration and economy of heat for developing the various changes of the brood.

154. Queens lay more or less according to, *1st*, The season; *2nd*, The number of bees that keep up the heat of the

brood-nest, and *3d,* The quantity of food which they eat.
When bees harvest honey or pollen, or when these necessa-
ries are provided artificially by the Apiarist, they feed the
queen as they pass by her, oftener than they would otherwise ; hence her laying increases in Spring, and decreases in
Summer or Fall.   It is certain that when the weather is un-
congenial, or the colony too feeble to maintain sufficient heat,
fewer eggs are matured, just as unfavorable circumstances
diminish the number of eggs laid by the hen ; and when the
weather is very cold, the queen stops laying, in weak colonies.

In the latitude of Northern Massachusetts, we have found
that the queen ordinarily ceases to lay some time in October ; and begins again, in strong stocks, in the latter part of
December.   On the 14th of January, 1857 (the previous
month having been very cold, the thermometer sometimes
sinking to 17° below zero), we examined three hives, and
found that the central combs in two contained eggs and unsealed brood ; there were a few cells with sealed brood in
the third.   Strong stocks, even in the coldest climates, usually contain some brood ten months in the year.

155. " Queens differ much as to the degree of their fertility.
Those are best which deposit their eggs with uniform regularity,
leaving no cells unsupplied—as the brood hatches at the same time
on the same range of comb, which can be again supplied ; the
queen thus losing no time in searching for empty cells."—(Dzierzon.)

In bee-life, as well as in human affairs, those who are
systematic, ordinarily accomplish the most.

To test the difference of fecundity between queens. Mr.
De Layens, while transferring bees (**574**), in middle April,
counted the eggs dropped on a black cloth (**577**), in forty
minutes, by the queens of four different colonies.   The
poorest queen dropped but one egg, the second twelve, the
third eighteen, and the fourth twenty.   On the fifteenth of

PLATE 6.

DZIERZON,

Discoverer of Parthogenesis in Queen-bees.

This writer is mentioned pages 51, 53, 55, 56, 58, 60, 62, 64, 66, 68, 72, 80,
120, 122, 137, 138, 142, 240, 284, 286, 287, 345, 346,
350, 359, 378, 514.

July the colony of the first queen was very poor, the second was of average strength, and both the others were very strong.

156. It is amusing to see how the supernumerary eggs of the queen are disposed of. If the workers are too few to take charge of all her eggs, or there is a deficiency of bee-bread to nourish the young; or if, for any reason, she does not judge best to deposit them in the cells, she stands upon a comb, and simply extrudes them from her oviduct, the workers devouring them as fast as they are laid.

One who carefully watches the habits of bees will often feel inclined to speak of his little favorites as having an intelligence almost if not quite akin to reason; and we have sometimes queried, whether the workers who are so fond of a tit-bit in the shape of a newly-laid egg, ever experienced a struggle between appetite and duty; so that they must practice self-denial to refrain from breakfasting on the eggs so temptingly deposited in the cells.

157. It is well known to breeders of poultry, that the fertility of a hen decreases with age, until at length she may become entirely barren. By the same law, the fecundity of the queen-bee ordinarily diminishes after she has entered her third year. An old queen sometimes ceases to lay worker-eggs; the contents of her spermatheca becoming exhausted, the eggs are no longer impregnated, and produce only drones.

The queen-bee usually dies of old age in her fourth year, although she has been known to live much longer. There is great advantage, therefore, in hives which allow her, when she has passed the period of her greatest fertility, to be easily removed.

5

The Worker-Bee.

158. The workers are the smallest inhabitants of a bee-hive, and compose the bulk of the population. A good swarm ought to contain at least 20,000; and in large hives, strong colonies which are not reduced by swarming, frequently number four or five times as many during the height of the breeding season.

Fig. 23.

159. Their functions are varied. *The young bees* work inside of the hive, prepare and distribute the food to the larvæ, take care of the queen, by brushing her with their tongue, nurse her, maintain the heat of the hive, or renew the air and evaporate the newly-gathered honey (**249**), by ventilating (**744**). They clean the hive of dirt or débris, close up all the cracks, and secrete the greater part of the wax which is produced in the hive.

The old bees may, if necessary, do a part of the same work; but, as we have seen, (**39**), old age renders many unfit to prepare the food of the larvæ. More alert than the young bees, they do the outside work, gather honey (**246**), pollen (**263**), and water (**271**), for the use of the family, and propolis (**236**) to cement the cracks.*

160. " Dzierzon states it as a fact, that worker-bees attend more exclusively to the domestic concerns of the colony in the early period of life; assuming the discharge of the more active

---

* Huber speaks of two kinds of workers: "One of these is, in general, destined for the elaboration of wax, and its size is considerably enlarged when full of honey; the other immediately imparts what it has collected, to its companions; its abdomen undergoes no sensible change, or it retains only the honey necessary for its own subsistence. The particular function of the bees of this kind is to take care of the young, for they are not charged with provisioning the hive. In opposition to the wax-workers, we shall call them small bees, or nurses."

"Although the external difference be inconsiderable, this is not an imaginary distinction. Anatomical observations prove that the stomach is not the

out-door duties only during the later periods of their existence. The Italian bees (**551**) furnished me with suitable means to test the correctness of this opinion.

"On the 18th of April, 1855, I introduced (**533**) an Italian queen into a colony of common bees; and on the 10th of May following, the first Italian workers emerged from the cells. On the ensuing day, they emerged in great numbers, as the colony had been kept in good condition by regular and plentiful feeding. I will arrange my observations under the following heads:

161. "*1.* On the 10th of May, the first Italian workers emerged; and on the 17th they made their first appearance outside of the hive. On the next day, and then daily till the 29th, they came forth about noon, disporting in front of the hive, in the rays of the sun. They, however, manifestly, did not issue for the purpose of gathering honey or pollen, for during that time none were noticed returning with pellets; none were seen alighting on any of the flowers in my garden; and I found no honey in the stomachs of such as I caught and killed for examination. The gathering was done exclusively by the old bees of the original stock, until the 29th of May, when the Italian bees began to labor in that vocation also—being then 19 days old.

162. "*2.* On the feeding troughs placed in my garden, and which were constantly crowded with common bees, I saw no Italian bees till the 27th of May, seventeen days after the first had emerged from the cells.

"From the 10th of May on, I daily presented to Italian bees, in the hive, a stick dipped in honey. The younger ones never attempted to lick any of it; the older occasionally seemed to sip a little, but immediately left it and moved away. The common bees always eagerly licked it up, never leaving it till they had filled their honey-bags. Not till the 25th of May did I see any Italian bee lick up honey eagerly, as the common bees did from the beginning.

"These repeated observations force me to conclude that, during

same: experiments have ascertained that one of the species cannot fulfill all the functions shared among the workers of a hive. We painted those of each class with different colors, in order to study their proceedings; and these were not interchanged. In another experiment, after supplying a hive, deprived of a queen, with brood and pollen, we saw the small bees quickly occupied in nutrition of the larvæ, while those of the wax-working class neglected them. Small bees also produce wax, but in a very inferior quantity to what is elaborated by the real wax-workers." The two kinds spoken of by Huber were bees at different stages of life.

the first two weeks of the worker-bee's life, the impulse for gathering honey and pollen does not exist, or at least is not developed: and that the development of this impulse proceeds slowly and gradually. At first the young bee will not even touch the honey presented to her ; some days later she will simply taste it, and only after a further lapse of time will she consume it eagerly. Two weeks elapse before she readily eats honey, and nearly three weeks pass, before the *gathering* impulse is sufficiently developed to impel her to fly abroad, and seek for honey and pollen among the flowers.

163. "I made, further, the following observations respecting the domestic employments of the young Italian bees :

" *1.* On the 20th of May, I took out of the hive all the combs it contained, and replaced them after examination. On inspecting them half an hour later, I was surprised to see that the edges of the combs, which had been cut on removal,* were covered by Italian bees exclusively. On closer examination, I found that they were busily engaged in re-attaching the combs to the sides of the hive. When I brushed them away, they instantly returned, in eager haste, to resume their labors.

" *2.* After making the foregoing observations, I inserted in the hive a bar from which a comb had been cut, to ascertain whether the rebuilding of comb would be undertaken by the Italian bees. I took it out a few hours subsequently, and found it covered almost exclusively by Italian workers, though the colony, at that time, still contained a large majority of common bees. I saw that they were sedulously engaged in building comb; and they prosecuted the work unremittingly, whilst I held the bar in my hand. I repeated this experiment several days in succession, and satisfied myself that the bees engaged in this work were always almost exclusively of the Italian race. Many of them had scales of wax visibly protruding between their abdominal rings (**201**). These observations show that, in the early stage of their existence, the impulse for comb-building is stronger than later in life.

164. "*3.* Whenever I examined the colony during the first three weeks after the Italian bees emerged, I found the brood-combs covered principally by bees of that race : and it is, hence, probable that the brood is chiefly attended to and nursed by the

---

* Mr. Donhoff, the writer of this quotation, used the Dzierzon hive, the combs of which are suspended in the hive by an upper bar only, and cannot be taken out unless their edges, that are built against the sides of the hive, are cut.

younger bees. The evidence, however, is not so conclusive as
in the case of comb-building, inasmuch as they *may* have con-
gregated on the brood-combs because these are warmer than the
others.

" I may add another interesting observation. The fæces in the
intestines of the young Italian bees was viscid and yellow; that
of the common or old bees was thin and limpid, like that of the
queen-bee. This is confirmatory of the opinion, that, for the
production of wax and jelly, the bees require pollen ; but do
not need any for their own sustenance."—(*B. Z.*, 1855, p. 163.
Dr. Donhoff, translated by the late S. Wagner.)

165. There are none but *gentlemen* of leisure in the com-
monwealth of bees, but assuredly there are no such *ladies*,
whether of high or low degree. The queen
herself has her full share of duties, the
royal office being no sinecure, when the
mother who fills it must daily superintend
the proper deposition of thousands of
eggs.

Fig. 24.
THE EGG IN THE CELL.
(Magnified.)

" The eggs of bees are of a lengthened, oval
shape with a slight curvature, and of a bluish
white color: being besmeared, at the time of
laying, with a glutinous substance, they ad-
here to the bases of the cells, and remain unchanged in figure or
situation for three or four days; they are then hatched, the bottom of each cell presenting to view a small white worm."—
(Bevan.)

166. For the first three days after their hatching, these
worms are fed with a jelly, thought to be prepared or secreted by the upper pair of glands of the worker-bees (**39**),
which are very large in the nurses. This milky food is a
whitish, transparent fluid, and is distributed to the larvæ,
as it is needed. After four or perhaps five days, the larva
is too large for the bottom of the cell, where it was coiled
up, to use the language of Swammerdam, like a dog when
going to sleep ; and stretches itself till it occupies the whole

length of the cell, lying on its back. Its food at this time, is different from that first given.

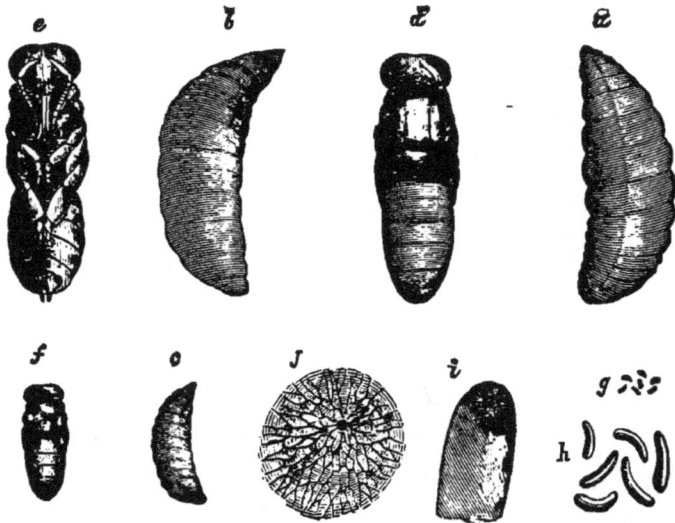

Fig. 25.
BROOD IN ALL STAGES.
(From Girard.)

a, *b*, magnified larvæ; *c*, the same, natural size; *d*, *e*, magnified nymph; *f*, natural size; *g*, eggs, natural size; *h*, magnified; *i*, egg, showing micropyle, magnified; *j*, micropyle, magnified.

"The mixture of honey and pollen given at the end of the nursing, is easily detected by its color, which is yellower, on account of the pollen, and can be seen through the skin of the larva."—(Dubini.)

167. "The larva, or grub, grows apace, but not without experiencing a difficulty to which the human family is, in some sort, subject in the period of youth. Its coat is in-

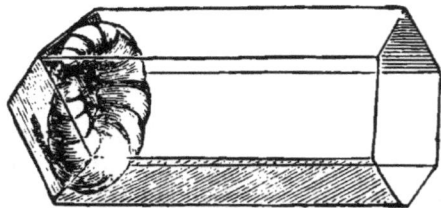

Fig. 26.
COILED IN THE CELL.
(Magnified. From Sartori and Rauschenfels.)

elastic and does not grow with the wearer, so that it soon, fitting badly, has to be thrown off; but, happily in the case of the larva, a new and larger one has already been formed beneath it, and the discarded garment, more delicate than gossamer, is pushed to the bottom of the cell."—(Cheshire.)

168. "The nursing-bees now seal over the cell with a light brown cover, externally more or less convex (the cap of a drone-cell being more convex than that of a worker), and thus differing from that of a honey-cell, which is paler and somewhat concave."—("Bevan on the Honey-Bee.")

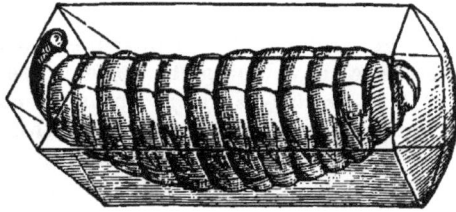

Fig. 27.
STRETCHED IN THE CELL.
(Magnified.)

The cap of the brood-cell is made not of pure wax, but of a mixture of bee-bread and wax; and appears under the microscope to be full of fine holes, to give air to the inclosed insect. From its texture and shape it is easily thrust off by the bee when mature, whereas if it consisted wholly of wax, the insect would either perish for lack of air, or be unable to force its way into the world. Both the material and shape of the lids which close the honey-cells are different: they are of pure wax, and are slightly concave, the better to resist the pressure of their contents. The bees sometimes neglect to cap the cells of some of the brood, and some persons have thought that this brood was diseased, but it hatches all the same. The larva is no sooner

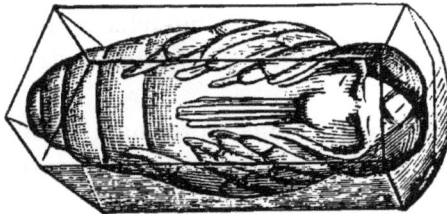

Fig. 28.
THE TRANSFORMATION IN THE SEALED CELL.
(Magnified.)

perfectly inclosed, than it begins to spin a cocoon after the manner of the silk-worm, and Cheshire teaches us that it does not encase the insect, but is only at the mouth of the cell, "and in no case extends far down the sides."

To return to Bevan:

169. "When it has undergone this change, it has usually borne the name of *nymph*, or *pupa*. It has now attained its full growth, and the large amount of nutriment which it has taken serves as a store for developing the perfect insect.

"The *working-bee nymph* spins its cocoon in thirty-six hours. After passing about three days in this state of preparation for a new existence, it gradually undergoes so great a change as not to wear a vestige of its previous form."

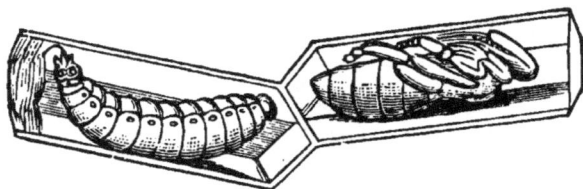

Fig, 29.

SPINNING OF THE COCOON AND TRANSFORMATION INTO NYMPH.
(Magnified. From Sartori and Rauschenfels.)

170. The last cast-off skin of the larva, "which, by the creature's movements within the cell, becomes plastered to the walls and joins the cocoon near the mouth end" (Cheshire), is left behind, and forms a closely-attached and exact lining to the cell; by this means the breeding-cells become smaller, and their partitions stronger, the oftener they change their tenants.

So thin is this lining, that brood combs more than twenty years old have been found to raise bees as large apparently as any other in the Apiary.

171. About twenty-one days are usually required for the transformations from the worker-egg to the perfect insect. But the time may be shortened or lengthened by the temperature, or the conditions of the colony. Dzierzon and others wrote that a worker-bee can hatch in nineteen to twenty-one days. Collin says nineteen to twenty-three. That the brood can remain even longer before hatching, is confirmed by the report of A. Saunier, in the South of France. Having deprived a hive of all its inhabitants, he found bees, hatching twenty-three days afterwards, that had not even been sealed in their cells, since there had been no nurses there to do this work. ("L'Apiculteur." Paris, 1870.) As these were already full-grown larvæ, when the hive was deprived of its bees, they must have been twenty-seven days old when hatching. In this experiment, the heat produced by the larvæ, coupled with that of the atmosphere, had been sufficient to keep them alive and help their slow development.

We have often noticed the brood of swarms, that had deserted their hives, still alive after a cold night, but in each case its development was delayed.

172. A newly hatched worker, like a newly hatched queen, is easily recognized by her small size, her pale gray color, and her weak appearance. After a few days, she has grown considerably larger. She is then in the bloom of health; her color is bright, she has not yet lost a single hair of the down which covers her body. These hairs fall gradually from age and work, and sometimes disappear almost entirely.

173. The first excursion of the young bee out of the hive takes place when she is about eight days old. (See Donhoff's experiment **160.**) The disturbing of the colony, or the lack of old bees may cause them to go out earlier.

The first flight of young worker-bees is easily remembered when once seen. It usually takes place in the afternoon of a sunny day. They first walk about on the platform in a hesitating manner and then take flight. Their humming, and joyous and peaceable circles to reconnoitre the location of their home, recalls to memory the gay playing of children in front of the school-house door. Their second trip is made about a week after the first; it is then that they bring in their first load. A young bee coming home is readily

recognized by the small size of the pellets of pollen she
carries, when compared with those of older bees, and by
the turns she makes before alighting,

174. The Apiarist should become acquainted with the
behavior of young bees, so as not to mistake their pleasant
flight for the disorderly and restless motions of robber-bees.
(**664.**)

175. Although the workers are of the same sex as the
queens (**4**), their sexual organs are undeveloped, owing to
the coarser food which they receive during the latter part of
their growth in the cell (**108**).   Yet they have rudimental
ovaries, containing a few undeveloped eggs (fig. 30). They
are incapable of fecundation.

176. Occasionally, some of them appear to be sufficiently developed to be capable of laying eggs ; but these eggs,
like those of queens whose impregnation has been retarded,
always produce drones.   Drone-laying, by worker-bees,
only takes place when a colony which has lost its queen
despairs of obtaining another, and seems to be caused by
their eagerness to raise brood.   Huber thought that fertile
workers were reared in the neighborhood of the young
queens, and that they received some particles of the peculiar food or jelly on which these queens are fed.*   But
microscopic examinations seem to prove that a large number of the workers, raised during a good honey harvest,
are capable of laying eggs.   The number of drone-laying

*An extract from Huber's preface will be interesting in this connection. After
speaking of his blindness, and praising the extraordinary taste for Natural History, of his assistant, Burnens, "who was born with the talents of an observer, " he says: " Every one of the facts.I now publish, we have seen, over
and over again, during the period of eight years, which we have employed in
making our observations on bees. It is impossible to form a just idea of the
patience and skill with which Burnens has carried out the experiments which
I am about to describe; he has often watched some of the working-bees of our
hives, which we had reason to think fertile, for the space of twenty-four hours,
without distraction * * * * and he counted fatigue and pain as nothing, compared with the great desire he felt to know the results."

workers is sometimes very large in a hopelessly queenless hive; we have seen at least a dozen laying on the same comb. Mr. Viallon, a noted bee-keeper of Louisiana, once had so many in one queenless colony, that he was able to send several dozen for dissection to bee-keepers in this country and Europe.

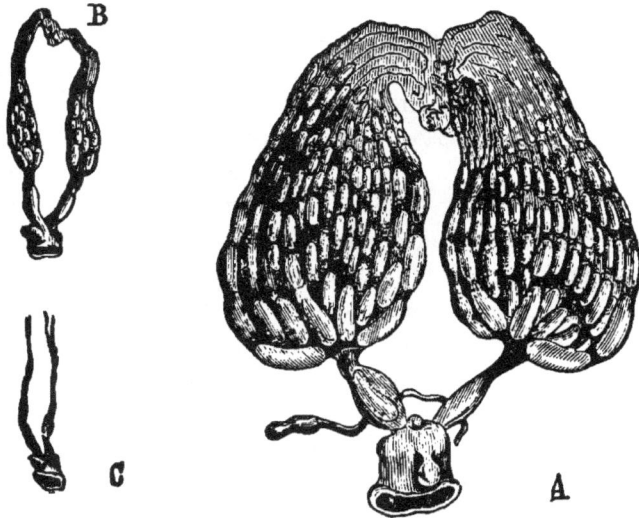

(Fig. 50.)

COMPARATIVE SIZES OF THE OVARIES OF QUEEN AND WORKER.
(All magnified. From Girard.)
A, queen ovaries; B, laying-worker ovaries; C, sterile-worker ovaries.

177. Some persons may question the wisdom of Nature in endowing the workers with the means of laying drone-eggs, when there is no queen in the colony to be fecundated by them. But Nature does nothing without purpose. The main cause of the loss of the queen, when there is no brood fit to raise others (**107**), and therefore, no hopes of survival for the colony, is usually the death of the young queen in her bridal flight (**122**). At some seasons, the drones are scarce, and a young queen may be compelled to make several trips before she finds one. If she gets lost, the hive

having remained queenless for at least eight or ten days
(**109**), the brood is too old to be used to raise another,
and the colony is doomed. That other colonies may not be
victims of similar accidents, owing to the scarcity of
drones, Nature endows this worthless colony with the faculty of drone-raising.

It is by the same provision of Nature that unhealthy
trees, on the eve of death, are seen covered with blossoms
and fruits. They make the strongest efforts to save their
race from extinction, and perish afterwards.

178. The drone-laying of worker-bees is easily discovered by the Apiarist. Their eggs are laid without order,
some cells containing grown larvæ, or sealed pupæ, by the
side of cells containing eggs; while the eggs of a queen are
very regularly laid. Huber states that the fertile workers
prefer large cells in which to deposit their drone eggs, re-
sorting to small ones, only when unable to find those of
greater diameter. A hive in our Apiary having much
worker-comb, but only a small piece of drone size, a fertile
worker filled the latter so entirely with eggs that some of
the cells contained three or four each.

179. Sometimes the bees do not seem to know that these
eggs are drone-eggs, and in their eagerness to raise a queen,
they treat some of them as such, by enlarging their cells
and feeding them on special food (**109**). The poor over-
fed drones, thus raised, usually perish in the cell (**136**).
The workers soon dwindle away, and the colony perishes.

180. They often even fail to raise any queen from brood,
which may be given them by the Apiarist, unless some
hatching bees are given at the same time. The latter, when
informed of the needs of the colony, usually succeed in
raising a queen. The introduction of a laying-queen in a
laying-worker colony, is the best remedy. (**533**.)

181. The bees of the same colony understand each other
very well for all their necessities, and they work with an

entrain which is truly admirable. They know each other, probably by smell, for it is very rare to see a bee of the hive treated as a robber (**664**). They never use their sting except to defend themselves, when hurt, or their home, when they think it is threatened.

182. Their life is short, but their age depends very much upon their greater or less exposure to injurious influences, and severe labors. Those reared in the Spring and early part of Summer, upon whom the heaviest labors of the hive devolve, appear to live not more than thirty-five days, on an average; while those bred at the close of Summer, and early in Autumn, being able to spend a large part of their time in repose, attain a much greater age. It is very evident that "the bee" (to use the words of a quaint old writer) "is a Summer bird;" and that, with the exception of the queen, none live to be a year old.

If an Italian queen be given, in the working season, to a swarm of common bees, in about three months none of the latter will be found in the colony, and as the black queen removed has left eggs in the cells, which take twenty-one days to hatch, it is evident that the bees all die from fatigue or accident in the remaining seventy days, making their average life thirty-five days *in the working season.*

The age which individual members of the community may attain, must not be confounded with that of the colony. Bees have been known to occupy the same domicile for a great number of years. We have seen flourishing colonies more than twenty years old; the Abbé Della Rocca speaks of some over forty years old; and Stoche says, that he saw a colony, which he was assured had swarmed annually for forty-six years! "Such cases have led to the erroneous opinion, that bees are a long-lived race. But this, as Dr. Evans* has observed, is just as wise as if a stranger, con-

---

* Dr. Evans was an English physician, and the author of a beautiful poem on bees.

templating a populous city, and personally unacquainted
with its inhabitants, should, on paying it a second visit,
many years after, and finding it equally populous, imagine
that it was peopled by the same individuals, not one of whom
might then be living.

> 'Like leaves on trees, the race of bees is found,
> Now green in youth, now withering on the ground,
> Another race the Spring or Fall supplies,
> They droop successive, and successive rise.' "
>
> EVANS.

Apiarists, unaware of the brevity of the bee's life, have
often constructed huge "bee-palaces" and large closets, vainly imagining that the bees would fill them, being unable to
see any reason why a colony should not increase until it
numbers its inhabitants by millions or billions.   But as the
bees can never at one time equal, still less exceed, the number which the queen is capable of producing in a season,
these spacious dwellings have always an abundance of spare
room.   It seems strange that men can be thus deceived,
when often in their own Apiary they have healthy stocks,
which, though they have not swarmed for a year or more,
are no more populous in the Spring, than those which have
regularly parted with vigorous colonies.

It is certain that the Creator has wisely set a limit to the
increase of numbers in a single colony ; and we shall venture
to assign a reason for this.   Suppose he had given to the
bee a length of life as great as the horse or the cow, and had
made each queen capable of laying daily some hundreds of
thousands of eggs ; or had given several hundred queens to
each hive ; then a colony must have gone on increasing, until it became a scourge rather than a benefit to man.   In the
warm climates of which the bee is a native, it would have
established itself in some cavern or capacious cleft in the
rocks, and would soon have become so powerful as to bid defiance to all attempts to appropriate the avails of its labors.

183. There is something cruel in the habits of the bee. Whenever one of them becomes unable to work from some cause or other, if she does not perish in her efforts to go to the fields, the other bees drag her out pitilessly; their love being concentrated on the whole family, not on a single individual. Yet, when one is hurt, and complains, hundreds of others resent the injury and are ready to avenge her.

184. Notched and ragged wings and shiny bodies, instead of gray hairs and wrinkled faces, are the signs of old age in the bee, indicating that its season of toil will soon be over. They appear to die rather suddenly; and often spend their last days, and even their last hours, in useful labors.

Place yourself before a hive, and see the indefatigable energy of these industrious veterans, toiling along with their heavy burdens, side by side with their more youthful compeers, and then judge if, while qualified for useful labor, you ought ever to surrender yourself to slothful indulgence.

Let the cheerful hum of their busy old age inspire you with better resolutions, and teach you how much nobler it is to die with harness on, in the active discharge of the duties of life.

THE DRONES.

185. The drones are the male bees. They are much larger and stouter than either the queen or workers; although

Fig. 31.

their bodies are not quite so long as that of the queen. They have no sting (**78**) with which to defend themselves; and no suitable proboscis (**48**) for gathering honey from the flowers; no baskets on their thighs (**59**) for holding bee-bread, and no pouches (**201**) on their abdomens for secreting wax. They are, therefore, physically disqualified for the ordinary work of the hive.

Their proper office is to impregnate the young queens.

> " Their short proboscis sips
> No luscious nectar from the wild thyme's lips,
> From the lime's leaf no amber drops they steal,
> Nor bear their grooveless thighs the foodful meal:
> On other's toils in pamper'd leisure thrive
> The lazy fathers of the industrious hive."
>
> EVANS.

186. The drones begin to make their appearance in April or May; earlier or later, according to the forwardness of the season, and the strength of the colony. Like the other inhabitants of the hive they cannot perform the work for which they are intended, till at least one week old. They go out of the hives only when the weather is warm, and at mid-day.

187. As we have seen (**122**), the mating of the queen with a drone always takes place in the air. Physiologists say that it cannot be otherwise, because the sexual organs of the drone cannot be extruded unless his abdomen is swelled by the filling of all the tracheæ with air. This happens only in swift flight (**74**).

Dzierzon supposes that the sound of the queen's wings, when she is in the air, excites the drones. Evidently their eyes (**11**) and ears (**25**) which are highly developed, as proven by Cheshire, help them also in the search of the queen, which is their sole occupation, when in the field. In the interior of the hive, they are never seen to notice her; so that she is not molested, even if thousands are members of the same colony with herself. But outside of the hive, they readily follow her, led, according to Dzierzon, by the peculiar hum of her flight, and certainly also, by the senses of smell and of sight, which are more perfect than those of the worker, most likely for this single purpose.

" When the queen flies abroad, the fleetest drone is more likely to succeed in his addresses than another, and thus he impresses upon posterity some part of his own superior activity and en-

PLATE 7.

F. R. CHESHIRE, F. L. S., F. R. M. S.

Author of " *Bees and Bee-keeping.*"

This writer is mentioned pages 2, 3, 4, 5, 7, 9, 10, 11, 12, 15, 16, 17, 20, 22, 23, 24, 27, 30, 37, 59, 71, 72, 80, 81, 89, 99, 118, 119, 124, 142, 186, 338, 341, 376, 415, 445, 446, 447, 450, 453, 454.

ergy. The slow and weak in the race die without heirs, so that the survival of the fittest is not an accident, but a predetermination. In previous chapters we have considered his highly-developed eyes, meeting at the vertex of his head, his multitudinous smell-hollows, and his strong large wings, the advantage of which now appears in a clearer light; his quickness in discovering a mate, whose neighborhood is to him filled with irresistible odours, and his ability in keeping her in view during pursuit, are no less helpful to his purpose than fleetness on the wing......."
—(Cheshire.)

188. The drone perishes in the act of impregnating the queen. Although, when cut into two pieces, each piece will retain its vitality for a long time, we accidentally ascertained, in the Summer of 1852, that if his abdomen is gently pressed, and sometimes if several are closely held in the warm hand, the male organ will often be permanently extruded, with a motion very like the popping of roasted popcorn ; and the insect, with a shiver, will curl up and die, as quickly as if blasted with the lightning's stroke. This singular provision is unquestionably intended to give additional security to the queen when she leaves her hive to have intercourse with the drone. Huber first discovered that she returned with the male organ torn from the drone, and still adhering to her body. If it were not for this arrangement, her *spermatheca* could not be filled, unless she remained so long in the air with the drone, as to incur a very great risk of being devoured by birds. In one instance, some days after the impregnation of a queen, we found the male organ,*

* We give, as interesting in this connection, the following extract from Mr. Langstroth's journal: ''*August 25th*, 1852.—Found the male organ protruding from a young queen; could not remove it without exerting so much force that I feared it would kill her. Dr. Joseph Leidy examined this queen-bee with the microscope, so as to demonstrate that—to use his words—'it was the penis and its appendages of a male, corresponding in all its anatomical peculiarities, with the same organs examined, at the same time, in other drones. The testicles and *vasa deferentia* of these drones were found to be full of the spermatic fluid. The *spermatheca* of the queen was distended with the same semi-fluid, spermatic matter.' This one examination *demonstrates* that the drones are males, and that they impregnate the queen by actual coition.''

6

in a dried state, adhering so firmly to her body, that it
could not be removed without tearing her to pieces.

Fig. 32.
THE ORGANS OF THE DRONE.
(Magnified. From Girard.)
a,a, testicles; *b,b*, mucous glands; *c*, seminal duct; *d*, part in which the
spermatophore is formed; *e*, hollow horns and penis; *f*, spermatozoids,
much magnified.

189. The number of drones in a hive is often very great,
amounting not merely to hundreds, but sometimes to thousands.   As a single one will impregnate a queen for life, it
would seem that only a few should be reared.   But as sexual intercourse always takes place high up in the air, the
young queens must necessarily leave the hive; and it is
very important to their safety that they should be sure to
find a drone without being compelled to make frequent excursions; for being larger than workers, and less active on
the wing, queens are more exposed to be caught by birds,
or destroyed by sudden gusts of wind.

In a large Apiary, a few drones in each hive, or the number usually found in one, would suffice. Under such circumstances bees are not in a state of nature, like a colony living in a forest, which often has no neighbors for miles.

A good stock, even in our climate, sometimes sends out three or more swarms, and in the tropical climates, of which the bee is probably a native, they increase with astonishing rapidity.* Every new swarm, except the first, is led off by a young queen; and as she is never impregnated until she has been established as the head of a separate family, it is important that each should be accompanied by a goodly number of drones: this requires the production of a large number in the parent-hive.

190. This necessity no longer exists when the bee is domesticated, since several colonies are kept in the same place, and the breeding of so many drones should be discouraged. Their brood takes useful space that might as well be occupied with worker-brood. One thousand good-for-nothing drones take up as much breeding-space as fifteen hundred workers (**224**), and require as much food, with negative results. Some hives, in a state of nature, produce so many drones that a great part of the surplus crop is disposed of by these voracious loafers. Besides, the comparatively large volume of the male organs, in connection with the gluttony of the drones, explains why they void their dejections in the hive, while workers retain them till they are on the wing (**73**), and why the cells of the combs of hives which have a large quantity of these gormands, become dark and thick sooner.

The importance of preventing the over-production of drones has been corroborated by the discovery of Mr. P. J. Mahan, that those *leaving* the hive have quite a large drop of honey in their stomachs — while those *returning* from

* At Sydney, in Australia, a single colony is stated to have multiplied to 300, in three years.

their pleasure excursions, having digested their dinners, are prepared for a new supply.

Aristotle (" History of Animals," Book IX, Chap. XI) speaks of the *irregular* and *thick* combs built by some colonies, and the superabundance of drones issuing from them. He describes their excursions as follows:

" The drones, when they go abroad, rise into the air with a circular flight, as though to take violent exercise, and when they have taken enough, return home, and gorge themselves with honey."

" The drone," says quaint old Butler (1609) " is a gross, stingless bee, that spendeth his time in gluttony and idleness. For howsoever he brave it with his round velvet cap, his side gown, his full paunch, and his loud voice, yet is he but an idle companion, living by the sweat of others' brows. He worketh not at all, either at home or abroad, and yet spendeth as much as two laborers: you shall never find his maw without a drop of the purest nectar. In the heat of the day he flieth abroad, aloft and about, and that with no small noise, as though he would do some great act; but it is only for his pleasure, and to get him a stomach, and then returns he presently to his cheer."

191. The bee-keepers in Aristotle's time were in the habit of destroying the excess of drones. They excluded them from the hive — when t a k i n g their accustomed airing — by contracting the entrances with a kind of basket work. Butler recommends a similar trap, which he calls a " *drone-pot*."

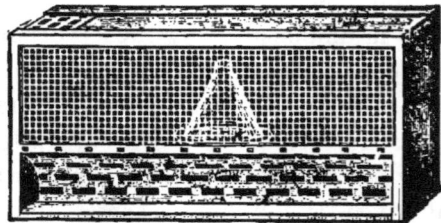

Fig. 33.

ALLEY'S DRONE-TRAP

One of the modern inventions to destroy them is Alley's drone-trap* improved by J. A. Batchelder; but it is much

* The perforated zinc, used in drone-traps, which we think was invented by Collin, ("Guide," p. 3. Paris, 1865), is so cut, that a worker-bee can pass through its openings, but neither queen nor drone can pass through them.

better to save the bees the labor and expense of rearing such a host of useless consumers. This can readily be done, when we have the control of the combs; for, by removing the drone-comb, and supplying its place with worker-cells, the over-production of drones may be easily prevented. Those who object to this, as interfering with nature, should remember that the bee is not in a state of nature; and that the same objection might, with equal force, be urged against killing off the supernumerary males of our domestic animals.

192. Soon after the harvest is over, or if there is a lull in the yield of honey, the drones are expelled from the hive. The worker-bees sting them, or gnaw the roots of their wings, so that when driven from the hive, they cannot return. If not ejected in either of these summary ways, they are so persecuted and starved, that they soon perish. At such times they often retreat from the comb, and keep by themselves upon the sides or bottom-board of the hive. The hatred of the bees extends even to the unhatched young, which are mercilessly pulled from the cells and destroyed with the rest.

Healthy colonies almost always destroy the drones, as soon as forage becomes scarce. In the vicinity of Philadelphia, there were only a few days in June, 1858, when it did not rain, and in that month the drones were destroyed in most of the hives. When the weather became more propitious, others were bred to take their place. In seasons when the honey-harvest has been abundant and long protracted, we have known the drones to be retained, in Northern Massachusetts, until the 1st of November. If bees could gather honey and could swarm the whole year, the drones would probably die a natural death.

How wonderful that instinct which, when there is no longer any occasion for their services, impels the bees to destroy those members of the colony reared with such devoted attention!

193. It is interesting to notice the actions of the drones when they are excluded from the hive. For a while they eagerly search for a wider entrance, or strive to force their bulky bodies through the narrow gateway. Finding this to be in vain, they solicit honey from the workers, and when refreshed, renew their efforts for admission, expressing, all the while, with plaintive notes, their deep sense of such a cruel exclusion. The bee-keeper, however, is deaf to their entreaties; it is better for him that they should stay without, and better for them—if they only knew it—to perish by his hands, than to be starved or butchered by the unfeeling workers. Towards dark, or early in the morning—when clustered, for warmth, in the portico—they may be brushed into a vessel of water, and given to chickens, which will soon learn to devour them.

194. Drones are sometimes raised in worker-cells (**150**). They are smaller in size, but apparently as perfect as the full-size drones, all their organs being well developed.

For the stages of development of drones, see the comparative table at the end of this chapter (**197**).

195. We have repeatedly queried, why impregnation might not have taken place *in the hive*, instead of in the open air. A few dozen drones would then have sufficed for the wants of any colony, even if it swarmed, as in warm climates, half a dozen times, or oftener, in the same season; and the young queens would have incurred no risks by leaving the hive for fecundation.

For a long time we could not perceive the wisdom of the existing arrangement; although we never doubted that there was a satisfactory reason for this seeming imperfection. To have supposed otherwise, would have been entirely unphilosophical, when we know that with the increase of knowledge many mysteries in nature, once inexplicable, have been fully cleared up.

The disposition cherished by many students of nature, to

reject some of the doctrines of revealed religion, is not prompted by a true philosophy. Neither our ignorance of all the facts necessary to their full elucidation, nor our inability to harmonize these facts in their mutual relations and dependencies, will justify us in rejecting any truth which God hath seen fit to reveal, either in the book of Nature, or in His holy Word. The man who would substitute his own speculations for the divine teachings, has embarked without rudder or chart, pilot or compass, on an uncertain ocean of theory and conjecture; unless he turns his prow from its fatal course, storms and whirlwinds will thicken in gloom on his "voyage of life;" no "Sun of Righteousness" will ever brighten for him the expanse of dreary waters; no favoring gales will waft his shattered bark to a peaceful haven.

The thoughtful reader will require no apology for this moralizing strain, nor blame a clergyman, if sometimes forgetting to speak as the mere naturalist, he endeavors to find

" Tongues in trees, books in running brooks,
" Sermons in ' bees,' and ' God ' in every thing."
(102.)

196. To return to the attempt to account for the existence of so many drones. If a farmer persists in what is called "breeding in and in," that is, without changing the blood, the ultimate degeneracy of his stock is the consequence.* This law extends, as far as we know, to all animal life, man himself not being exempt from its influence. Have we any reason to suppose that the bee is an exception? or that degeneracy would not ensue, unless some provision were made to counteract the tendency to "in-and-in breeding?" If fecundation had taken place in the hive, the queen would have been impregnated by drones from a com-

---

* In the above, Mr. Langstroth refers to indiscriminate breeding. In-and-in breeding, by selection, intensifies certain qualities, such as the development of fat, or of muscle, but it also intensifies the defects, generally causing a decrease of vitality or of health in the race.

mon parent; and the same result must have taken place in each successive generation, until the whole species would eventually have "run out." By the present arrangement, the young queens, when they leave the hive, often find the air swarming with drones, many of which belong to other colonies, and thus, by crossing the breed, provision is constantly made to prevent deterioration.

Experience has proved that impregnation may be effected not only when there are no drones in the colony of the young queen, but even when there are none in her immediate neighborhood. Intercourse takes place very high in the air (perhaps that less risk may be incurred from birds), and this favors the crossing of stocks.

197. "COMPARATIVE TABLE OF THE NORMAL DURATION OF THE BEE'S TRANSFORMATIONS FROM EGGS TO WINGED INSECTS.

		Queen.	Worker.	Drone.
Eggs . . . . days .	. 3	3	3
Growth of larva . . . . . "	. 5½	6	6½
Spinning of cocoon . . . . "	. 1	2	1½
Period of rest . . . . . "	. 2	2	3
Metamorphosis into pupa . . "	. 1	1	1
Duration of this stage . . . "	. 3½	7	9
Average time from egg to winged insect	16	21	24

PLATE 8.

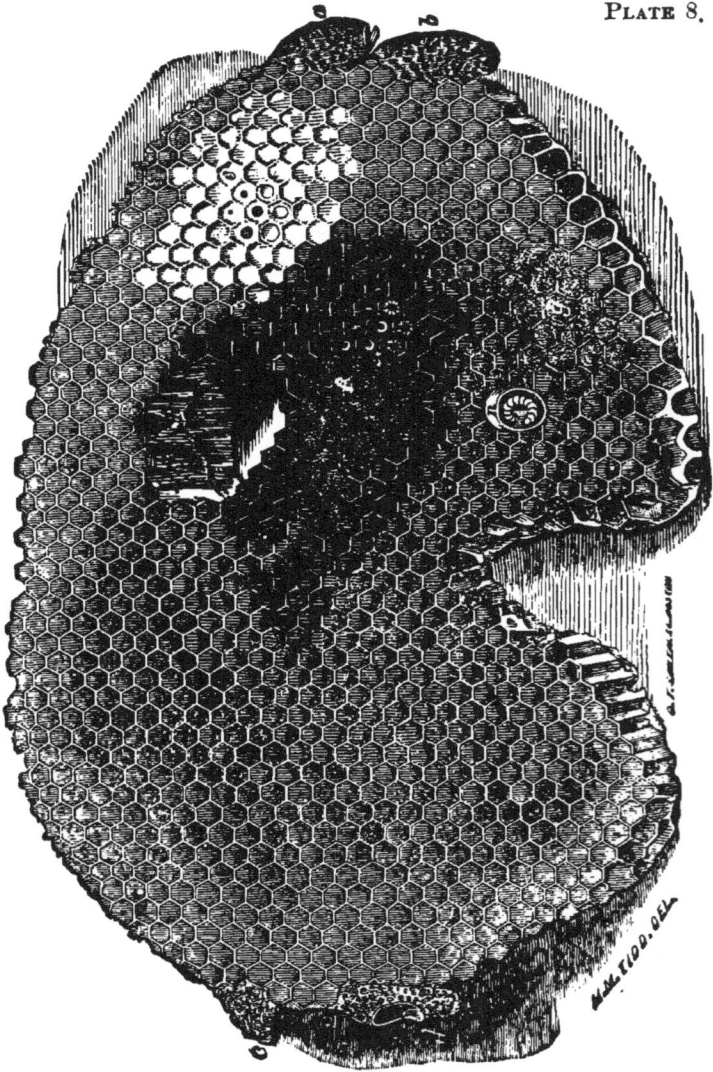

HONEY-COMB AND HOOD.

A little under natural size.

a, queen-cell with lid, as it often appears after the queen has hatched out of it; *b*, unhatched queen-cell; *c*, remains of a queen-cell; *d*, queen-cell whose inmate has been destroyed; *n*, queen-cell in the midst of a comb; *e*, cells containing honey; *f*, cells with worker brood; *g*, cells with drone brood.

CHAPTER II.

THE BUILDING OF BEES.—COMB.

198. When a swarm (**406**) has found a suitable habitation, some of the bees clean it of its rubbish, if necessary, while others, at once, prepare to build the furniture, which is intended as cradles for the young bees, and as a store-room for the provisions, and is called comb.

According to Webster, this word is probably taken from the Anglo-Saxon " comb," which means a hollow; the combs being hollow structures, with exceedingly light walls.

199. The combs are usually begun at the highest point of the hive and built downwards, yet, when some breaking happens, the bees sometimes build them upwards; but they are far from having the usual regularity. Combs are made of wax, a natural secretion which is produced by bees as cattle produce fat, by eating.

200. " Wax is not chemically a fat or glyceride, yet it is nearly allied to the fats in atomic constitution, and the physiological conditions favouring the formation of one are curiously similar to those aiding in the production of the other. We put our poultry up to fat in confinement, with partial light, to secure bodily inactivity, we keep warm and feed highly. Our bees, under Nature's teaching, put themselves up to yield wax under conditions so parallel, that the suitability of the fatting coops is vindicated."—(Cheshire.)

Fig. 34.
WAX SCALES.
(Magnified.)

201. If they remain quietly clustered together, when

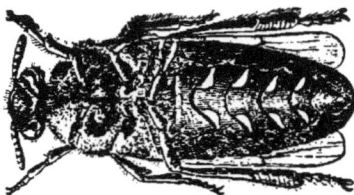

Fig. 35.
SECRETION OF WAX SCALES.
(Magnified.)
(From the "*Illustrierte Bienenzeitung.*")

gorged with honey, or any liquid sweet, the wax is secreted in the shape of delicate scales in four small pouches, on each side of the abdomen of worker-bees.

"These scales, of an irregular pentagonal shape, are so thin and light, that one hundred of them hardly weigh as much as a kernel of wheat."—(Dubini, "L'Ape.")

202. In the young bees, which are endowed with a great appetite, they form, probably, without their knowledge, during the honey season; and if there is no place to use them, they are gathered in small knots here and there. This

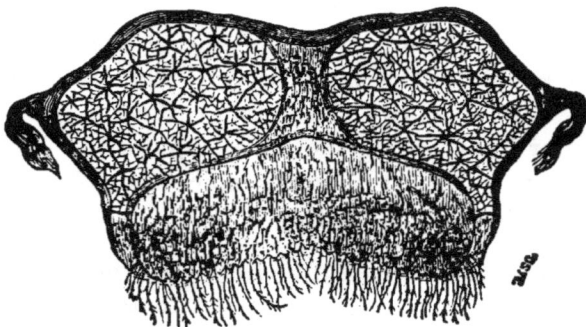

Fig. 36.
THE WAX-PRODUCING ORGAN OF THE WORKER-BEE.
(Magnified. From Girard.)

only happens when the combs are entirely filled and sealed. It has been noticed, most especially, in hives in which a comb had been broken down by heat. (**333.**) In such cases, many of the bees gorge themselves with the wasting honey, and cluster on the outside, until the heat has subsided, and the running honey has been gathered up.

Scales of wax, in lumps, can then be found where they have clustered.

203. Although the faculty of producing wax is diminished in old bees, who are subject to the natural law which makes it more difficult to fatten an old animal, it is proved that they also produce small scales of wax.

" During the active storing of the past season, especially when comb building was in rapid progress, 1 found that nearly every bee taken from the flowers contained wax scales of varying sizes in the wax-pockets."—(A. J. Cook.)

204. The first condition indispensable for bees to produce wax, is to have the stomach well filled.

It is an interesting fact that honey-gathering and comb-building go on simultaneously ; so that when one stops, the other ceases also. As soon as the honey harvest begins to fail, so that consumption is in advance of production, the bees cease to build new comb, even though large portions of their hive are unfilled. When honey no longer abounds in the fields, it is wisely ordered that they should not consume, in comb-building, the treasures which may be needed for Winter use. What safer rule could have been given them ?

It takes about twenty-four hours, for a bee's food to become transformed into wax.

205. " Having filled themselves with honey, they gather in chains; not in a single group, but in a number of groups, hanging in a parallel curtain, in the direction of the comb to be constructed. Thus a bee clings to the ceiling with her claws, or the sticky rubber of her feet, her posterior limbs hanging down; another bee grapples the claws of these posterior feet, with the claws of her anterior limbs, letting her hind limbs hang also, to be grappled by a third, and so on, till the first chain meets another, and both united form an arch, top downward. (fig. 37.) This single chain becomes compound when several are in the same line (fig. 38), and grouped near one another."—(Sartori and Rauschenfels, " L'Apicoltura in Italia," Milan, 1878.)

206. " If we examine the bees closely during the season of
comb-building and honey-gathering, we shall find many of them
with the wax scales protruding between the rings that form the
body, and these scales are either picked from their bodies, or
from the bottom of the hive or honey boxes in which they are
building. If a bee is obliged to carry one of these wax scales
but a short distance, he takes it in his mandibles, and looks as
business-like with it thus, as a carpenter with a board on his
shoulder. If he has to carry it from the bottom of the honey box,
he takes it in a way that I cannot explain any better than to say
he slips it under his chin, in the mandibles or jaws. When thus
equipped, you would never know he was encumbered with anything, unless it chanced to slip out, when he will very dextrously
tuck it back with one of his forefeet. The little plate of wax

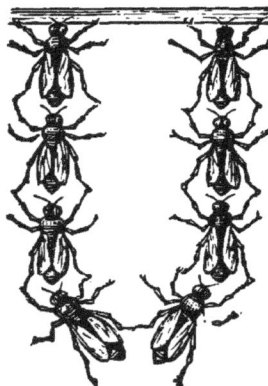
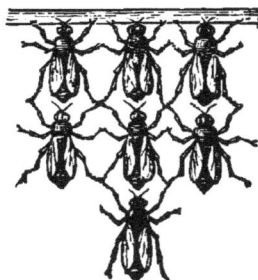

Fig. 37.   Fig. 38.

is so warm, from being kept under his chin, as to be quite soft
when it gets back; and as he takes it out, and gives it a pinch
against the comb where the building is going on, one would
think he might stop a while and put it into place; but not he;
for off he scampers and twists around so many different ways,
you might think he was not one of the working kind at all. Another follows after him sooner or later, and gives the wax a pinch,
or a little scraping or burnishing with his polished mandibles,
then another, and so on, and the sum total of all these manœu-
vres is that the comb seems almost to grow out of nothing; yet
no bee ever makes a cell himself, and no comb building is ever
done by any bee while standing in a cell; neither do the bees ever
stand in rows and ' excavate,' or any thing of the kind."

"The finished comb is the result of the united efforts of the moving, restless mass, and the great mystery is, that anything so wonderful can ever result at all, from such a mixed-up, skipping about way of working, as they seem to have."

"When the cells are built out only part way, they are filled with honey or eggs, and the length is increased when they feel disposed, or 'get around to it, perhaps; as a thick rim is left around the upper edge of the cell, they have the material at hand, to lengthen it at any time. This thick rim is also very necessary to give the bees a secure foothold, for the sides of the cells are so thin, they would be very apt to break down with even the light weight of a bee. When honey is coming in rapidly, and the bees are crowded for room to store it, their eagerness is so plainly apparent, as they push the work along, that they fairly seem to quiver with excitement; but, for all that, they skip about from one cell to another in the same way, no one bee working in the same spot to exceed a minute or two, at the very outside. Very frequently, after one has bent a piece of wax a certain way, the next tips it in the opposite direction, and so on until completion; but after all have given it a twist or a pull, it is found in pretty nearly the right spot. As near as I can discover, they moisten the thin ribbons of wax, with some sort of fluid or saliva (41). As the bee always preserves the thick rib* or rim of the comb he is working, the looker-on would suppose he was making the walls of a considerable thickness, but if we drive him away, and break this rim, we will find that his mandibles have come so nearly together, that the wax between them, beyond the rim, is almost as thin as a tissue paper."—(A. I. Root, " A. B. C. of Bee Culture.")

207. It is very difficult to ascertain who first discovered these scales of wax. According to Mr. S. Wagner, J. A. Overbeck, in his " Glossarium Melliturgium," p. 89, Bremen, 1765, claims that a Hanoverian pastor, named Herman C. Hornbostel, described them in the *Hamburg Library*, about the year 1745.

---

* The constant preserving of this rib or heavy edge of the comb while the work progresses, explains why old comb lengthened and sealed with new wax, sometimes retains a part of its dark color throughout. Some of the old wax is undoubtedly mixed with the new, in the constant remodeling of this heavier edge, till the comb is sealed.

They were also discovered, in Germany, by a farmer. This discovery was communicated to the naturalist Bonnet by Willelmi, under the date of August 22, 1765. (Huber.)

In 1779, Thos. Wildman had noticed the scales of wax on the abdomen of the workers; and he was so thoroughly convinced that wax was secreted from honey, that he recommended feeding new swarms, when the weather is stormy, that they may sooner *build comb* for the eggs of the queen.

From the books written in the French language, it seems that it was Duchet, who, in his "Culture des Abeilles," printed in Friburg in 1771, wrote first that beeswax is produced from honey, of which they eat a large quantity, "*which is cooked in their bodies, as in a stove,*" increasing thereby the warmth of the hive, and that beeswax "*exudes out of this stove*" through the rings of their body which are near the corselet. This idea of Duchet led Beaunier to examine bees, and he discovered that they produce, at one time, not two scales of wax only, but nine, the last ring having seemed to produce one. He adds:

208. "To employ this material, bees use their jaws, their tongues, and their antennæ. In favorable years you can see a great quantity of these pieces of wax which have fallen on the bottom of the hives."—("Traité sur l'Éducation des Abeilles," Vendôme, 1808.)

209. When bees are building combs, some scales of wax are often found on the bottom board, the bees having been unable to use them before they became too tough. Sometimes they pick them up afterwards and use them; some races of bees, the Italian (**551**), for instance, often use also pieces of old combs, which may be within their reach.

The comb, thus built, is easily detected on account of its darker color. Queen-cells seem to be always built of particles, taken from the comb on which they hang, and are never of pure wax (**104**).

" Thus, filtered through yon flutterer's folded mail,
Clings the cooled wax, and hardens to a scale.
Swift, at the well-known call, the ready train
(For not a buz boon Nature breathes in vain)
Spring to each falling flake, and bear along
Their glossy burdens to the builder throng.
These with sharp sickle, or with sharper tooth,
Pare each excrescence, and each angle smoothe,
Till now, in finish'd pride, two radiant rows
Of snow white cells one mutual base disclose.
Six shining panels gird each polish'd round;
The door's fine rim, with waxen fillet bound;
While walls so thin, with sister walls combined,
Weak in themselves, a sure dependence find."

EVANS.

210. The cells of bees are found to fulfill perfectly the most subtle conditions of an intricate mathematical problem.

Let it be required to find what shape a given quantity of matter must take, in order to have the greatest *capacity* and *strength*, occupying, at the same time, the least *space*, and consuming the least *labor* in its construction. When this problem is solved by the most refined mathematical processes, the answer is the hexagonal or six-sided cell of the honey-bee, with its three four-sided figures at the base!

The shape of these figures cannot be altered ever so little, except for the worse.

211. The bottom of each cell is formed of three lozenges, the latter forming one third of the base of three opposite cells.

" If the little lozenge plates were square, we should have the same arrangement, but the bottom would be too sharp pointed as it were, to use wax with the best economy, or to best accommodate the body of the infantile bee. Should we, on the contrary, make the lozenge a little longer, we should have the bottom of the cell too nearly flat to use wax with most economy, or for the comfort of the young bee." (A. I. Root, "A. B. C. of Bee Culture.")

212. "There are only three possible figures of the cells," says Dr. Reid, " which can make them all equal and similar, without any useless spaces between them. These are the equilateral triangle, the square, and the regular hexagon. It is well known to mathematicians, that there is not a fourth way possible in which a plane may be cut into little spaces that shall be equal, similar, and regular, without leaving any interstices.''

An equilateral triangle would have been impossible for an insect with a round body to build. A circle seems to be the best shape for the development of the larvæ; but such a figure would have caused a needless sacrifice of space, materials, and strength. The body of the immature insect, as it undergoes its changes, is charged with a superabundance of moisture, which passes off through the reticulated cover of its cell; may not a hexagon, therefore, while approaching so nearly to the shape of a circle, as not to incommode the young bee, furnish, in its six corners, the necessary vacancies for a more thorough ventilation?

Is it credible that these little insects can unite so many requisites in the construction of their cells?

213. The fact is that the hexagonal shape of the cells is naturally produced, and wihout any calculation, by the bee. She wants to build each cell round; but as every cell touches the next ones, and as she does not wish to leave any space between, each one of the cells flattens at the contact, as would soap bubbles if all of the same diameter. It is the same for the lozenges of the bottom. The bee, wanting the bottom of the cell concave inside, makes it, naturally, convex outside. As this convexity projects on the opposite side of the median line, the bee who builds the opposite cells begins, naturally, on the tip of the convexity, the walls of cells just begun, since she wants also to make their bottom concave. The final result is that one-third of the bottom of each of three cells makes the bottom of the one cell opposite, and each one of the lozenges is flattened, so as not to encroach on the opposite cells.

214. The cells are not horizontal, but inclined from the orifice to the bottom (fig. 39), so as to be filled with honey

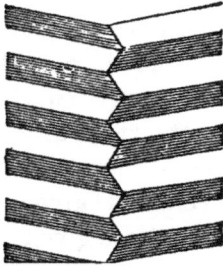

Fig. 39.

SHOWING THE SLOPE OF THE CELLS AND SHAPE OF THE BASE.

(From Sartori and Rauschenfels.)

more easily. The thickness of worker-brood comb is about one inch, with cells opening on each side. The distance between combs is about $\frac{7}{16}$ of an inch. This space is not always exact, but is never under $\frac{5}{16}$, that being necessary for the bees to travel between the combs without interfering with one another. These distances can be a little increased without troubling the bees, and we place the combs in our hives one and a half inches from center to center, for easier manipulation.

215. When the combs are newly built, they are white, but they get color shortly afterwards, especially during the harvest of yellow honey. When used for breeding, the cast skins and residues from the larvæ (**167**) give them a dark color, which becomes nearly black with age, especially if bees have suffered with diarrhœa (**784**), or raised a great many drones (**73–190**).

As wax is a bad conductor, the combs aid in keeping the bees warm, and there is less risk of the honey candying in the cells.

216. Is the size of the cells mathematically exact? When the first Republic of France inaugurated the decimal system of weights and measures, Réaumur proposed to take the cells of the bees as a standard to establish the basis of the system, but it was ascertained that cells are not uniform in size.

217. The cells in which workers are reared are the smallest. Those in which the drones are reared are larger. It is generally admitted that five worker-cells measure about

7

a linear inch, or twenty-five to the square inch, but this is incorrect. If five worker-cells measured exactly an inch, the number contained in a square .inch would be about twenty-nine. As they are usually somewhat larger, the average number in a square inch is a trifle over twenty-seven. Drone-cells number about eighteen, in the same area.

L'Abbé Collin measured the average dimensions of the cells very carefully, and the measurements given in his work (Paris, 1865) are about the same as those given above.

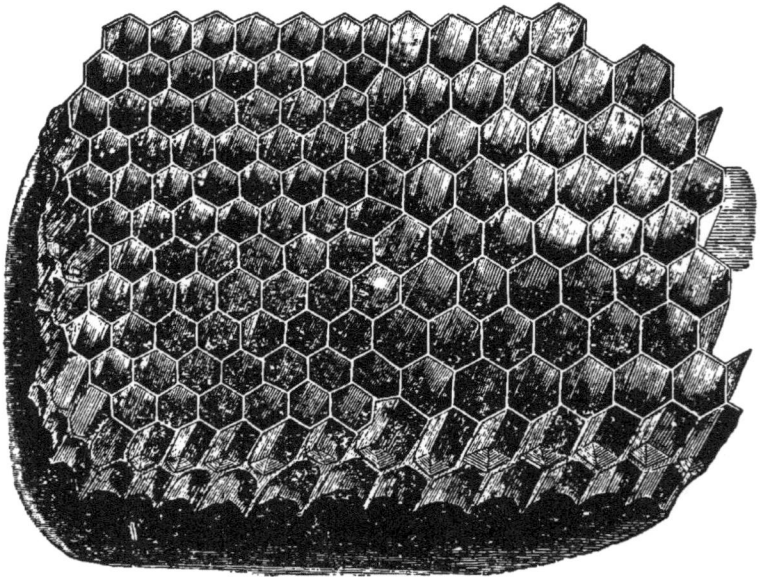

Fig. 40.

218. The queen-cells have already been described. (**104.**)

As bees, in building their cells, cannot pass immediately from one size to another, they display an admirable sagacity in making the transition by a set of irregular intermediate cells. Fig. 40 exhibits an accurate and beautiful representation of comb, drawn for this work from nature, by M. M. Tidd, and engraved by D. T. Smith, both of Boston Mass. The cells are of the size of nature. The large ones are drone-cells, and the small ones, worker-cells. The irregular, five-sided cells between them, show how bees pass from one size to another.

Mr. Cheshire, in his book, has criticized this engraving, on account of the acuteness of the cells of transition, or as he terms them, of accommodation. He writes: "The head of a bee could not reach the bottom of the acute angles as they are represented." Our first impression, on reading the criticism, was that Mr. Cheshire was right. Then the thought that Mr. Langstroth had his engravings made from nature led us to inspect some combs, when we found several cells of *accommodation* with angles at least as acute as in the cut. But we noticed also that this acuity exists only on the rims of the cells and not inside; the bees, inside the cells, having pushed out the walls, to be enabled to reach the bottom of the angles which were thus rounded inside.*

219. The combs are built with such economy, that the entire construction of a hive of a capacity of nine gallons does not yield more than two pounds of bees-wax when melted.

According to Dr. Donhoff, the thickness of the sides of a cell in a new comb is only the one hundred and eightieth part of an inch! Cheshire states that he found some that measured only the four hundreth of an inch.

220. Most Apiarists before Huber's time supposed that wax was made from pollen, either in a crude or digested state. Confining a new swarm of bees to a hive in a dark and cool room, at the end of five days he found several beautiful white combs in their tenement; these being taken from them, and the bees supplied with honey and water, new

---

* Mr. Langstroth wrote to us, in regard to this criticism of Mr. Cheshire: "This piece of comb was actually copied from nature by a man of extraordinary accuracy."

L. OF C.

combs were again constructed. Seven times in succession their combs were removed, and were in each instance replaced, the bees being all the time prevented from ranging the fields to supply themselves with pollen. By subsequent experiments, he proved that sugar-syrup answered the same end with honey. Giving an imprisoned swarm an abundance of fruit and pollen, he found that they subsisted on the fruit, but refused to touch the pollen; and that no combs were constructed, nor any wax-scales formed in their pouches.

Notwithstanding Huber's extreme caution and unwearied patience in conducting these experiments, he did not discover the whole truth on this important subject. Though he demonstrated that bees can construct comb from honey or sugar, without the aid of pollen, and that they cannot make it from pollen, without honey or sugar, he did not prove that when *permanently* deprived of it they can continue to work in wax, or if they can, that the pollen does not aid in its elaboration.

Some pollen is always found in the stomach of wax-producing workers, and they never build comb so rapidly as when they have free access to this article. It must, therefore, in some way, assist the bee in producing it.

221. The experiments made by Berlepsch show that bees, which are deprived of pollen when they construct combs, consume from sixteen to nineteen pounds of honey to produce a pound of comb, while, if provided with it, the amount of honey is reduced to ten or twelve pounds. If the experiment is continued without pollen for some time, the bees become exhausted and begin to perish. It is therefore demonstrated that although nitrogen, which is one of the elements of pollen, does not enter into the composition of bees-wax (**222**), yet it is indispensable as food to sustain the strength of bees during their work in comb making.

222. Honey and sugar contain by weight about eight pounds of oxygen to one of carbon and hydrogen. When converted into wax, these proportions are remarkably changed, the wax containing only one pound of oxygen to more than sixteen of hydrogen and carbon. Now as oxygen is the grand supporter of animal heat, the large quantity consumed in secreting wax aids in generating that extraordinary heat which always accompanies comb-building, and which enables the bees to mould the softened wax into such exquisitely delicate and beautiful forms. This interesting instance of adaptation, so clearly pointing to the Divine Wisdom, seems to have escaped the notice of previous writers.

223. Careful experiments prove that from ten to sixteen pounds of honey are usually required to make a single pound of wax. As wax is an animal oil, secreted chiefly from honey, this fact will not appear incredible to those who are aware how many pounds of corn or hay must be fed to cattle to have them gain a single pound of fat. From experiments made by Mr. P. Viallon here, and by Mr. De Layens in France, it seems that in good circumstances bees use only about seven pounds of honey to produce a pound of wax.

Many bee-keepers are unaware of the value of empty comb. Suppose honey to be worth only ten cents per pound, and comb, when rendered into wax, to be worth thirty cents, the Apiarist who melts a pound of comb loses largely by the operation, even without estimating the time his bees have consumed in building it. It is, therefore, considered a first principle in bee-culture never to melt good worker-combs. A strong colony of bees, in the height of the honey-harvest, will fill them with very great rapidity.

With the box hives (**274**), but little use can be made of empty comb, unless it is new and can be put into the surplus honey-boxes (**728**), but by the use of movable frames, every good piece of worker-comb may be given to the bees (286).

224. As we have seen before (217), while the small cells are designated as worker-cells, the large ones, which vary greatly in depth and are more especially prepared to store honey, and in which the drones are raised, are known as store or drone-cells.

225. Generally, bees build a larger number of worker than of store-cells; yet they do not follow any regulation as to the relative proportion in the quantity of each kind. Not two colonies, in the same Apiary, will show the same number of large cells, even when the hives are of equal capacity, and even if the building was done in circumstances seemingly identical. You will find a colony whose comb will consist of two-thirds worker and one-third store cells, the adjacent colony will have but one-sixth of the latter, another a few square inches only. In a hive all the large cells are together, in another they are scattered. Some of these drone-combs are built from top to bottom of the hive, others are at the top only, others at the side, or at the bottom, or scattered, etc.

226. These facts, not explainable by themselves, when added to the wonderful habits of bees, have led to the theory that it was with foresight, with perfect knowledge and for a special purpose, that bees construct such a varied proportion of the two kinds of cells. Bees are represented as knowing the sex of the eggs which each kind of cells will receive; and foreseeing that their queen may not live long and that the young queens have to be fecundated (120), they build large cells in which drones could be raised.

227. We have demonstrated (213) that bees construct their cells without any geometrical calculation. We had previously (142) established that the queen does not know the sex of the eggs she is laying, and although regretting to decrease the charm with which bees were surrounded by

the imagination of bee-keepers, we will try to demonstrate that, in the building of cells, they simply follow their inclination ; as do all other beings, in the acts that they perform. But we have first to put forward a few facts, which are generally accepted, on which we will ground our reasoning.

228. *1st*, A swarm (**406**), hived on empty frames, always begins its constructions by worker or small cells :

2d, If the queen of a swarm is very prolific (**97**), very little of large, or store-comb, will generally be built by her bees :

3d, If, on the contrary, from old age, or from some other cause, the fecundity of a queen is deficient (**155**), her bees will fill the hive with a large quantity of store-combs :

4th, If the queen of a swarm is removed, or dies while the bees are building, all the combs, made during her absence, will consist of store-cells :

5th, If all or part of the store-combs of a hive are removed, the bees will rebuild large cells, at least three times out of four.

229. Besides these five propositions, we will remember that queens prefer to lay in small cells (**145**), and that they seem to know how to ask the workers to narrow the orifices of the store-cells, when there are no others in the hive to receive their impregnated eggs (**146 to 148**).

We have to remark also that, while the queen prefers the narrow cells, the workers prefer to build the wide ones, since they cease to construct worker-cells when the queen is gone, or when she is not on the spot, to remind them, by her presence, that she needs narrow cells for her impregnated eggs (**146**), and we will find out the cause of such differences, in the number and in the position of each kind of combs, by following the work of the bees, in some of the circumstances in which they may have to build.

230. (*a*) The queen of a swarm is very prolific, the crop

is abundant, and the building goes on very fast. The queen
lays in all the cells, as soon as begun, disputing for them
with the workers, who want to fill them with honey. As
she follows the builders, waiting for cells, no large cells
are made. After about three weeks, the bees of the first
laid eggs begin to leave their cells (**171**); the queen
goes back to fill these empty cells, and the workers, hence-
forth free from restraint, follow their preferences by building store-combs. Result: A few large cells, placed on the
side or at the back of the hive.

231. (*b*) This other swarm has a queen as prolific as
the one above. For two weeks she follows the builders as
the first did, laying in the cells as soon as built. But, the
crop stopping suddenly, both the building and the laying
slacken, when only two-thirds of the constructions are
made. After three weeks of scarcity, abundance comes
again, and the building is resumed. But the queen is no
longer among the workers, waiting for cells; she is at the
other end of the hive, where she lays in the cells which
were left empty when the larvæ that they harbored were
born. Result: About one-third of store-combs.

232. (*c*) This third swarm has a queen whose prolificness is deficient, yet she has been able to follow the builders for a few days. She is at last left behind, and the
workers begin combs with large cells. On reaching these
cells, one or two days later, she passes over them without
laying (**149**), and rejoins the builders, who hasten to com-
ply with her desire to have worker-cells. But she is soon
left behind for the second time, and the workers, unre-
strained again, build large cells till she again rejoins them,
to be again left behind, and so on. Result: Parts of store-
combs mixed, here and there, with worker-combs.

233. (*d*) We have removed from a hive all its drone-
combs; but as the queen is occupied in filling empty worker-
cells in another part of the hive, the builders, following their

preference, reconstruct large cells, thus annulling our work of removal.

234. (*e*) We have given one or two combs to a swarm as soon as it was hived (**422**), and we wonder why its bees have built so much drone-comb. The cause is obvious: the queen, finding empty cells to fill, remained a long time far from the builders, who, following their inclination, constructed drone-cells.

235. We have to utilize the facts just enunciated. If we desire to prevent a swarm from building too many store-combs, we should watch the builders, and remove the large cells as soon as built; these combs, if worth saving, may be used in the surplus sections (**728**). We must remember that, to succeed, it is indispensable that no other cells but the ones to be rebuilt be left at the disposal of the queen. The same rule applies also to the removal of drone-combs at any time; and as the fulfilling of this condition is not always possible, it is better to replace the removed combs with worker comb or comb foundation (**674**).

The above rules are not without exceptions, for unnoticed circumstances may have some influence on the building of combs; but we think that we have stated the main causes of variation.

PROPOLIS.

236. This substance, which is used by the bees to coat the inside of the bee-hive, and make it water and air tight, is obtained from the resinous buds and limbs of trees; the different varieties of poplar yield a rich supply. When first gathered, it is usually of a bright golden color, and so sticky that the bees never store it in cells, but apply it at once to the purposes for which they procured it. If a bee is caught while bringing in a load, it will be found to adhere very firmly to her legs.

Huber planted in Spring some branches of the wild poplar, before the leaves were developed, and placed them in pots near his Apiary; the bees alighted on them, separated the folds of the large buds with their forceps, extracted the varnish in threads, and loaded with it, first one thigh and then the other; for they convey it like pollen, from one leg to the other. We have seen them thus remove the warm propolis from old bottom-boards standing in the sun.

Propolis is frequently gathered from the alder, horse-chestnut, birch, and willow; and as some think, from pines and other trees of the fir kind. Bees will often enter varnishing shops, attracted evidently by their smell; and in the vicinity of Matamoras, Mexico, where propolis seems to be scarce, we saw them using green paint from window-blinds, and pitch from the rigging of a vessel. Bevan mentions the fact of their carrying off a composition of wax and turpentine from the trees to which it had been applied. Dr. Evans says he has seen them collect the balsamic varnish which coats the young blossom-buds of the holly-hock, and has known them to rest at least ten minutes on the same bud, moulding the balsam with their fore-feet, and transferring it to the hinder legs, as described by Huber.

> " With merry hum the Willow's copse they scale,
> The Fir's dark pyramid, or Poplar pale;
> Scoop from the Alder's leaf its oozy flood,
> Or strip the Chestnut's resin-coated bud;
> Skim the light tear that tips Narcissus' ray,
> Or round the Hollyhock's hoar fragrance play;
> Then waft their nut-brown loads exulting home,
> That form a fret-work for the future comb;
> Caulk every chink where rushing winds may roar,
> And seal their circling ramparts to the floor."
>
> EVANS.

237. A mixture of wax and propolis being much more adhesive than wax alone, serves admirably to strengthen the attachments of the combs to the top and sides of the hive.

If the combs are not filled with honey or brood soon after they are built, they are varnished with a delicate coating of propolis, which adds greatly to their strength; but as this natural varnish impairs their snowy whiteness, the bees ought not to be allowed access to the surplus honey-receptacles, except when about ready to store them with honey. (**734.**)

238. Bees make a very liberal use of propolis to fill any crevices about their premises; and as the natural summer-heat of the hive keeps it soft, the bee-moth (**802**) selects it as a place of deposit for her eggs. Hives ought, therefore, to be made of lumber entirely free from cracks. The corners, which the bees usually fill with propolis, may have a melted mixture run into them, consisting of three parts of resin and one of bees-wax; this remaining hard during the hottest weather, will bid defiance to the moth.

239. Bees gather propolis, especially when they can find neither honey nor pollen in the fields. Thus, during the honey-crop, very little of it is taken. In some countries, they use it much more plentifully, owing to its being found more readily.

240. Propolis is hard and brittle in the Winter, and its use by the bees to glue up all parts of the hive, has created the greatest objection to drawers, close-fitting frames, hinged doors, etc., with which some patent hives are provided, and which become entirely immovable, when once coated with it. It is, at all times, the greatest hindrance to the neat handling of the combs, and in warm weather daubs the hands of the Apiarist. It can only be cleaned from the fingers by the use, in place of soap, of a few drops of turpentine, alcohol, spirits of hartshorn, or ether.

241. Propolis is sometimes put to a very curious use by the bees.

"A snail, having crept into one of M. Réaumur's hives early in the morning, after crawling about for some time, adhered, by means of its own slime, to one of the glass panes. The bees having discovered the snail, surrounded it, and formed a border of propolis round the verge of its shell, and fastened it so securely to the glass that it became immovable."—(Bevan.)

> " Forever closed the impenetrable door ;
> It naught avails that in its torpid veins
> Year after year, life's loitering spark remains."
>
> EVANS.

" Maraldi, another eminent Apiarist, states that a snail without a shell having entered one of his hives, the bees, as soon as they observed it, stung it to death ; after which, being unable to dislodge it, they covered it all over with an impervious coat of propolis."

> "For soon in fearless ire, their wonder lost,
> Spring fiercely from the comb the indignant host,
> Lay the pierced monster breathless on the ground,
> And clap in joy their victor pinions round :
> While all in vain concurrent numbers strive
> To heave the slime-girt giant from the hive—
> Sure not alone by force instinctive swayed,
> But blest with reason's soul-directing aid,
> Alike in man or bee, they haste to pour,
> Thick, hard'ning as it falls, the flaky shower ;
> Embalmed in shroud of glue the mummy lies,
> No worms invade, no foul miasmas rise."
>
> EVANS

242. In these instances, who can withhold his admiration of the ingenuity and judgment of the bees? *In the first case*, a troublesome creature gained admission to the hive, which, from its unwieldiness, they could not remove, and which, from the impenetrability of its shell, they could not destroy ; here, then, their only source was to deprive it of locomotion, and to obviate putrefaction ; both which objects they accomplished most skillfully and securely, and, as is usual with these sagacious creatures, at the least possible expense of labor and materials. They applied their cement where alone it was required—round the verge of the shell. *In the latter case*, to obviate the evil of decay, by the total exclusion of air, they were obliged to be more lavish in the use of their embalming material, and to case over the "slime-girt giant," so as to guard themselves from his noisome smell. What means more effectual could human wisdom have devised, under similar circumstances?

243. In bygone days, it was a prevalent belief, that when any member of a family died, the bees knew what had happened; and some were superstitious enough to put the hives in mourning, to pacify their sorrowing occupants; imagining that, unless this was done, the bees would never afterwards prosper!* It was frequently asserted that they sometimes took their loss so much to heart, as to alight upon the coffin whenever it was exposed. A clergyman told the writer that he attended a funeral, where, as soon as the coffin was brought from the house, the bees gathered upon it so as to excite much alarm. Some years after this occurrence, being engaged in varnishing a table, the bees alighted upon it in such numbers, as to convince him, that love of varnish, rather than sorrow or respect for the dead, was the occasion of their conduct at the funeral. How many superstitions, believed even by intelligent persons, might be as easily explained, if it were possible to ascertain as fully all the facts connected with them!

244. COMMERCIAL USES OF PROPOLIS.—"Dissolved in alcohol and filtered, it is used as a varnish, and gives a polish to wood, and a golden color to tin. A preparation made with finely-ground propolis, gum arabic, incense, storax, benzoin, sugar, nitre, and charcoal, in quantities varied at will, is moulded into fumigating cones, for perfuming rooms or halls."—(Dubini, Milan, 1881.)

245. The following letter from a noted Russian Apiarist, to Mr. E. Bertrand, editor of the *Revue Internationale d'Apiculture*, of Nyon, Switzerland, one of the most progressive bee-publications, will be found of interest:

---

* Whittier has written a little poem entitled "Telling the Bees," *apropos* of their knowing of some one's death.

"During my pleasant stay at your pretty villa, I spoke to you of the utilization of propolis in the varnish of our wooden ware, which resists the dissolving power of hot water so well. I have just found a description of the process, and will communicate it to you.

"Propolis is purchased by hucksters, who pay five copecks—a little over two cents—and sometimes even less, for permission to scrape or plane the propolis from the walls of a hive that has lost its bees. The shavings, covered with propolis, are heated, put into a wax-press, and subjected to the treatment used in the extraction of beeswax; the propolis is then purified in hot water, to which sulphuric acid is added. About fifty per cent. of propolis is thus obtained, which sells at forty cents per pound.

"This propolis is poured into hot linseed-oil and beeswax, in the following proportions: Propolis 1, beeswax ½, oil 2. Previously, the oil should 'linger,' as we say, on the stove, for fifteen or twenty days, that is, remain hot without boiling, to give it the property of drying. The wooden ware is dipped into the above mentioned preparation, and must remain in it ten or fifteen minutes, after which it is cooled, and rubbed and polished with woolen rags."—(A. Zoubareff, St. Petersburgh, Sept. 26, 1882.)

We would suggest to manufacturers of supplies, that the soaking or painting of wooden feeders, and of queen-cages, with a similar preparation, would prevent the warm feed from soaking into the wood.

CHAPTER III.

FOOD OF BEES.—HONEY.

246. The main food of bees is the honey or nectar, produced by plants and flowers. That honey is a vegetable product was known to the ancient Jews, one of whose Rabbins asks: "Since we may not eat bees, which are *unclean*, why are we allowed to eat honey?" and replies: "Because bees do not *make* honey, but only *gather* it from plants and flowers."

247. Yet during its sojourn in the honey-sack, the nectar undergoes a chemical change. Most of its cane-sugar, or saccharose, is changed into grape-sugar, or glucose.* This change is due to its being mixed with saliva and gastric juice in the honey-sack **(63)**. "But the cane-sugar yet remains in large proportion in honey gathered on the mountains."—(Girard.)

248. The nectar is produced by the plants in *nectariferous tissues*, in which accumulations of sugar can be found, and exudes most frequently through small apertures, named *stomatœ*.

249. It contains more or less water, according to the kind of flowers, and the conditions in which it is produced. Some flowers give nectar which is almost completely deprived of water. Such is the *Fuschia* (fig. 41). When the nectar of this flower is produced in very dry weather, it sometimes crystallizes in the blossom, as it comes in contact with the air.

---

* What is *chemically* known as glucose should not be confounded with the impure *glucose* of commerce.

In some other flowers, as in the *Fritillaria imperialis*, the nectar contains as much as ninety-five per cent of water. If we except dry and warm days, we can safely assert that, in most cases, the proportion of water in the nectar varies between sixty and eighty per cent.

250. The quantity of nectar produced by the flowers decreases during drought, and increases on the first or second day after a rain. But it is then more watery. In some seasons the saccharine juices abound, while in others they are so deficient that bees can obtain scarcely any food from fields all white with clover. A change in the secretion of honey will often take place so suddenly, that the bees will, in a few hours, pass from idleness to great activity.

Fig. 41.

FUSCHIA.

As a rule, the quantity of nectar, exuded by the plants, varies according to the time of day and atmospheric conditions. Usually, it is most abundant in the morning. Its quantity decreases as the sun rises higher. At three o'clock in the afternoon, the flowers give the least nectar. Then the yield again increases till dark. In Algeria, Africa, in the neighborhood of Blidah, bees cannot find honey later than eight in the morning.

251. It is when the blossom is ready for fertilization, that the nectar is most abundant in it; if it is not gathered by insects, it is re-absorbed by the plant and serves, together with the sugar accumulated in the ovaries, to nourish the seeds.

252. The accumulations of sugar in the tissues, may exist, not only in the flower, but in different parts of plants, in the cotyledons, in the leaves, in the stipules, in the bracteas, and between the leaves and twigs. They help the development of the tissues.

Sometimes the nectariferous tissues are destitute of stomatæ or openings. Then the accumulated nectar may force itself through the cuticle or skin of the plant.

The water of the sap, which runs incessantly in the plants, goes out through the different tissues in unequal quantities; as some tissues are more porous than others. Generally, water escapes in the form of steam; but, in some circumstances, when the air is moist, the water is emitted in liquid form, and may carry with it, to the outside, a part of the accumulations of sugar through which it has passed, thus producing honey-dew. The more sugar this water contains, the slower its evaporation will be.

253. The dampness of the soil and of the air, and a temperature producing a profuse transpiration in plants, then a sudden stop of transpiration, are the best conditions to produce the maximum of nectar in the nectariferous tissues and of liquid exudations on the outside.

254. Most of the above statements are taken, or rather abridged, from "Les Nectaires," of Gaston Bonnier, a professor at the École Normale Supérieure of Paris (1879). This work was awarded a medal by the Academy of Science of Paris. Bonnier backs his statements with one hundred and thirty engravings made from microscopic researches.

255. He explains, not only how the nectar is formed in the blossoms, but also how the extra floral nectar, the so-called *honey-dew*, is produced on different parts of plants, o˙ trees.

He has noticed and described the production of nectar

8

(honey-dew without *aphides*),* on many herbaceous plants, and on the following trees or shrubs: Two kinds of oak,

Fig. 42.

FIELD APIARY, TEMPORARY LOCATION.

(From the German of Gravenhorst.)

* Honey-dew without *aphides* was noticed in this country, on wheat, and even on wheat stubble. (J. O. Shearman, *American Bee-Journal*, 1887, page 503.) See also, O. W. Bellemey, "Gleanings," March, 1882, page 398, and others.

the ash, two kinds of linden, the sorb, the barberry, two kinds of raspberry, the poplar, the birch, two kinds of maple, and the hazel brush. In some parts of Europe, this *honey-dew* is so plentiful, that some Apiarists transport their bees to the districts in which it is produced, during its yield. (Fig. 42.)

256. Bees also harvest, in some seasons, a sweet substance of poorer quality, which is a discharge from the bodies of small *aphides* or " plant lice."*

Messrs. Kirby and Spence, in their interesting work on Entomology, have given a description of the honey-dew furnished by the *aphides:*

"The loves of the ants and the *aphides* have long been celebrated; you will always find the former very busy on those trees and plants on which the latter abound; and, if you examine somewhat more closely, you will discover that the object of the ants in thus attending upon the *aphides*, is to obtain the saccharine fluid secreted by them, which may well be denominated their milk. This fluid, which is scarcely inferior to honey in its sweetness, issues in limpid drops from the abdomen of these insects, not only by the ordinary passage, but also, by two setiform tubes, placed one on each side, just above it. Their sucker being inserted in the tender bark is, without intermission, employed in absorbing the sap, which, after it has passed through these organs, they keep continually discharging. When no ants attend them, by a certain jerk of the body, which takes place at regular intervals, they ejaculate it to a distance."

257. "Mr. Knight once observed a shower of honey-dew descending in innumerable small globules, near one of his oak trees. He cut off one of the branches, took it into the house, and, holding it in a stream of light admitted through a small opening, distinctly saw the *aphides* ejecting the fluid from their bodies with considerable force, and this accounts for its being

---

* The Abbé Boissier de Sauvages, in 1763, described two species of honeydew. The first kind, he says, has the same origin with the *manna* on the ash and maple trees of Calabria and Briançon, where it flows plentifully from their leaves and trunks, and thickens in the form in which it is usually seen. —(" Observations sur l'Origine du Miel.") We have received specimens of a honey-dew from California, which is said to fall from the oak trees in stalactites of considerable size.

frequently found in situations where it could not have arrived
by the mere influence of gravitation. The drops that are thus
spurted out, unless interrupted by the surrounding foliage, or
some other interposing body, fall upon the ground; and the
spots may often be observed, for some time, beneath and around
the trees, affected with honey-dew, till washed away by the rain.
The power which these insects possess of ejecting the fluid from
their bodies, seems to have been wisely instituted to preserve
cleanliness in each individual fly, and, indeed, for the preservation of the whole family; for, pressing as they do upon one another, they would otherwise soon be glued together, and rendered
incapable of stirring. On looking steadfastly at a group of these
insects (*Aphides salicis*) while feeding on the bark of the willow,
their superior size enabled us to perceive some of them elevating
their bodies and emitting a transparent substance in the form of
a small shower:

> " Nor scorn ye now, fond elves, the foliage sear,
> When the light aphids, arm'd with puny spear,
> Probe each emulgent vein, till bright below,
> Like falling stars, clear drops of nectar glow."
>
> EVANS.

258. " Honey-dew usually appears upon the leaves as a viscid transparent substance, as sweet as honey itself, sometimes in
the form of globules, at others resembling a syrup. It is generally most abundant from the middle of June to the middle of
July—sometimes as late as September.

" It is found chiefly upon the oak, the elm, the maple, the plane,
the sycamore, the lime, the hazel, and the blackberry; occasionally also the cherry, currant, and other fruit trees. Sometimes
only one species of trees is affected at a time. The oak generally affords the largest quantity. At the season of its greatest
abundance, the happy, humming noise of the bees may be heard
at a considerable distance, sometimes nearly equalling in loud-
ness the united hum of swarming."—(Bevan.)

In some seasons, bees gather large supplies from these
honey-dews, but it is abundant only once in three or four
years. The honey obtained from this source is usually of a
dark color, and seldom of a very good quality.

259. It is very difficult to ascertain, at all times, the
special source of honey-dew, whether from the trees or from

the *aphides*. In order to give all sides a hearing, we will cite a letter from Mr. Bonnier on this subject, and leave the reader to draw his own conclusions:

"Plant lice are seen even on trees, that have no extra floral nectaries. They do not produce exudations (properly speaking), but bore the tissues to eat the contents. Their presence on the plant has no connection with that of the nectar. The excremental liquid of *aphides* is not equally sweet in all the species, and the bees harvest only that which is very sweet. They generally prefer the true honey-dew (*miellée*), which exudes from the leaves at certain times, and contains mannite and saccharine matter.

" I have seen bees, however, harvesting the sweet liquid of the *aphides* and the true *miellée* at the same time, on the aspen, maple, and sycamore.

" I have rarely seen the extra floral nectar of the special nectaries overflow and run in drops, but the true *miellée* of trees may fall in small drops, and some observers conclude from this fact, that it is produced by *aphides*. I have often seen some trees, and even all the trees, of a timber, covered with an abundant *miellée*, falling in small drops, although there was not a single louse on the higher limbs.

" To sum up, we must not confound the three kinds of sweet liquid, which may be produced outside the flowers : *1st*, The extra-floral nectar proper, produced, like the nectar of flowers, from special sugar tissues ; *2d*, The true *miellée*, produced on the surface of the leaves of trees or shrubs, without the action of *aphides ; 3d*, The excretion, more or less sweet, sometimes containing very little sugar, abundantly produced by a great number of *aphides*."

260. In some blossoms, as in the red clover, the corolla is so deep and narrow, that the nectar is out of reach of the honey-bee. Larger insects, such as the bumble-bee, or smaller ones, as some wasps, enjoy it to the exclusion of our favorites. Yet in some seasons, we have seen bees working on red-clover bloom, and have attributed this to the corollas being shorter, owing to drouth, or scant growth. Mr. Bonnier has discovered that, in some such flowers, the nectar is sometimes so abundant, that the bees can reach it. It is true that insects, and even bees, can tear the tender corollas of some blossoms, opposite the honey receptacle, to reach the nectar, but this is of such rare instance, in the honey-bee, that it cannot be considered of any practical value.

261. The honey, when harvested, is stored in the rear of the hive, above the brood, and as near it as possible.

When just gathered, it is too watery to be preserved for the use of the bees. To evaporate this water, they force a strong current of air through the hive, and the bee-keeper can ascertain the days of large honey-yield, by the greater roar of the bees in front of their hive during the night following. If a strong colony is put on a platform scale, it will be found, during the height of the honey-harvest, to gain a number of pounds on a pleasant day. Much of this weight will be lost in the night, from the evaporation of the newly-gathered honey. A thorough upward ventilation, in hot weather, will therefore contribute to increase the ripening of honey.

When the cell is about full, the bees seal it with a flat cover or capping made of wax. This capping is begun at the lower edge of the cell, and is raised gradually, as the honey is deposited within, till the cell is entirely sealed. These cappings being flat, depressed, or uneven, are easily distinguished from the caps of the brood, which are convex and of a darker color.

262. Are the caps of the honey-cells air-tight? This much-debated question is not yet satisfactorily answered.

The caps of the brood-cells, made of pollen and wax, are undoubtedly porous enough to allow the air to reach the larva; and some Apiarists question the imperviousness of the sealing of honey-comb. Mr. Cheshire himself, while of opinion that "the bee aims at compact coverings for her honey," says that "not more than ten per cent. of these are absolutely impervious to air." Yet his own description of the cause of the well-known whiteness of the cappings, owing to the air which is left behind and "cannot escape," would prove that these cappings are originally made as airtight as a thin coat of wax can make them. The fact that honey shrinks and swells inside of the cell, is only a proof that, like many other things, its volume depends on the temperature. Again, its fermenting in sealed cells, proves only that it contains the elements of fermentation, and these can be developed at certain degrees of temperature, even in air-tight vessels. Mr. Cheshire's tests of honey-combs, steeped in water, to ascertain whether the honey in sealed cells would absorb moisture and expand, have been tried by us with altogether contrary results. The difference of opinion on this subject may be due to the fact that the cappings are very fragile, and crack imperceptibly, when exposed to variations of temperature outside of the hive.

Would it be possible that the thin coat of wax, though evidently air-tight, be, *in some circumstances*, porous enough to allow moisture to soak through it slowly, like water through leather?

POLLEN.

263. The pollen, or fertilizing dust of flowers, is gathered by the bees from blossoms, and is indispensable to the nourishment of their young—repeated experiments having proved that brood cannot be raised without it. It is very rich in the nitrogenous substances which are not contained in honey, and without which ample nourishment could not be furnished for the development of the growing bee. Dr. Hunter, on dissecting some immature bees, found that their stomachs contained pollen, but not a particle of honey.

We are indebted to Huber for the discovery, that pollen is the principal food of the young bees. As large supplies were often found in hives whose inmates had starved, it was

evident that, without honey, it could not support the mature
bees; and this led former observers to conclude that it
served for the building of comb.   Huber, after demonstrating that wax can be secreted from an entirely different substance, soon   ascertained   that   pollen   was   used   for   the
nourishment of the embryo bees.   Confining some bees to
their hive without any pollen, he supplied them with honey,
eggs, and larvæ.   In a short time, the young all perished.
A fresh supply of brood being given to them, with an ample
allowance   of   pollen,   the   development   of   the   larvæ   pro-
ceeded in the natural way.

264. We had an excellent opportunity of testing the
value of this substance, in the backward Spring of 1852.   On
the 5th of February, we opened a hive containing an artificial swarm of the previous year, and found many of the cells
filled with brood.   The combs being examined on the 23d,
contained neither eggs, brood nor bee-bread; and the colony was supplied with pollen from another hive; the next
day, a large number of eggs were found in the cells.   When
this supply was exhausted, laying again ceased, and was
only resumed when more was furnished.   During the time
of these experiments, the weather was so unpromising, that
the bees were unable to leave the hive.

Dzierzon is of opinion that bees can furnish food for their
young, without pollen; although he admits that they can do
it only for a short time, and at a great expense of vital energy; just as the strength of an animal nursing its young is
rapidly reduced, if, for want of proper food, the very substance of the mother's body must be converted into milk.
The experiment just described does not corroborate this
theory, but confirms Huber's view, that pollen is indispensable to the development of brood.

Gundelach, an able German Apiarist, says that if a colony with a fertile queen be confined to an empty hive, and
supplied with honey, comb will be rapidly built, and the

cells filled with eggs, which in due time will be hatched; but the worms will all die within twenty-four hours.

Sometimes bees, unable to feed their brood for lack of pollen, desert their hives (**407**).

265. In September, 1856, we put a very large colony of bees into a new hive, to determine some points on which we were then experimenting. The weather was fine, and they gathered pollen, and built comb very rapidly; still for ten days, the queen-bee deposited no eggs in the cells. During all that time, these bees stored very little pollen in the combs. One of the days being so stormy that they could not go abroad, they were supplied with rye flour (**267**), none of which, although very greedily appropriated, could be found in the cells. During all this time, as there was no brood to be fed, the pollen must have been used by the bees either for nourishment, or to assist them in secreting wax; or, as we believe, for both these purposes.

266. Bees prefer to gather *fresh* pollen, even when there are large accumulations of old stores in the cells. With hives giving the control of the combs, the surplus of old colonies may be made to supply the deficiency of young ones; the latter, in Spring, being often destitute of this important article.*

If honey and pollen can both be obtained from the same blossom, the industrious insect usually gathers a load of each. To prove this, let a few pollen-gatherers be dissected when honey is plenty; and their honey-sacks will ordinarily be full.

When the bee brings home a load of pollen, she stores it away, by inserting her body in a cell, and brushing it from her legs; it is then carefully packed down, being often covered with honey, and sealed over with wax. Pollen is seldom deposited in any except worker-cells. This fact

---

* Although the bees of queenless colonies do not usually go in quest of pollen, some occasionally harvest it, and as it is not used, it accumulates in the hive.

supports the idea that large cells are not built to raise brood
(**224**).

Aristotle observed, that a bee, in gathering pollen, con-
fines herself to the kind of blossom on which she begins, even
if it is not so abundant as some others; thus a ball of this
substance taken from her thigh, is found to be of a uniform
color throughout; the load of one insect being yellow, of
another, red, and of a third, brown; the color varying with
that of the plant from which the supply was obtained. They
may prefer to gather a load from a single species of plant,
because the pollen of different kinds does not pack so well
together. Réaumur has estimated, that a good colony may
gather and use as much as one hundred pounds of it in a
year.

267. When bees cannot find pollen, in early Spring, they
will gather flour, or meal, or even fine sawdust, as a substitute. This was noticed by Hartlib, as early as 1655.

Dzierzon, early in the Spring, observed his bees bringing
rye-meal to their hives from a neighboring mill, before they
could procure any pollen from natural supplies. The hint
was not lost; and it is now a common practice, wherever
bee-keeping is extensively carried on, to supply the bees
early in the season with this article. Shallow troughs or
boxes are set not far from the Apiaries, filled about two
inches deep with *finely-ground, dry, unbolted rye-meal, oat-
meal or even with flour*. Where bolted flour, or meal, is
given, it should be tightly pressed with the hands, to prevent the bees from drowning in it. To attract them to it,
we bait them with a few old combs, or a little honey.

The boxes must be placed in a warm spot sheltered from
the wind. Thousands of bees, when the weather is favorable, resort eagerly to them, and return heavily laden to
their hives.

This artificial pollen or bee-bread, is kneaded by them
with saliva, or honey brought from the hive. This is easily

ascertained by tasting the little pellets, which in the hurry are loosened from their baskets, and fall to the bottom of the flour box. In fine, mild weather, they labor at this work with great industry; preferring the meal to the *old* pollen stored in their combs. They thus breed early, and rapidly recruit their numbers. The feeding is continued till, the blossoms furnishing a preferable article, they cease to carry off the meal.

We will here add that, as a rule, colonies that do not carry in meal or pollen, at the opening of Spring, are without brood, either because they are queenless, or from want of honey, or from some other cause.

The discovery of flour, as a substitute for pollen, removes a very serious obstacle to the culture of bees. In many districts, there is for a short time such an abundant supply of honey, that almost any number of strong colonies will, in a good season, lay up enough for themselves, and a large surplus for their owners. In many of these districts, however, the supply of pollen is often quite insufficient, and in Spring, the swarms of the previous year are so destitute, that unless the season is early, the production of brood is seriously checked, and the colony cannot avail itself properly of the superabundant harvest of honey.

268. As bees carry on their bodies the pollen, or fertilizing substance, they aid most powerfully in the impregnation of plants, while prying into the blossoms in search of honey or bee-bread. In genial seasons, fruit will often set abundantly, even if no bees are kept in its vicinity; but many Springs are so unpropitious, that often during the critical period of blossoming, the sun shines for only a few hours, so that those only can reasonably expect a remunerating crop whose trees are all murmuring with the pleasant hum of bees.

269. One of the laws of Nature is that the crossing of the races produces offspring with greater vigor, endurance, and faculty of reproduction. Fruits succeed better, when the pollen, which fertilizes the pistil, comes from some other blossom; and the insects are intrusted with the mission of transporting this pollen from one blossom to another, while gathering it for their own use. In some plants, fertilization would have been impossible, without the help of insects. For instance, some plants, such as the willows, are diecious, having their male organs on one tree, and their female organs on another. The bees after visiting the one for pollen, go to the other for honey, and the fecundation is effected.

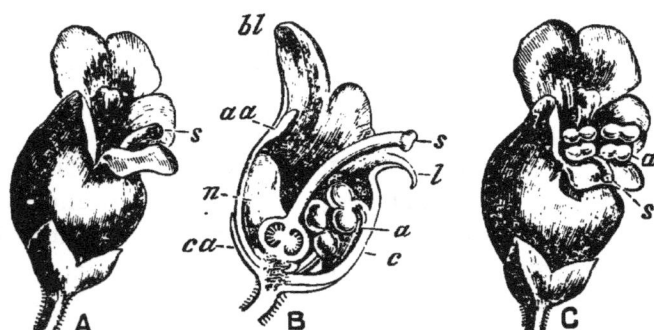

Fig, 43.

SCROPHULARIA NODOSA.

(Magnified. From Cheshire.)

A, young blossom. *s*, stigma.

B, section of blossom. *ca*, calyx; *c*, corolla; *aa*, aborted anthers; *s*, stigma; *l*, lip; *a*, anthers; *n*, nectar; *bl*, black lip.

C, older blossom. *s*, dropping stigma; *a*, anthers.

In some other plants, such as the *Scrophularia Nodosa* (Simpson honey plant—Fig. 43), the female organs are ready for fecundation earlier than the male. But as the flower secretes a large quantity of honey, which is replaced in its nectaries as fast as the bees gather it, the bees, in traveling from one blossom to another, carry the pollen of an old blossom to the pistil of a younger one, and fertilization is accomplished. Some plants, corn, for instance, produce such quantities of pollen, that the agency of insects is less indispensable to the fertilization of their blossoms.

270. To determine the advantages which flowers derive from insect fertilization, any one can wrap a few flowers in gauze, just before the opening of the bud, and compare the number of fertile seeds, from flowers thus treated, with those of other blossoms.

We have heard farmers mention the fact that the first crop of red clover furnishes but little seed, compared with the second crop. This is because the bumble-bees, which help its fertilization, are very scarce in Spring, while they are much more plentiful in Summer. " In Australia it was found impossible to obtain seed from red clover until the bumble-bees were imported into that country " (Darwin).

A large fruit-grower told us that his cherries were a very uncertain crop, a cold northeast storm frequently prevailing when they were in blossom. He had noticed that, if the sun shone only for a couple of hours, the bees secured him a crop.

If those horticulturists, who regard the bee as an enemy, could exterminate the race, they would act with as little wisdom as those who attempt to banish from their inhospitable premises every insectivorous bird, which helps itself to a small part of the abundance it has aided in producing. By making judicious efforts early in the Spring, to entrap the mother-wasps and hornets, which alone survive the Winter, an effectual blow may be struck at some of the worst pests of the orchard and garden. In Europe, those engaged extensively in the cultivation of fruit, often pay a small sum in the Spring for all wasps and hornets destroyed in their vicinity.

WATER.

271. Water is necessary to bees to dissolve the honey, which sometimes granulates in the cells, and to raise brood. They can raise a certain amount of brood without water, but they always seem to suffer more or less in consequence (**662**). In the Winter, they breed but little, and the moisture which condenses on the walls of the hive is generally sufficient. Yet we have noticed that as soon as bees are brought out of the cellar (**653**), if the temperature is sufficiently warm,'a great many will be seen sucking water. This fact shows that Berlepsch was right when he advised bee-keepers to give water to bees during Winter, to avoid what he called disease of the thirst. Besides, every one may notice that bees take advantage of any warm Winter day to bring it to their hives; and, in early Spring, may be seen busily drinking around pumps, drains, and other moist places. Later in the season, they sip the dew from the grass and leaves.

272. Every careful bee-keeper will see that his bees are well supplied with water. If he has not some sunny spot, close at hand, where they can safely obtain it, he will furnish them with shallow wooden troughs, or vessels filled with floats or straw, from which — sheltered from cold winds, and warmed by the genial rays of the sun—they can drink without risk of drowning.

A barrel half filled with earth and then filled with water, in which some water-

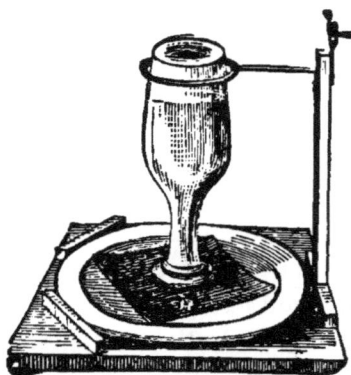

Fig. 44.

WATER SUPPLY BOTTLE.

(From Sartori & Rauscheufels, ot Milan, Italy.)

cress or other aquatic plants are kept, to preserve it from putrefaction, and to prevent the bees from drowning, will do very well. For a small Apiary, a jug or bottle (fig. 44), filled with water, and inverted on a plate, covered with a small piece of carpet, will be sufficient. It can also be given in the combs. Mr. Vogel, editor of the *Bienen-Zeitung*, on the 19th of March gave to a colony a comb containing crystallized honey, and another containing about three-fourths of a pound of water. Within sixteen hours, both combs were altogether emptied by the bees.

273. A learned French bee-keeper, Mr. De Layens, made many experiments in regard to this matter.

"In the month of May, 1878, I put a lump of sugar near a spot where a great many bees came for water; they paid no attention to it. The sugar was then moistened and covered with honey. The bees, attracted by the honey, came in great numbers, and sucked up most of the moist sugar. After they became accustomed to this, I decreased the moistening, till I gave them nothing but dry sugar, when they brought water to dissolve the sugar, and removed all except the parts which were too hard to be dissolved easily."—(*Bulletin de la Suisse*, Nov. 1880.)

The same writer has noticed that, in Spring, if the bees are compelled to go very far for water, many of them perish. He found a loss of three hundred and fifty grammes of bees—four-fifths of a pound—from a hive, during a sudden Spring storm.

From the 10th of April to the 31st of July, forty colonies consumed 187 litres of water, about fifty gallons; the greatest quantity used in a day being seven litres, or about fifteen pints.

That bees do not need water, in circumstances other than those above named, is evidenced from the fact that, in importing bees from Italy, we did not succeed in receiving them alive, until our shippers reluctantly consented to send them without water (**595**).

SALT.

274. Bees seem to be so fond of salt, that they will alight upon our hands to lick up the saline perspiration.

" During the early part of the breeding season," said Dr. Bevan, " till the beginning of May, I keep a constant supply of salt and water near my Apiary, and find it thronged with bees from early morn till late in the evening. About this period the quantity they consume is considerable, but afterwards they seem indifferent to it. The eagerness they evince for it at one period of the season, and their indifference at another, may account for the opposite opinions entertained respecting it."

CHAPTER IV.

THE BEE–HIVES. — HIVES WITH IMMOVABLE COMBS.

275. The first hives that were provided for bees were as rude as their natural abodes. We do not need to look back very far to remember the " bee-gum," so called, probably, because it had often been made out of the gum tree, with two sticks crossing in the middle, and a rough board nailed on top, while a notch in the lower end formed the entrance. In the Old World, they manufactured straw or willow " skeps " and pottery hives, which are still used in Asia and Africa. The earthen hive was simply a tube, laid on its side, and closed at each end with a movable wooden disk. This disk was removed to take the honey, which is always located at the back part of the hives.

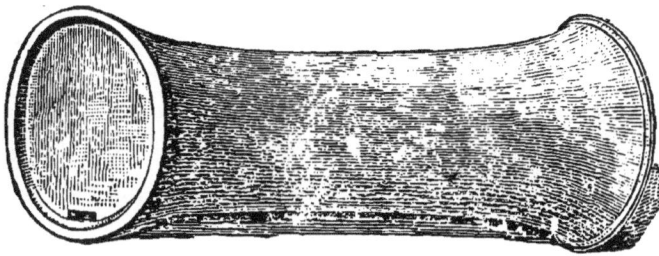

Fig. 45.
EARTHEN HIVE OF AFRICA AND CYPRUS.
(From "L'Apicoltore," Milan.)

These earthen hives were, unquestionably, the most sensible of those old kinds. In the Islands of Greece they were set in thick stone walls, built on purpose with the entrance on one side of the wall. Sometimes they were located in the walls of the houses, and the honey was removed from the inside of the house, or, if in walls, from behind, out of the flight of bees.

9

276. To get the honey from the gums, or boxes, the
bee-keepers used at first to drive the bees to another hive
(**574**) and take all the contents.   But most of the thus
impoverished colonies perished.   This led to the thought
that killing bees would be more facile, and the brim-
stone-pit was invented.   This killing of bees was so customary that, about one hundred years ago, Joseph II,
Emperor of Austria, decreed that every bee-keeper who
would cut the combs in Spring, instead of brimstoning
the bees, would receive one florin (about forty cents) per
colony.

Fig. 46.
STRAW HIVE, WITH CAP.
(From Hamet.)

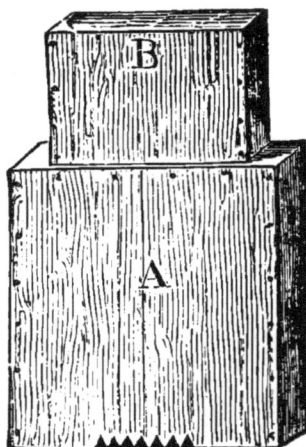

Fig. 47.
BOX HIVE, WITH CAP.
(From Hamet.)

277. Nearly sixty years ago, our senior, then a boy,
saw this harvesting of combs for the first time.   Clothed with
a heavy linen frock, equipped with a mask of wire,
strong enough to be sword-proof, and sweating under a
scorching sun in this heavy garment, he helped (?) the old
priest of his village to prune about twenty colonies, removing
the back combs with a curved knife, from the upturned
hives.   It was in April; and, while the crop thus harvested

was light, the damage inflicted to the bees was immense, for they had to rebuild their combs at a time when queens begin their greatest laying. But the bee-keepers of old were persuaded that this crop of beeswax was beneficial to bees, since it compelled them to make new combs, which were considered better than older ones (**676**).

Fig. 48.
STRAW EKE HIVE.
(From Hamet.)
B, body; *A*, hole to connect the stories with the surplus cap.

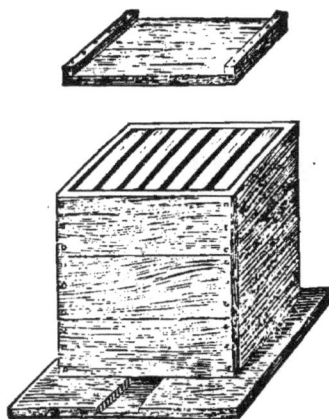

Fig. 49.
THE RADOUAN EKE HIVE.
(From Hamet.)

278. Some bee-keepers, having noticed that bees place their honey at the highest part of the hive, added a cap or upper story, which communicated with the hive through a hole in the top of the latter (figs. 46 and 47). Still later, Apiarists found out that when the hive was very deep and the connecting hole small, the bees refused to store their honey in the cap, and they made their hives with open ceilings, replacing the top board of the breeding-story with slats or bars. The hives were afterwards divided into several horizontal sections, called "ekes" (figs. 48 and 49). Instead of using a cap, some Apiarists removed the upper story, when full of honey, and placed a new story under the others.

The bees then continued their constructions downwards.
To separate the sections from one another, they used a wire
that cut the combs. Butler, in his "Feminine Monarchy,"
1634, shows hives composed of four sections, piled upon one
another. Palteau, in 1750, advises bee-keepers to use a
perforated ceiling at the top of each section. Radouan, in
1821, instead of a perforated ceiling, uses triangular bars,
to which the bees attach their combs (fig. 49). Chas.
Soria, in 1845, used these bars at the bottom of each
story as well as at the top, with bee space between, so that
they can be removed, exchanged, or reversed, without
crushing any bees, or damaging a single cell (fig. 50).

Fig. 50.
EKE OF CHAS. SORIA.
(From Hamet.)

Fig. 51.
DIVIDING HIVE OF JONAS.
DE GELIEU.
(From Hamet.)

279. Other Apiarists divided their hives vertically, con-
formably with the shape of the combs of the bees, which
hang vertically. If we are correctly informed, it was Jonas
de Gelieu who inaugurated this style (fig. 51). He made
his hive divisible into only two parts. Œttl, towards the
middle of this century, made a straw hive divided into three
vertical parts. The main advantage of these hives resides
in the facility of dividing them for artificial swarming. But
as this method of making artificial swarms is defective. as
will be shown further, (**471**), and as all these contri-

vances did not allow a close study of the habits of the bee, or permit the needed manipulations, it became necessary to invent a hive whose every comb, and every part, the Apiarist could promptly and easily control ; a hive which, to employ the forcible expression of Mr. Hamet, could " *se démonter comme un jeu de marionettes;*" (be taken to pieces like a puppet-show).

REQUISITES OF A COMPLETE HIVE.

280. *1.* A complete hive should give the Apiarist such perfect control of all the combs, that they may be easily taken out without cutting them, or exciting the anger of the· bees.

2. It should permit all necessary operations to be performed without hurting or killing a single bee.

Some hives are so constructed, that they cannot be used without injuring or destroying some of the bees ; and the destruction of even a few materially increases the difficulty of managing them (**399**).

3. It should afford suitable protection against extremes of heat and cold, sudden changes of temperature, and the injurious effects of dampness.

The interior of a hive should be dry in Winter, and free in Summer from a pent and almost suffocating heat.

4. Not one unnecessary motion should be required of a single bee.

As the honey-harvest, in most locations, is of short continuance, all the arrangements of the hive should facilitate, to the utmost, the work of the busy gatherers. Hives which compel them to travel with their heavy burdens through densely crowded combs, are very objectionable. Bees instead of forcing their way through thick clusters, must easily pass into the top surplus honey-boxes of the hives, from any comb in the hive, and into every box, without traveling much over the combs.

5. It should be capable of being readily adjusted to the wants of either large or small colonies (**349**).

6. It should allow every good piece of worker-comb to be given to the bees, instead of melting it into wax, and should permit of the use of comb-foundation (**674**).

7. It should prevent the over-production of drones, by permitting the removal of drone-comb from the hive.

A hive containing too much comb suitable only for storing honey, or raising drones, cannot be expected to prosper.

8. It should allow the bottom board to be loosened or fastened at will, for ventilation, or to clear out the dead bees in Winter. If suffered to remain, they often become mouldy, and injure the health of the colony. In dragging them out, when the weather moderates, the bees often fall with them on the snow, and are so chilled, that they never rise again; for a bee, in flying away with the dead, frequently retains its hold until both fall to the ground.

9. No part of the interior of the hive should be below the level of the place of exit.

If this principle is violated, the bees must, at great disadvantage, drag, *up hill*, their dead, and all the refuse of the hive.

10. It should afford facilities for feeding bees, both in warm and cool weather, in case of need.

11. It should furnish facilities for enlarging, contracting, and closing the entrance, to protect the bees against robbers; and when the entrance is altered, the bees ought not, as in some hives, to lose valuable time in searching for it.

12. It should furnish facilities for admitting at once a large body of air, that the bees may be tempted to fly out and discharge their fæces, on warm days in Winter, or early Spring (**344**).

If such a free admission of air cannot be given, the bees,

by losing a favorable opportunity of emptying themselves, may suffer from diseases resulting from too long confinement.

13. It should allow the bees, together with the heat and odor of the main hive, to pass in the freest manner, to the surplus honey-receptacles.

In this respect, many hives with which we are acquainted are more or less deficient; the bees being forced to work in receptacles difficult of access, and in which, in cool nights, they find it impossible to maintain the requisite heat for comb-building, or, in which, in hot days, they cannot send air enough to make the place habitable.

14. Each of the parts of every hive in an Apiary should be so made, as to be interchangeable from one hive to another. In this way, the Apiarist can readily make the exchanges of brood, honey, or pollen, which circumstances demand.

15. The hive should permit the surplus honey to be taken away in the most convenient, beautiful and salable forms.

16. It should be equally well adapted to be used as a swarmer, or non-swarmer.

17. It should enable the Apiarist to multiply his colonies with a certainty and rapidity which are impossible if he depends on natural swarming.

18. It should enable the Apiarist to supply destitute colonies with the means of obtaining a new queen.

19. It should enable him to catch the queen, for any purpose; especially to remove an old one whose fertility is impaired by age.

20. It should enable a single bee-keeper to superintend several hundred colonies for different individuals.

Many persons would keep bees, if an Apiary, like a garden, could be superintended by a competent individual. No person can agree to do this with the common hives. If the bees are allowed to swarm, he may be called in a dozen different directions at once, and if any accident, such as the loss of a queen, happens to the colonies of his customers, he can usually apply no remedy.

21. All the joints of the hive should be water-tight and moth-proof (**804**), and there should be no doors or shutters liable to shrink, swell, or get out of order.

22. A complete hive should be protected against the destructive ravages of mice in Winter (**348**).

23. It should permit the honey, after the gathering season is over, to be concentrated where the bees will most need it.

24. It should permit the space for spare honey receptacles to be enlarged or contracted at will, without any alteration or destruction of existing parts of the hive.

Without the power to do this, the productive force of a colony is in some seasons greatly diminished.

25. Its surplus honey receptacle should be as close to the brood as possible.

26. A complete hive, while possessing *all* these requisites, should, if possible, *combine* them in a *cheap* and *simple* form, adapted to the wants of all who are competent to cultivate bees.

281. There are a few desirables to which a hive, even if it were perfect, could make no pretensions!

It could not promise splendid results to those who are too ignorant or too careless to be entrusted with the management of bees. In bee-keeping, as in all other pursuits, man must first understand his business, and then proceed upon the good old maxim, that "the hand of the diligent maketh rich." "*In a word, to succeed it is indispensable to know what to do, and to do it just in time.*"—(S. Wagner).

It could not have the talismanic influence to convert a bad situation for honey into a good one; or give the Apiarist an abundant harvest, whether the season was productive or otherwise.   As well might the farmer seek for some kind of wheat which will yield an enormous crop, in any soil, and in every season.

It could not enable the cultivator, while rapidly multiplying his stocks, to secure the largest yield of honey from his bees.   As well might the breeder of poultry pretend, that in the same year, and from the same stock, he can both raise the greatest number of chickens, and sell the largest number of eggs.

Movable-Comb Hives.

282. The bee-keepers of Greece and of Candia seem to have been the first to provide their hives with movable bars, under which bees suspended their combs.   Della-Rocca mentions these and gives engravings of them in his work, published in 1790.   In 1838, Dzierzon revived this hive and improved it.   In spite of the difficulty of its management, since the combs not being attached to movable-frames, but

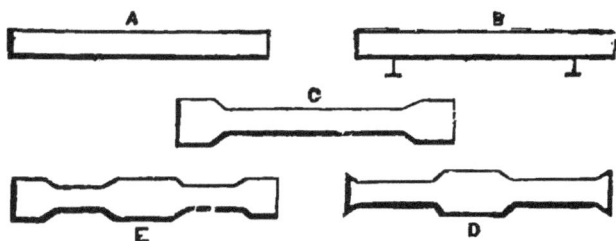

Fig. 52.

DIVERS MOVABLE BARS TO SUPPORT THE COMBS.

to top bars (fig. 52), cannot be removed without cutting them loose from the sides of the hive, Dzierzon succeeded in making discoveries, in bee physiology, which rank among the most important (**132**).   His success was marvelous for the epoch.   Mr. Wagner wrote of him in 1852:

283. "As the best test of the value of Mr. Dzierzon's system is the *results* which have been made to flow from it, a brief account of its rise and progress may be found interesting. In 1835, he commenced bee-keeping in the common way, with twelve colonies, and after various mishaps which taught him the defects of the common hives and the old mode of management, his stock was so reduced, that, in 1838, he had virtually to begin anew. At this period he contrived his improved hive, in its ruder form, which gave him the command over all the combs, and he began to experiment on the theory which observation and study had enabled him to devise. Thenceforward his progress was as rapid, as his success was complete and triumphant. Though he met with frequent reverses, about seventy colonies having been stolen from him, sixty destroyed by fire, and twenty-four by a flood, yet, in 1846, his stock had increased to three hundred and sixty colonies, and he realized from them that year six thousand pounds of honey, besides several hundred weight of wax. At the same time, most of the cultivators in his vicinity, who pursued the common methods, had fewer hives than they had when he commenced.

"In the year 1848, a fatal pestilence, known by the name of 'foul brood' (**787**), prevailed among his bees, and destroyed nearly all his colonies before it could be subdued, only about ten having escaped the malady which attacked alike the old stocks and his artificial swarms. (**469**). He estimates his entire loss that year at over five hundred colonies. Nevertheless, he succeeded so well in multiplying by artificial swarms, the few that remained healthy, that, in the Fall of 1851, his stock consisted of nearly four hundred colonies. He must therefore have multiplied his stocks more than three-fold each year."

But in the Dzierzon hive, it is often necessary to cut and remove many combs to get access to a particular one ; thus if the tenth from the end is to be removed, nine must be taken out. This hive cannot furnish the surplus honey in a form the most salable in our markets, or admitting of safe transportation in the comb. Notwithstanding these disadvantages, it has achieved a great triumph in Germany, and given a new impulse to the cultivation of bees.

Dzierzon builds hives in structures of two, four and even more colonies, piled upon one another. On the frontispiece to the first edition of this work, Mr. Langstroth gave a representation of a triple hive. The little that can be saved in the first cost of such hives, he found to be more than lost by the great inconvenience of handling them.

MOVABLE–FRAME HIVES.

284. About one hundred years ago, Huber invented the leaf-hive, which enabled him to make his discoveries. It consisted of twelve frames, each an inch and a quarter in width, which were connected together by hinges, so that they could be opened or shut at pleasure, like the leaves of a book.

(Fig. 53.)
THE HUBER LEAF HIVE.
(From Hamet.)

285. This hive was lately improved upon by several bee-keepers in Europe and America, the most noted of whom are the late Mr. Quinby, and his son-in-law, L. C. Root, author and publisher of one of the most progressive bee-books, "Quinby's New Bee-keeping." This style of hive

is generally known as the closed-end standing-frame hive.
Mr. Armstrong of Illinois, seems to be successful with a
hive almost entirely similar to the Huber leaf-hive in its
principles.   Mr. Heddon, of Michigan, has also patented a
closed-end frame hive, which is praised by some bee-keepers
of note.   The reader will understand that, in these hives,
the combs hang separately in frames, which, when joined
together, make a body, enclosed in an outer covering.   Their
being used by a number of Apiarists, shows that these
hives have some advantages, the greatest objection to them
being the difficulty of fitting the frames together, after inspection, without crushing some bees, unless they have been
previously shaken out.

286. Several attempts were made, in the first half of
this century, to invent a practical hanging-frame hive; that
is, a hive in which each comb, hanging in a separate frame,
could be readily taken out and replaced without jarring the
hive, or removing the other frames.   Propokovitsch, in
Russia, Munn, in England, Debeauvoys, in France, tried
and failed.   At last, in October, 1851, Mr. Langstroth
invented the top-opening movable-frame hive, now used
the world over, in which the combs are attached to movable
frames so suspended in the hives as to touch neither the
top, bottom, nor sides; leaving, between the frames and
the hive walls, a space of from one-fourth to three-eighths
of an inch, called bee-space.   (Fig. 54.)

287. By this device the combs can be removed at pleasure, without any cutting, and speedily transferred to another hive.   Our congenial friend, Prof. A. J. Cook, of
the Michigan State Agricultural College, and author of
"The Bee-keeper's Guide," says of it: "It is this hive,
the greatest Apiarian invention ever made, that has placed
American Apiculture in advance of that of all other countries."   And no one knows, better than the revisers of this
work, that such is the plain truth, as they have watched

the progress of bee-keeping in Europe, through its French, Italian, Swiss, and German bee-papers, for twenty years past.

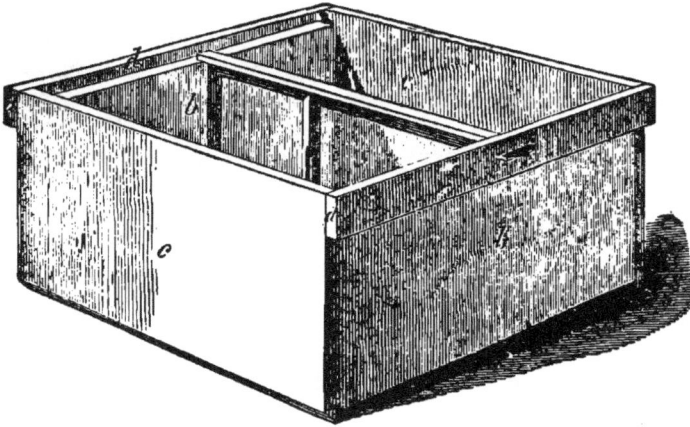

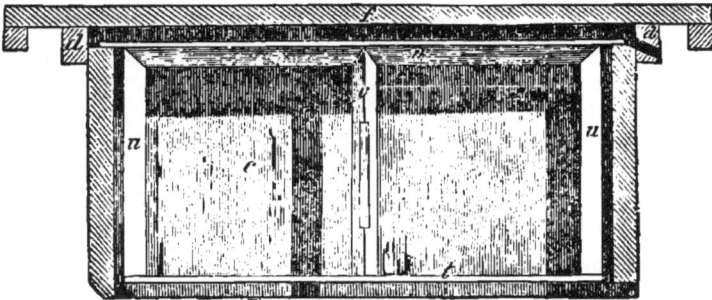

Fig. 54.

ORIGINAL LANGSTROTH HIVE.

b,b, front and rear of hive; *d,d*, pieces forming the rabbets for the frames to rest upon; *c,c*, sides of hive; *f*, movable cover; *u,u,t*, movable frame.

288. Mr. Langstroth, however, modestly disclaimed the idea of having attained perfection in his hive. He wrote:

"Having carefully studied the nature of the honey-bee, for many years, and compared my observations with those of writers and cultivators who have spent their lives in extending the sphere of Apiarian knowledge, I have endeavored to remedy the many difficulties with which bee-culture is beset, by adapting my invention to the actual habits and wants of the insect. I have also tested the merits of this hive by long continued experiments, made on a large scale, so that I might not, by deceiving both myself and others, add another to the useless contrivances which have deluded and disgusted a too credulous public. I would, however, utterly repudiate all claims to having devised even a perfect bee-hive. Perfection belongs only to the works of Him, to whose omniscient eye were present all causes and effects, with all their relations, when He spake, and from nothing formed the Universe. For man to stamp the label of perfection upon any work of his own, is to show both his folly and presumption."

289. A short time after the issuing of the Langstroth patent, the Baron Von Berlepsch, of Seebach, Thuringia, invented frames of a somewhat similar character. Carl T. E. Von Siebold, Professor of Zoölogy and Comparative Anatomy, in the University of Munich, thus speaks of these frames:

"As the lateral adhesion of the combs built down from the bars frequently rendered their removal difficult, Berlepsch tried to avoid this inconvenience, in a very ingenious way, by suspending in his hives, instead of the bars, small quadrangular frames, the vacuity of which the bees fill up with their comb, by which the removal and suspension of the combs are greatly facilitated, and altogether such a convenient arrangement is given to the Dzierzon-hive, *that nothing more remains to be desired.*" (???)

Mr. Cheshire (2d vol. page 46) was mistaken in attributing to Dzierzon the invention of the frame-hive, for Dzierzon has not even invented, but only perfected the movable-comb hive (**282–283**), having always, to this day, been opposed to frames. So the German hive is known as the Berlepsch hive.

290. For years, both of these inventions shared equally the attention of bee-keepers in Europe. Berlepsch's hive is used principally in Germany, Italy, and part of Switzerland; Langstroth's in England, France, and the French-speaking part of Switzerland; but it is to be noted, that hives made on the principle of the Langstroth invention, are steadily gaining ground wherever both styles are used.*

291. And this is not to be wondered at. The Berlepsch

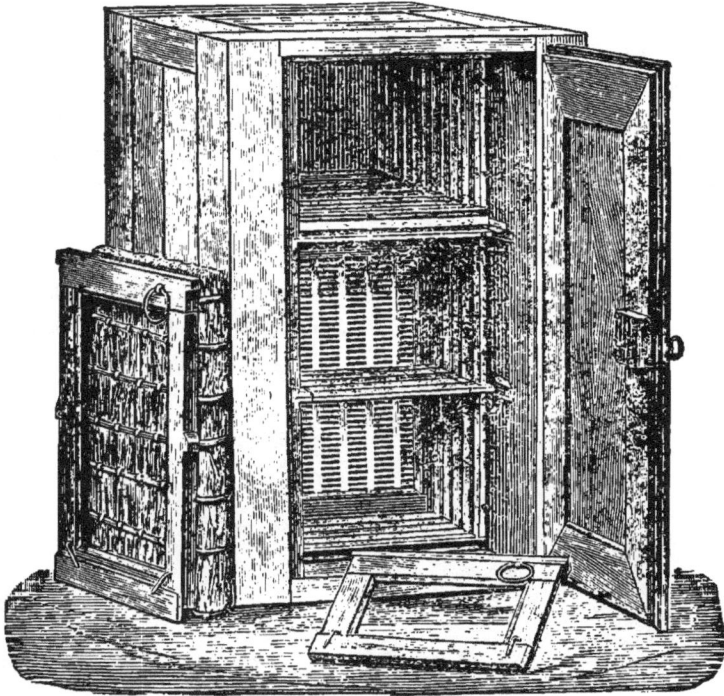

Fig. 55.
BERLEPSCH HIVE WITH BACK CUSHION.
(From the "*Illustrierte Bienenzeitung.*"

hive opens from the rear, like a cupboard. Two stories are used for the brood, and the third for surplus honey. This is sometimes separated from the main apartment by perforated zinc (**467**), to exclude the queen, or by a board

* At the Italian Bee-keepers' Convention, held in Milan, in September 1885, several Apiarists exhibited hives of this style, and yet none could be found in Italy, sixteen years ago. The first Langstroth hive which appeared in Italy was introduced by us, in 1872.

with a square hole in the center. The frames are suspended, in grooves, by the ends of their upper bars, and have to be taken out with pincers.

292. The worst feature of this hive is that, if it is necessary to reach the last frame, every one of the others has to be taken out. There are twenty combs in the brood-chamber. It is safe to say, that a hive built on the Langstroth principle, can be visited five times more rapidly, than a hive built on the Berlepsch idea. These inconveniences, coupled with the fact that the brood apartment of the Berlepsch hive is divided into two stories, and that the surplus apartment cannot be enlarged, *ad infinitum*, make the Berlepsch hive inferior; and we can safely predict that hives with movable ceiling will, some day, be exclusively used throughout the world.

Fig. 56.

SHOWING SOME OF THE EARLY IMPROVEMENTS OF THE LANGSTROTH HIVE, STILL IN USE IN SOME SECTIONS.

293. The superiority of the Langstroth hive is so evident that we were not surprised to read in the *Revue Internationale d'Apiculture*, Sept. 1885 :

Plate 9.

M. QUINBY,

Author of "*The Mysteries of Bee-Keeping.*"

This writer is mentioned pages 139, 147, 148, 150, 151, 152, 153, 157, 162, 168, 189, 363, 471.

"The question of the mobility of the ceiling was discussed at length at the Bee-keepers' Meeting held in Milan, Italy, in September 1885. Mr. Cowan and I were unable to conceal from the Italian bee-keepers our wonder that it was not solved for them, as it has been, for a long time, in the countries of large production.

"We can predict, and without any fear of mistake, that the principles on which the Langstroth hive is based will be admitted sooner or later by the most progressive bee-keepers of the world."—(Ed. Bertrand.)

294. The introduction of the Langstroth hive in Italy, and especially in Germany, has been hindered, so far, by the premature adoption of a standard frame, which "shuts the door to progress."—(Ed. Bertrand.)

295. The success of American bee-culture, in the last twenty years, was first attributed, by European bee-keepers, to the honey-producing power of the country; but the most intelligent Apiarists, who have tried the American methods, with the Langstroth hive, now recognize that success is principally due to the manipulations that it permits.

296. Nay, if the student will but refer to the former revision of this very book (1859), the first words of it will show him the progress accomplished since then:

"Practical bee-keeping in this country is in a very depressed condition, being entirely neglected by the mass of those most favorably situated for its pursuit. Notwithstanding the numerous hives which have been introduced, the ravages of the bee-moth have increased, and success is becoming more and more precarious. While multitudes have abandoned the pursuit in disgust, many even of the most experienced are beginning to suspect that all the so-called 'Improved Hives' are delusions or impostures; and that they must return to the simple box or hollow log, and 'take up' their bees with sulphur in the old-fashioned way."

297. Mr. Gravenhorst, also a German, invented a movable-frame hive made of straw. We give a cut of his hive

10

and Apiary, not that they have any practical importance for us, but because his system is peculiar. The frames are removed from the bottom, so that, in order to open one of

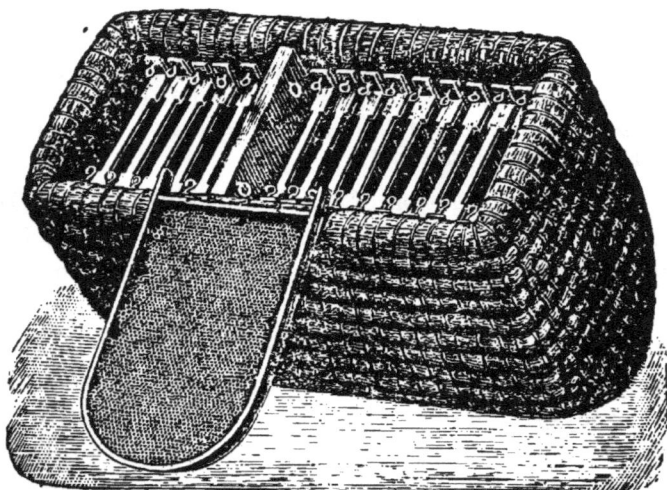

Fig. 7.
THE GRAVENHORST HIVE.
(From the "*Illustrierte Bienenzeitung.*")

these hives, it requires the strength of a strong man to invert it, especially if it is full of honey.

The Gravenhorst hive is not intended for ladies.

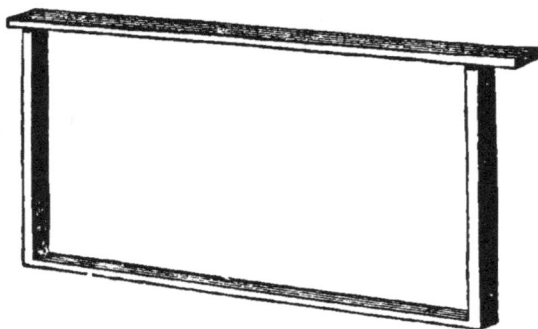

Fig. 58.
OLD STANDARD LANGSTROTH FRAME

298. Although the movable frame, hanging in the hive, by projections of the top bar (figs. 54, 58), as invented by Mr. Langstroth, is the style now almost universally adopted, there is a great diversity of opinions as to the proper size and shape of the frames, and the number, which a hive should contain. Hundreds of different sizes are used with success, from Maine to California, and from Canada to Texas. We herewith give a diagram of the principal frames

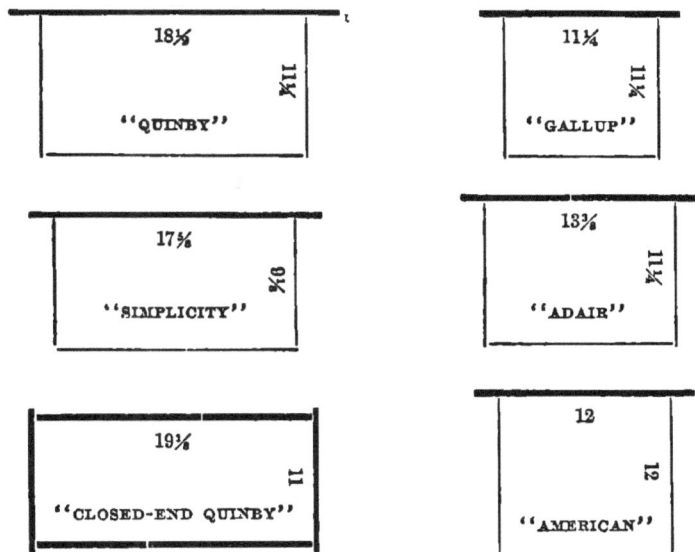

Fig. 59.

DIAGRAM OF PRINCIPAL FRAMES IN USE.

Figures given are outside dimensions in inches. Suspended frames have ¾-inch supporting arms, or an equal prolongation of top bar.

in use. The "Simplicity" is almost exactly similar to the original Langstroth frame: so much so, in fact, that they are interchangeable. This style of frame has been manufactured and sold, by the most prominent dealers, to such an extent, that it may be called the *Standard Frame of America*.

299. The "Hanging Quinby" is the frame preferred by

the writers. The "Gallup" frame is used with success by such practical Apiarists as G. M. Doolittle and O. Clute, author of a charming little novel entitled "The Blessed Bees," under the *nom de plume* of "John Allen." The "American" and "Adair" frames are somewhat in use also. The "Closed-End Quinby" (**285**) is not a hanging frame, but it is nevertheless used by such bee-keepers as Messrs. L. C. Root, Hetherington, Bingham, etc.

300. It is evident that profit can be derived from bee-culture with almost any style of frame; but it is certain also, that, in every pursuit, some conditions produce better effects than others, under the same circumstances.

In Apiculture, as in everything else, we should try to obtain the best results with the least labor and expense, and these can only be attained by studying the habits of the bee, and complying with them, as far as is practicable.

The combs of the brood-chamber, or main apartment of the hive, are used by the bees to raise their young, and to store their food for Winter. The size of frames must be considered, with reference to this.

301. We have seen (**153**) that the queen lays her eggs in a circle. In fact, it is necessary that she should do so, in order to lose no time in hunting for cells; else how could she lay three thousand eggs, or more, per day? A very shallow frame will break the circle, and compel her to lose time. In a comb five inches deep, for instance, and fifteen or sixteen inches long, the largest circular area contains less than twenty square inches, or five hundred and fifty worker-cells on each side. When these are occupied with eggs, the queen, while hunting for empty cells, will find wood above and below, instead of comb, at every half turn, and will lose not only time, but eggs; for, in the busy season, her eggs have to drop, like mature fruit, if not laid in the cells. *Loss of eggs is loss of bees; loss of bees at the proper time is loss of honey.*

302. A two-story shallow brood-chamber is objectionable for the same reason.   Besides, the bees which
cover the brood and keep it warm, must also keep warm
the lower bar of the top frame, the upper bar of the lower
frame, and the space between the two, without deriving any
benefit from such an arrangement.   This division of the
brood-combs into two shallow stories, is one of the causes,
which prevent the bee-keepers of Germany from raising as
many bees, in their hives, as we do here in the ordinary
Langstroth hives.   This disadvantage was so evident that
the bee-keepers of Switzerland, who had adopted, as a
standard, the Berlepsch hive (fig. 55), decided to replace
the double story by a single one of the same dimension, as
the Italian bee-keepers had done before, but for half the
hive only.

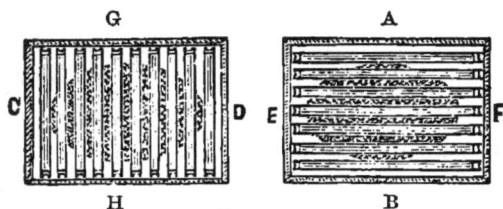

Fig. 60.

DIAGRAMS OF GALLUP AND LANGSTROTH HIVES.
(From the "A. B. C. of Bee-Culture.")

A small frame like the Gallup (fig. 59), presents another
objection, the cluster being divided among a greater number of frames.

"For Winter, it is evident that the sides of the clusters
A. B. and C. D. (fig. 60) are better protected than the ends G. H.
and E. F., and also that the long frames protect the center of the
brood-nest much better than the short ones."—(A. I. Root, "A.
B. C.")

Even a cross-bar through a frame (fig. 54) will hinder
the laying of the queen, so that brood will often be raised
only on one side of it.   Any one can easily try this.

303. From the foregoing, it appears that a square frame is the best for breeding. But square frames are objectionable. If they are small, they do not have enough space in each frame for Winter supplies, above or behind the brood. If they are large, they are unhandy, and their depth makes them difficult to take out without crushing bees. We have used some sixty hives, American frames, 12 by 12, for eighteen years or more, and this is our greatest objection to them.

304. A deeper frame is still more objectionable for the same reason,* and because the surplus cases on top are too remote from the brood. (**278.**) In early Spring, the bees have more difficulty in keeping the lower end of such frames warm, as the heat always rises, and a part of it is wasted, warming up the stores, which in this hive are all above the brood. In hot weather, the combs are also more apt to break down from heat and weight combined. Such a hive is deficient in top-surface for the storing of honey in boxes.

305. It is thus evident, that Mr. Langstroth and Mr. Quinby† were right in using frames of greater length than depth, especially as these frames allow of more surplus room above the brood, a matter of some importance.

306. But we must beware of excess in anything. A

---

* The *deeper* the frames, the more difficult it is to make them hang *true* on the rabbets, and the greater the difficulty of handling them without crushing the bees or breaking the combs.

† The late Mr. M. Quinby, of St. Johnsville, New York, in calling my attention to some stocks, which he had purchased in box hives of this shape, informed me that bees wintered in them about as well as in tall hives, the bees drawing *back* among their stores in cold weather, just as in tall hives they draw *up* among them. My hive, as at first constructed, was fourteen and one-eighth inches from front to rear, eighteen and one-eighth inches from side to side, and nine inches deep, holding twelve frames. After Mr. Quinby called my attention to the wintering of bees in his long box-hives, I constructed one that measured twenty-four inches from front to rear, twelve inches from side to side, and ten inches deep, holding eight frames. I have since preferred to make my hives eighteen and one-eighth inches from front to rear, fourteen and one-eighth inches from side to side, and ten inches deep. Mr. Quinby preferred to make my movable frames longer and deeper.—L. L. L.

shallow frame has too little honey above the cluster in Winter, and in long cold Winters, like that of 1884–5, a great many bees die for want of food above them, in hives containing plenty of honey (**630**), the combs, *back of the cluster*, being too cold.

The Langstroth-Simplicity frame is long enough, but hardly deep enough. The Quinby frame is deep enough, but would be better if a little shorter.

307. We have used on a large scale Quinby, American and Standard Langstroth-sized frames for years, and have obtained better results from the Quinby, both for wintering out of doors, and for honey producing. Yet, the Langstroth-Simplicity being the standard frame of America, we would hesitate to advise any Apiarist to change from this size ; knowing, by practical experience, how annoying it is, not to have all frames and all hives in one Apiary uniform in size.

But we would counsel beginners to use the Quinby size, —especially if they intend to winter out-of-doors,—or at least to use a frame as long as the standard Langstroth and as deep as the Quinby.

308. The number of frames to be used in a hive depends on their size ; for we should manage our bees, as we do our other domestic animals, and give them as much space as is necessary to obtain the best results. What would we think of a farmer who would build a barn without first considering the number of animals and the amount of feed which he intended to shelter in it?

309. Many hives cannot hold one-quarter of the bees, comb, and honey which, in a good season, may be found in large ones ; while their owners wonder that they obtain so little profit from their bees. A good swarm of bees, put, in a good season, into a diminutive hive, may be compared to a powerful team of horses harnessed to a baby wagon, or a noble fall of water wasted in turning a petty water-wheel.

As the harvest of honey is always in proportion to the number of bees in the hive, and as a large colony requires no more labor from the Apiarist than a small one, the hive should afford the queen sufficient space to deposit all the eggs, which she is able to lay* during twenty-one days, the average time for an egg to be transformed into a worker. Besides, it should contain a certain amount of food, honey and pollen.

310. We have seen before (**97**) that a good queen can lay 3,500 eggs per day in the good season, so that 73,500 cells may be occupied with brood at one time. If we add to this number about 20,000 cells for the provisions needed in the breeding season, we have about 94,000 cells as the number required for a strong colony. As every square inch of comb contains about 55 cells (**217**), 27 to 28 on each side, the combs of a hive should measure over 1,700 square inches. This space must, of course, allow of contraction, according to the needs of the colony, by what is called movable division boards. (**349.**)

311. If the reader will refer to the dimensions of frames given (**298**), he will ascertain that as a Quinby frame measures 189 square inches inside, a hive should contain at least 9 of these frames.

As the Standard Langstroth-Simplicity frame measures about 149 square inches, the hive must contain 12 frames. The American frames must number 13, and the Gallup 14.

312. We know that many Apiarists object† to these figures, because they succeed, and harvest good crops, with

---

* It is unquestionable that the quality of a queen depends on the quantity of eggs that she is able to lay. Then why limit her, by using hives so narrow that she cannot develop her fertility?

† It is perhaps necessary to say here, that we have found more opposition on this subject than on any other, especially in the bee-papers. But we take this opportunity of again *energetically* asserting that our preference for large hives is based on a successful practice of more than twenty years, with several hundred colonies in different sized hives, while our opponents could bring forward nothing but their preconceived ideas.

AN APIARY I

PLATE 10.

smaller hives. But figures, based on facts, cannot lie. Smaller hives will do only in localities, where late Springs and short honey crops make it impossible for the queen to lay to the utmost of her capacity, before the time when her bees would be useful.

313. It is only by testing different sizes of hives and frames side by side, for years, on a large scale, and with the same management, as we have done, that the comparison can be made serviceable. Our experiments prove also that small frames impede the laying of the queen. *The brood-chamber of a large hive can easily be reduced in size, if need be; but a small hive cannot be enlarged at will, except by the addition of upper stories, which should properly be devoted to the storing of honey.*

314. In addition to the disadvantages of small frames and small hives already enumerated, another — and the greatest of all—is the excess of natural swarming which they cause. The leading advocates of small hives, some of whom are large honey producers, invariably acknowledge that they have *too much natural swarming;* nor is it to be wondered at, since swarming is mainly caused by the lack of breeding room for the queen. (**406.**)

315. The main criterion of a good farmer, is the care that he takes to improve his stock, by selecting the best animals as reproducers. If we use hives so narrow that we cannot discern which are our most prolific queens, and that they incite natural swarming, we are unable to improve our bees by selection. (**452, 511.**)

316. The distance, between frames from center to center, can be varied, as we have seen before (**214**), from 1⅜ inches to 1½, in the breeding apartment, of which we are now treating. In the surplus cases, it may be made much greater.

317. The distance of 1½ inches, advised by Mr. Quinby, is preferable two for reasons:

1st, It facilitates the taking out of the combs, giving a little more room to handle them, and thus aids in interchanging combs, which may have slight irregularities; when such changes are necessary to help weak colonies with brood or honey from stronger ones.

2nd, It gives more room between brood-combs for the bees to cluster in Winter, and a greater thickness of honey above them, thereby placing the bees in better condition for Winter.

318. The frames must be properly distanced in the hive, and the combs must be built straight in them; for a movable-frame hive, with crooked combs, is worse than a hive without any frames.

319. The building of straight combs in the frames was formerly tolerably secured by the use of a triangular wooden guide fastened to the underside of the top bar of the frame, and which the bees follow in most instances. Something of this kind was mentioned by Della Rocca as early as 1790. (" Traité Complet sur les Abeilles.")

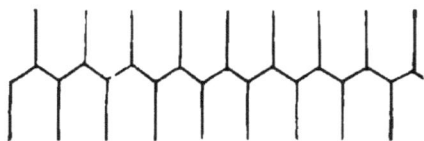

Fig. 61.

The figure 61 shows the form of a metallic stamp, invented by Mr. Mehring, of Bavaria, Germany, *for printing or stamping the shape of the combs* upon the under side of the top bar of the frames. After the outlines were made he rubbed melted wax over them, and scraped off all that did not sink into the depressions. Mr. Mehring represented this device as enabling him to dispense with guide combs, the bees appearing to be delighted to have their work thus accurately sketched out for them.* In practice it

* This invention should not be confused with that of comb-foundation, made a few years later by the same distinguished Apiarist. (677)

was found to be inferior to the triangular comb guides.

Pieces of worker-comb, glued to the under side of the top bar with melted wax, were used successfully. But the introduction of comb-foundation (**674**) has finally given us the means of securing straight combs at all times, and it may be used, for this purpose, in such narrow strips, that its cost cannot be an objection.

320. STANDARD L. MOVABLE FRAME. — Top bar, $19\frac{1}{4}$ long $\times$ $\frac{7}{8}$ wide $\times$ $\frac{5}{8}$ thick. In each end a notch $\frac{5}{16} \times 1\frac{3}{16}$ is made in the thickness of it, leaving a projecting or supporting shoulder which is to rest in a rabbet in the upper ends of the hive, and by which the frame is suspended (fig. 54). Ends or vertical pieces: two pieces $8\frac{7}{8}$ long $\times$ $\frac{7}{8}$ wide $\times$ $\frac{5}{16}$ thick. Bottom bar $16\frac{3}{4}$ long $\times$ $\frac{7}{8}$ wide $\times$ $\frac{1}{4}$. We will call the attention of manufacturers to the fact, that this makes a much stronger frame than the former style, given in previous editions, and preserves the exact outside measurements. The ends, or vertical pieces, are nailed both ways to the top bar (fig. 71), and the bottom bar is nailed inside of them, instead of under them as formerly.

321. We must not forget that these bottom bars sometimes have to support the weight of heavy combs, as in transferring (**574**), and that the bees may glue them fast to lumps, which happen to be on the bottom board. Hence the necessity of having them nailed, so that they will not pull out.*

All the parts of the movable frames should be cut out by circular saws, and the measurement should be exact, so that the frames when nailed together may be square. If they are not *strong* and *perfectly square*, the proper working of the hive will be greatly interfered with.

322. The under side of the top bar may be cut to a tri-

---

* As a rule, manufacturers make the top bar of the frames too weak; some have remedied this by excessive wiring, and a tin brace in the center. Such contrivances are costly and worse than useless.

angular edge, but where comb foundation is used, the flat top bar will be found much better (**693**). Above all, the outside measurements of the frame must be carefully preserved.

323. The width of the *top bar* has something to do with the amount of *bridges and brace combs* (**397**), built by the bees, between the brood-chamber and the upper stories. A wide top bar, leaving but a narrow space for passage above, will almost altogether prevent the building of bridges, but it has other disadvantages that have rendered it unpopular, although some bee-keepers of note—Col. Camm of Illinois, among others — use it. In producing extracted honey (**749**) these bridges and brace combs do not annoy much.

324. L. SIMPLICITY FRAME (fig. 59). —This frame has been made and sold so largely by A. I. Root, and other dealers, that it is established now. The length of the top bar and the height of the frame are the same as those of the Standard L. Frame, the frame itself being one-fourth inch longer outside. They are sometimes made with metal corners invented by A. I. Root (fig. 62).

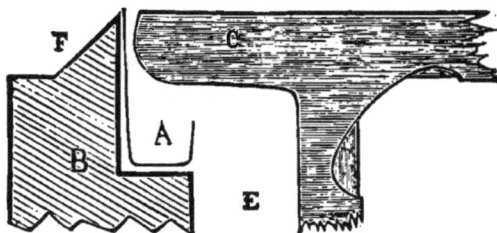

Fig. 62.

METAL CORNER AND ITS POSITION IN THE HIVE.

The engraving is full size. The ⅞ board *B* is supposed to be the end of the hive. *A* is a section of the metal rabbet, and *C* is the corner. *E* is the space between the hive and the frame; *F* is the beveled edge to receive the upper story.—("A. B. C. of Bee-Culture.")

325. These tin corners have the advantage of making the frames very strong; and as the tin shoulder rests by a "knife edge," C., on another tin edge, at right angles with it, A., nailed in the rabbet of the hive, the bees cannot glue the frames fast. But these frames have the disadvantage of getting out of place easily, too easily in fact, and their sharp edges make them very inconvenient to handle.

326. For the L. Quinby suspended frame, see diagram (fig. 68). This frame is one-fourth inch deeper than that originally given by Mr. Quinby in his "Mysteries of Bee-keeping." Mr. Quinby had too much space in the hive, under the frame.

327. It is necessary that the hive should always slant forward, toward the entrance, when occupied by bees, to facilitate the carrying out of dead bees, and other useless substances; to aid the colony in protecting itself against robbers, to carry off moisture, and prevent rain from beating into the hive.

328. For this, and other reasons, the combs should run from front to rear,—so as to hang perpendicularly —and not from side to side as they do in the Berlepsch hive.

329. The Langstroth hive, from the simple form given in fig. 54, was improved upon in many different ways. The Standard Langstroth hive has been, for a long time (fig. 63), a hive with portico, honey-board, permanent bottom-board, and ten frames.

Fig. 63.

330. In this hive, the "*observing-glass*," in the rear, was first discarded, and replaced by a board, making the hive more simple and cheaper.   The glass in the rear is of no use, in practical bee-keeping, and for experimenting, the observing hives such as described (**375**), with only one comb, and both sides of glass, are to be preferred (fig. 80).

331. The *movable honey-board*, between the brood-chamber and the upper stories, has been also discarded of late years, the great objection to honey-boards being that the bees glue them, and build small pieces of comb or *bridges*, in the space between them and the frames ; the jar of their breaking, when the honey-board is removed, anger-ing the bees.

332. The *permanent bottom-board* has lost favor with the

Fig. 64.
VAN DEUSEN CLAMP.

great majority of bee-keepers, and is now replaced by mov-able bottom-boards adjustable at will.   The Van Deusen hive-clamp (fig. 64), is used by many Apiarists for fastening movable bottoms or additional stories.   We have discarded the permanent bottom-board, owing to the difficulty of prompt-ly cleaning it of dead bees and rubbish, when removing bees from the cellar in Spring, or after a hard winter passed out of doors.

333. In the ventilation of the hive, we should endeavor, as far as possible, to meet the necessities of the bees, under all the varying circumstances to which they are exposed in our uncertain climate, whose severe extremes of temperature forcibly impress upon the bee-keeper, the maxim of Virgil,

" Utraque vis pariter apibus metuenda."
" Extremes of heat or cold, alike are hurtful to the bees."

To be useful to the majority of bee-keepers, *artificial* ventilation must be simple, and not as in Nutt's hive, and other labored contrivances, so complicated as to require almost as close supervision as a hot-bed or green-house.

Fig. 65.

HIVE, WITH EXTRACTING SUPERS SET BACK FOR VENTILATION IN VERY HOT WEATHER.

The cap is thrown back to show the straw mat.

334. With an independent bottom-board, ventilation can

be given to any amount by raising the hive, as in fig. 65, or even more. By furnishing ventilation independent of the entrance, above the brood-chamber, or between the different surplus apartments, if necessary, we improve upon the method which bees, in a state of nature, are compelled to adopt, when the openings in their hollow trees are so small, that they must employ, in hot weather, a larger force in ventilation, than would otherwise be necessary.

335. The bees, finding their home more pleasant, will cease to cluster on the outside, as long as there will be honey to gather, and room to store it in.

336. On the other hand, by the use of movable blocks, the entrance may be kept so small, in cool weather, that only a single bee can go in at once, or it may be entirely closed.

While sufficient airing must be given, the supply should be controlled, so as not to injure the brood by admitting too strong a current of chilly air. In the chapter on wintering bees, directions are given for ventilating the hives in cold weather, so as to carry off all superfluous moisture. **(636.)**

337. For the benefit of beginners, it may be necessary to add, that the bees will glue up with propolis **(236)**, and sooner or later entirely close any ventilating holes through which they cannot pass Hence air holes, covered with wire cloth, miss their purpose altogether. In the same manner, and with a great deal of labor, bees will try to close any upper entrances, such as that of figs. 65 and 54*d*, if these remain open, when not needed for the welfare of the colony.

338. The *portico* of the Langstroth hive has advantages, and disadvantages, which about balance one another. Its advantages are, that it shelters the bees from rain in Summer, and from cold and snow in Winter. Its disadvantages are, that it sometimes harbors enemies of bees, moths, spiders, etc., etc., and sometimes helps to hide the queen from the Apiarist's diligent search. It hinders the bee-keeper when he wants to watch closely the sport of bees before the entrance.

Fig. 66.
DOUBLE–STORY LANGSTROTH "SIMPLICITY," WITH PORTICO.

339. When the portico-hive is used, two entrance blocks are provided, as per accompanying diagram. By changing

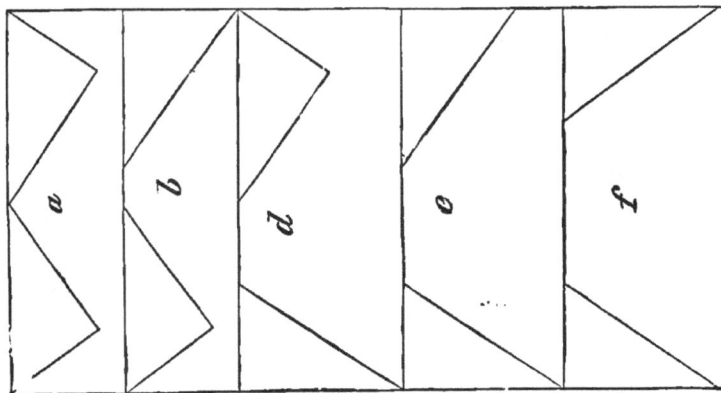

Fig. 67.
ENTRANCE BLOCKS.
a, hive closed; *b, c, d, e, f*, gradually enlarged openings.

11

the position of these blocks on the alighting-board (see fig. 67, in which some of the positions are shown), the size of the entrance to the hive may be varied in a great many ways, and the bees always directed to it by the shape of the block, without any loss of time in searching for it.

The Hive We Prefer.

340. The diagram we give (fig. 68), of the hive we prefer to all others, can be taken as a pattern for any other size, by changing the size of the pieces and retaining only the exact distances between the frames and the body, and the height of the entrance. Its details can be varied *ad infinitum*. It can hold eleven frames, but generally we use only nine frames and two contracting, or division-boards, or ten frames and one division-board. (**349.**)

This hive, in the dimensions given, is *not* a new, untried pattern. We have used several hundreds of them for years, with the best of success. It is used extensively by several large producers.

341. In consequence of our writings in the Swiss and French bee-papers, it was adopted, under the name of the Quinby-Dadant hive, by several progressive bee-keepers on the other side of the Atlantic, where it gets new partisans every year.

The publisher of the *Revue Internationale d'Apiculture*, Mr. Ed. Bertrand, in the number of October 1887, writes:

"These wide hives, several bee-keepers find that they are too small; for some have increased them to thirteen frames instead of eleven, and I have seen such large hives, last Summer, filled with bees and honey, besides two upper stories of thirteen half-frames each, the whole containing 120 quarts, all occupied by the daughters of the same queen."

In the same number, a German bee-keeper, Mr. Chas.

Regnier, of Sarrelouis, gives the result of a comparison of the Standard German (**289**) with these hives. He

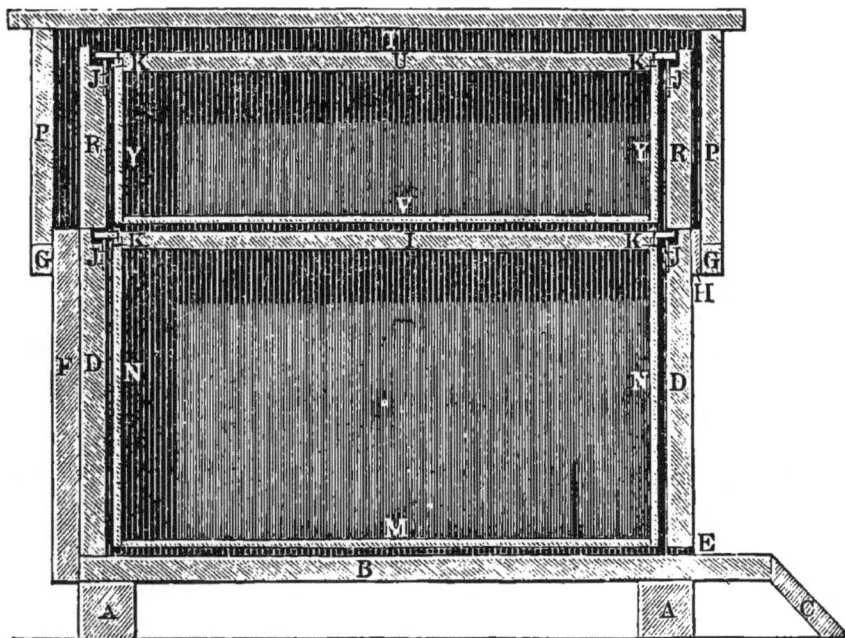

Fig. 68.

DIAGRAM OF OUR HIVE.

AA, cross-pieces to support the bottom, 18x2x4. *B*, bottom, 25x17½x⅞. *C*, apron, 10x17½x⅞. *DD*, front and rear of the hive, 16½x12¼x⅞. *E*, entrance, 8x⅜. *F*, double board nailed at the rear, 17¼x13x⅞. *GG*, square slats to support the cover. *H*, lath, ½x1¼ to widen the top edge of the front board. *I*, top bar of the frame, 20¼x¾x⅞. *JJJJ*, rabbets ⅝ wide x⅝ high, dug in front and rear boards, and furnished with sheets of iron ¾ inches wide, projecting ¼ of an inch, on which the frame-shoulders are supported. If the grooves are not provided with the sheets of iron, their size should be ½x⅜. *KKKK*, show how the uprights *NN* of the frames are nailed to the top bar. *M*, bottom bar of the frame, 17⅞x½x⅞. *NN*, sides of the frame, 11¼x5-16x⅞. *PP*, front and rear of the cap, 18½x9x⅞. *RR*, front and rear of the surplus-box, 16½x6¾x⅞. *T*, empty space on top of the surplus-box, 1¼. *U*, top bar of the surplus-frame, same as top-bar *I*. *V*, bottom bar of the surplus frames, same as *M*. *YY*, sides of the surplus frames, 6x¼x⅞.

The space between *M* and *B* is about ½ inch; between *DN*, *ND*, *VI*, *RY*, *YR*, should be ⅜ of an inch. Hives of every size can be constructed on this diagram, with the only caution to preserve the spaces of the width indicated.

writes that the crop of his German averaged about twenty-one pounds, while his Dadant hives averaged about forty-eight pounds, adding that, at the start, his German were full of combs, while the Dadant had several combs to build.

(Fig. 69.)

a, front of the hive; *b*, slanting board; *c*, movable block; *d*, cap; *e*, straw mat; *f*, enamel cloth; *g*, frames with combs.

342. Its *movable bottom board* (fig. 69), is adjusted or encased in the body of the main hive, on all sides but the front, to shed the rain and better protect the colony against ants and moths. It projects forward three inches, at least, to support an adjustable entrance-block. Some Apiarists use a tin slide, instead of an *entrance-block*. We object to it, because, if glued by bees it may be bent in handling, and if it is mislaid, it cannot always be promptly replaced ; while any square wooden-block can take the place of the entrance-block, if necessary.

Fig. 70.
HIVE, SETTING FLAT ON THE BOTTOM.

343. The *apron*, or slanting-board, helps overladen workers to reach the entrance, when they have fallen to the ground. The blocks that support the bottom, may be made of unequal height, so as to give the hive the proper forward slant, on level ground. If the grain of the lumber in the bottom-board runs from front to rear, it will shed water more readily, and rot less. If the bottom is nailed on the cross-blocks, it will not be in danger of warping.

Our Swiss friends make the bottom-board with the grain running from side to side. They say that in this way they can make it fit exactly in the lower rabbet of the hive, without swelling or shrinking. They also make the apron, with hinges fastened on the bottom-board, and in snowy or cold weather, they raise it and lean it against the hive, to protect the entrance.

344. The *adjustable bottom board* is convenient in many instances. If in taking the bees from a winter repository, it is found wet and mouldy, you can at once exchange it for a dry one, and wipe the wet board at leisure. Or, if a comb breaks down in Summer, by weight and heat, the hive can be lifted off its bottom, and placed on a clean stand, so that the leaking honey and broken combs can be instantly removed, and robbing or daubing of bees avoided. Moreover, the bottom-board is the first part of the hive to decay, and a hive-body and cover will usually outlast two bottom-boards. The movable bottom allowing the raising of the hive for ventilation, in extremely hot weather, enables us also to discard the back ventilator, of the old hive (fig. 63.)

345. The *body of the hive* is made double on the back, which should always be the North side of the hive. (**567.**) This, with the division-board inside, on the West, shelters

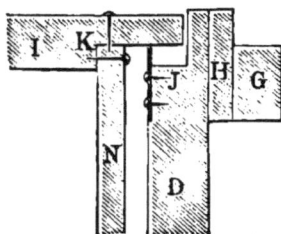

Fig. 71.

the colony more efficiently than a single board against the cold North-West winds of Winter. If the bees are to be wintered indoors, the double back may be dispensed with. A more simple form of body, setting flat on the bottom, as in fig. 70, can also be made.

The rabbet in which the frames hang, is made with a

(fig. 71), sheet-iron shoulder, supporting the frame, similar to Root's tin edge. This can be dispensed with altogether, but in such cases, the rabbet should be only deep enough for the frame to hang as represented in fig. 54. The plain

Fig. 72.
SHOWING HOW THE SPACING WIRE IS FIXED.

wooden rabbet is objectionable, because the bees glue the frame shoulders with propolis. Yet we use it in our hives almost altogether, because of the difficulty of fitting the division-board closely otherwise.

346. In any style of hanging-frame hives, it is indispensable for the frames to be so suspended, that a bee can pass between them and the body, bottom, and upper story, to prevent the gluing of them with propolis. (See bee-space, **286.**)

In our hives, we give only one-eighth of an inch of space, above the frames, below the top edge of the hive, and give one-fourth inch under the frames of the upper-story, which preserves the three-eighths bee-space, between each story (**286**). We found, in practice, that there was danger of crushing bees, in handling the upper stories, when they were made so that the frames were flush with their lower edge.

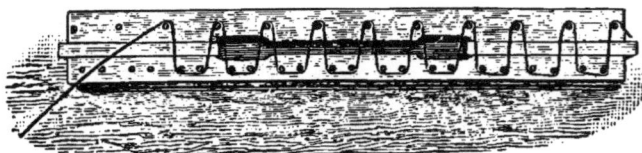

(Fig. 73.)
SHOWING THE TOOL USED TO BEND THE WIRE BRACES.

347. The *Spacing-wire*, an improvement on Quinby's wire brace, to space the frames at the bottom, is found very convenient in hives as deep as this. It is also useful in indicating to novices the number of frames to be placed in

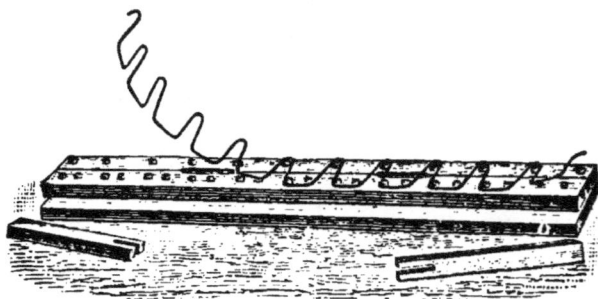

FIG. 74.
SHOWING HOW THE WIRE IS REMOVED.

the hive. Even a practical bee-keeper will sometimes make the mistake of putting eleven or thirteen frames, in a hive that should hold twelve. With this wire, mistakes are impossible, as they will at once be detected. Besides, if the hive has to be transported some distance, it keeps the frames from jarring. Its cost is insignificant. Some Swiss Apiarists use two of these, one in each end.

348. The *entrance* should not be less than five-sixteenths, or more than three-eighths of an inch in depth, in order to give easy passage to the bees, and at the same time, keep out mice. Round holes are objectionable. Each hive is furnished with an entrance-block, somewhat heavy, and cut as in fig 69, to reduce, or close the entrance, according to the emergencies.

349. The *division board,* also called contractor or dummy,

Fig. 75.
DIVISION BOARD.

is an *indispensable* feature of all good hives. With its help, the hive may be adjusted to the size of the weakest swarm, and in Winter, the space behind it can be filled with warm and absorbing material (**636**). The constant use of a division board, even in the strongest colonies, renders the handling of combs much easier. All Apiarists know that the first comb is the hardest to remove. By removing the board first, the combs are at once free and can be easily taken out.

350. This board is made of the same depth as the frames, with a similar top-bar. Some Apiarists use a division-board the full depth of the hive, but in moving it, bees are crushed under it, and if any bees happen to be on the outside of it, they cannot escape, and die there. On the other hand, this bee-passage is not objectionable, since heat, having a tendency to rise, does not escape through it. The board is made one-fourth inch shorter than the inside of the hive, and a strip of oil-cloth or enamel cloth, one and a half inches wide, is tacked on, to fill the spaces at each end. In this way, the board fits well against the ends, and is never glued so as to make it difficult to remove. A small half-round pine-strip, laid . against the end of the board, while tacking on the cloth, and pulled out afterwards, helps to tack the cloth properly. To prevent the bees from tearing or gnawing the edge of the cloth, some Apiarists nail a small strip of tin over it.

351. In the diagram (fig. 68), the reader will notice the strip H, used to widen the upper surface of the rabbeted end of the hive. This wide surface is very convenient, to make the cloth and straw-mat fit closely, as they can thus be cut a little longer.

352. *The oil-cloth* or enamel-cloth, first applied to hive purposes by R. Bickford, is used over the brood-frames in Spring. It fits closely, concentrates the heat, and can be removed without jar or effort. When the surplus arrangement, or upper story, is put on, this cloth is removed and placed at the top. (**759**) All Apiarists, or nearly all, who have tried the oil-cloth and

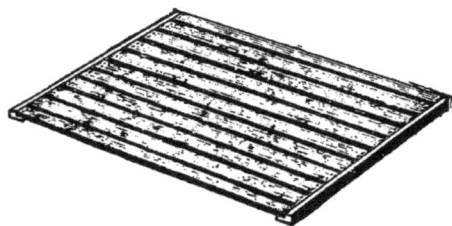

Fig. 76.
HEDDON'S SKELETON HONEY-BOARD.

honey-board simultaneously, have discarded the latter forever, except in some cases of comb-honey production, when a *skeleton* honey-board (fig. 76) is used between the stories. The oil-cloth is sometimes gnawed, or rather pulled to pieces by the bees in a few years, but its cost is so small, and its use so great, that it is worth while to replace it as often as necessary.

Fig. 77.
FRAME TO MAKE STRAW MATS.

353. The *straw-mat* is one of the most useful and necessary implements of the bee-hive. It is far superior to the wooden-mat described by one or two writers. It is flexible and porous, warm in Winter, cool in Summer. It may be made of rye straw, or of what is called slough-grass, a tough and coarse grass growing in marshy places, and abounding on the bottoms of the Mississippi Valley. The mat shown in fig. 69 is only about one inch thick. Mr. C. F. Muth manufactures mats much thicker and stronger; they are equal to a cushion.

In fig. 77 we present to our readers an engraving of a frame, for making these mats. They are very simple in construction. It is well, in making them, to use strong twine, soaked in linseed-oil; for the moisture, which escapes from the bees in Winter, would soon rot the string.

The enamel-cloth is removed before Winter (**635**), and the mat placed immediately over the frames. A good mat will last as long as the hive.

354. The upper story or cover may be a half-story cap, in one piece (fig. 65), or in two pieces (fig. 70), or, if only full stories are used for surplus, it may be a shallow cover (fig. 78), which will fit over either the first or the second story. We prefer the half-story cap, which can be readily filled

Fig. 78.

BLANTON'S TWO-STORY HIVE.

with absorbents for Winter, and is adapted to any style of supers.*

355. The caps must fit freely so as to be easily removed. They may be made of lighter lumber than the body of the hive, to save fatigue to the Apiarist in handling them. The top of the hive must be water-tight. Cracks, knots and seams should be avoided, or should be thoroughly painted with roof-cement. Before putting together the boards which form the top of the cap of our hives, we make, along both sides of the joints, a rounded groove, three-eighths of an inch wide and one-fourth of an inch deep, in which the rain-water runs, instead of leaking inside. Mr. McCord of Oxford, O., makes the covers of his hives water-tight, by covering them with strong muslin, tacked on with a strip nailed to the edges, and thoroughly painted. Mr. G. M. Doolittle of Borodino, N. Y., and Dr. C. C. Miller, of Marengo, Ill., both among the leading bee-writers and successful producers of honey, use tin, painted white, on the tops of their hives. The Swiss and French bee-keepers do the same.

356. The hives should always be painted, not only to make them last, but to give them a neat appearance. No

* This term is used by Apiarists to designate any upper box placed over the main lower hive.

dark colors should be used, as they absorb the sun's heat, nor should all the hives be of the same tint (**503**). If the joints are painted when they are put together, they will last much longer. Every old Apiarist well knows that the joints are the first to decay.

357. Each hive, in an Apiary, should bear a number, on the back of the brood apartment; and this should be printed in black characters, large enough to be seen at a distance. In small Apiaries bee-keepers use a slate, on each hive; but in large ones, where many operations are performed, it is better to keep a record of the condition of the colonies, and of all the operations, in a special book.

We will add, that a hive which does not furnish a thorough control over every comb cannot allow of the manipulations which the bee-keeper's necessities demand. Of such hives, the best are those which best unite *cheapness* and *simplicity*, with *protection in Winter*, and *ready access* to the spare honey-boxes.

358. In closing this chapter on hives, we cannot refrain from advising the beginners in bee-culture to be very cautious in buying patent hives. More than eight hundred patents on bee-hives and implements have been issued in the United States since January, 1873. Not ten of these have proved to be of any use to bee-keepers. The mention of this fact will suffice to show the small value of these 790 patents, and the loss incurred by those who have bought them, before they were able to judge of their merits.

MATERIALS FOR BEE-HIVES.

359. The variety of opinions respecting the best *materials* for hives, has been almost as great as on the subject of their proper size and shape. Columella* and Virgil rec-

---

* Columella, about the middle of the first century of the Christian Era, wrote twelve books on husbandry—"*De re rustica.*"

ommend the hollowed trunk of the *cork tree*, than which no material would be more admirable if it could only be cheaply procured. Straw hives have been used for ages, and are warm in Winter and cool in Summer. The difficulty of making them take and retain the proper shape for improved bee-keeping, is an objection to their use. Hives made of wood are, at the present time, fast superseding all others. The *lighter* and more *spongy* the wood, the poorer will be its power of conducting heat, and the warmer the hive in Winter and the cooler in Summer. Cedar, bass-wood, poplar, tulip-tree, and especially *soft pine*, afford excellent materials for bee-hives. The Apiarist must be governed, in his choice of lumber, by the cheapness with which any suitable kind can be obtained in his own immediate vicinity.

Scholz, a German Apiarist, recommends hives made of *adobe*—in which frames or slats may be used—as cheaply constructed, and admirable for Summer and Winter. Such structures, however, cannot be moved. But in many parts of our country, where both lumber and saw-mills are scarce, and where people are accustomed to build adobe houses, they might prove desirable. The material is plastic clay, mixed with cut straw, waste tow, etc.

360. To make the movable-frame hives to the best advantage, the lumber should be cut out by a circular saw, driven by steam, water, or horse-power, or even by foot-power. We have used the foot and hand circular-saws made by W. F. & J. Barnes, for years, and could not do without them in our shops. In buildings where such saws are used, the frames may be made from the small pieces of lumber, seldom of any use, except for fuel, and may be packed almost solid in a box, or in a hive which will afterwards serve for a pattern. One frame in such a box, properly nailed together, will serve as a guide for the rest. The parts of the hive can easily and cheaply be made by any one who can handle tools, but cannot be profitably manufactured to be sent far, unless made where lumber is cheap, and the parts closely packed,—in the flat,—to be put together after reaching their destination.

361. If the Apiarist desires minute instructions, on how to file his saws and keep them in order, select his lumber, and make his hives, with pleasure and profit, let him send to A. I. Root, of Medina, Ohio, for his "A. B. C., of Bee-Culture." He will be repaid a hundred-fold, by the many good points he will find in it.

362. We here cite, with illustration, his explanation of " why boards warp " :

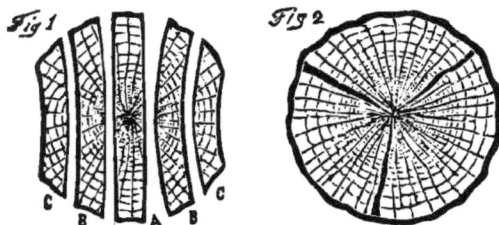

Fig. 79.

" Before going further, you are to sort the boards so as to have the heart side of the lumber come on the outside of the hive. If you look at the end of each board, you can see by the circles of growth, which is the heart side, as is shown in the cuts. At B, you see a board cut off just at one side of the heart of the tree; at C, near the bark; at A, the heart is in the centre of the board. You all know, almost without being told, that boards always warp like C; that is, the heart side becomes convex. The reason is connected with the shrinkage of boards in seasoning. When a log lies until it is perfectly seasoned, it often checks as in fig. 2. You will observe that the wood shortens in the direction of the circles, and but very little, if any, along the lines that run from the bark to the centre. To allow this shrinkage in one direction, the log splits or checks in the direction shown. Now to go back to our boards, you will see that B shrinks more than A, because A has the heart of the tree in its centre; that C will shrink, in seasoning, much more on the bark side, than on the heart side; that this cannot fail to bring the board out of a level; and that the heart side will always be convex. You have all seen bee-hives, probably, with the corners separated and gaping open, while the middle of the board was tight up in place. The reason was that the mechanic had put the boards on, wrong side out. If the heart side had been outward, the corners of the hive would have curled inwardly, and if the middle had been nailed securely, the whole hive would have been likely to have close, tight joints, even if exposed to the sun, wind, and rain."—("A. B. C. of Bee-Culture," page 103.)

363. Double-walled hives, chaff hives, and Winter covers, will be described in the chapter on "Wintering" (**619**). The upper-stories, half-stories, wide frames, sections, etc., for comb, or extracted honey, will be discussed in the chapter on honey producing (**716**).

VENTILATION OF THE BEE-HIVE.

364. If a populous colony is examined on a warm day, a number of bees may be seen standing upon the alighting-board, with their heads turned towards the entrance of the hive, their abdomens slightly elevated, and their wings in such rapid motion, that they are almost as indistinct as the spokes of a wheel, in swift rotation on its axis. A brisk current of air may be felt proceeding from the hive; and if a small piece of down be suspended at its entrance, by a thread, it will be drawn out from one part, and drawn in at another. Why are these bees so deeply absorbed in their fanning occupation, that they pay no attention to the busy numbers constantly crowding in and out of the hive? and what is the meaning of this double current of air? To Huber, we owe the satisfactory explanation of these curious phenomena. The bees, thus singularly plying their rapid wings, are *ventilating* the hive; and this double current is caused by pure air rushing in, to supply the place of the foul air which

PLATE 11.

A. I. ROOT ("NOVICE"),

Author of "*The A. B. C. of Bee Culture;*" Editor of "*Gleanings in Bee Culture.*"

This writer is mentioned, pages 61, 62, 92, 93, 95, 149, 156, 167, 175, 176, 272, 285, 288, 307, 312, 314, 315, 319, 320, 342, 368, 370, 422, 429, 438, 485, 486.

is forced out. By a series of beautiful experiments, Huber ascertained that the air of a crowded hive is almost as pure as the surrounding atmosphere. Now, as the entrance to such a hive is often very small, the air within cannot be renewed, without resort to artificial means. If a lamp is put into a close vessel, with only one small orifice, it will soon exhaust the oxygen, and cease to burn. If another small orifice is made, the same result will follow; but if a current of air is by some device drawn out from one opening, an equal current will force its way into the other, and the lamp will burn until the oil is exhausted.

365. It is on this principle of maintaining a double current by *artificial means*, that bees ventilate their crowded habitations. A file of ventilating bees stands inside and outside of the hive, each with head turned to its entrance, and while, by the rapid fanning of their "many twinkling" wings, a brisk current of air is blown out of the hive, an equal current is drawn in. As this important office demands unusual physical exertion, the exhausted laborers are, from time to time, relieved by fresh detachments. If the interior of the hive permits inspection, many ventilators will be found scattered through it, in very hot weather, all busily engaged in their laborious employment. If its entrance is contracted, speedy accessions will be made to their numbers, both inside and outside of the hive; and if it is closed entirely, the heat and impurity quickly increasing, the whole colony will attempt to renew the air by rapidly vibrating their wings, and in a short time, if unrelieved, will die of suffocation.

366. Careful experiments show that pure air is necessary not only for the respiration of the mature bees, but for hatching the eggs, and developing the larvæ; a fine netting of air-vessels enveloping the eggs, and the cells of the larvæ being closed with a covering filled with air-holes **(168).**

In Winter, if bees are kept in a dark place, which is

12

neither too warm nor too cold, they are almost dormant, and require very little air; but even under such circumstances, they cannot live entirely without it; and if they are excited by atmospheric changes, or in any way disturbed, a loud humming may be heard in the interior of their hives, and they need almost as much air as in warm weather. (**621.**)

367. If bees are greatly disturbed, it will be unsafe, especially in warm weather, to confine them, unless they have a very free admission of air; and even then, unless it is admitted above, as well as below the mass of bees, the ventilators may become clogged with dead bees, and the colony perish. Bees under close confinement become excessively heated, and their combs are often melted; if dampness is added to the injurious influence of bad air, they become diseased; and large numbers, if not the whole colony, may perish from diarrhœa. Is it not under precisely such circumstances that cholera and dysentery prove most fatal to human beings? the filthy, damp, and unventilated abodes of the abject poor, becoming perfect lazar-houses to their wretched inmates.

368. We have several times examined the bees of new swarms which were brought to our Apiary, so closely confined, that they had died of suffocation. In each instance, their bodies were distended with a yellow and noisome substance, as though they had perished from diarrhœa. A few were still alive, and although the colony had been shut up only a few hours, the bodies of both the living and the dead were filled with this same disgusting fluid, instead of the honey they had when they swarmed.

In a medical point of view, these facts are highly interesting; showing as they do, under what circumstances, and how speedily, diseases may be produced resembling dysentery or cholera.

369. In very hot weather, if thin hives are exposed to

the sun's direct rays, the bees are excessively annoyed by the intense heat, and have recourse to the most powerful ventilation, not merely to keep the air of the hive pure, but to lower its temperature.

Bees, in such weather, often leave, almost in a body, the interior of the hive, and cluster on the outside, not merely to escape the close heat within, but to guard their combs against the danger of being melted.

370. Few novices have an adequate idea of the danger to heavily laden combs from heat, especially if the cluster of bees, outside, happens to obstruct the entrance, by hanging in front of it. In the Summer of 1877, we have seen whole rows of hives, which were exposed to the sun's rays, in a large Apiary, "melt down" almost simultaneously,— causing a loss of hundreds of dollars,—for lack of sufficient ventilation, owing to the clustering of the bees in front of the entrance.

371. After one comb breaks down, the leaking honey spreads over the bottom board, runs out of the entrance, daubs the bees, and prevents further ventilation; then the rest of the combs fall pell-mell on one another, crushing the brood, the queen, and the remaining bees. It is utter destruction.

372. In very hot weather, the bees are specially careful not to cluster on new combs containing sealed honey, which, from not being lined with cocoons, and from the extra amount of wax used for their covers, melt more readily than the breeding-cells.

Apiarists have noticed that bees often leave their honey-cells almost bare, as soon as they are sealed; but it seems to have escaped their observation, that this is absolutely necessary in very hot weather. In cool weather, they may frequently be found clustered among the sealed honey-combs, because there is then no danger of their melting.

Few things are so well fitted to impress the mind with

their admirable sagacity, as the truly scientific device by
which they ventilate their dwellings.   In this important matter, the bee is immensely in advance of the great mass of
those who are called rational beings.   It has, to be sure,
no ability to decide, from an elaborate analysis of the chemical constituents of the atmosphere, how large a proportion·
of oxygen is essential to the support of life, and how rapidly
the process of breathing converts it into a deadly poison.
It cannot, like Liebig, demonstrate that God, by setting the
animal and the vegetable world, the one over against the
other, has provided that the atmosphere shall, through all
ages, be as pure as when it first came from His creating
hand.   But shame upon us! that with all our boasted intelligence, most of us live as though pure air was of little or
no importance ; while the bee ventilates with a philosophical
precision that should put to the blush our criminal neglect.
373. It is said that ventilation cannot, in one case, be had
without cost.   Can it then be had for nothing, by the industrious bees?   Those ranks of bees, so indefatigably plying
their busy wings, are not engaged in idle amusement ; nor
might they, as some shallow utilitarian may imagine, be
better employed in gathering honey, or superintending some
other department in the economy of the hive.   At great expense of time and labor, they are supplying the rest of the
colony with the pure air so conducive to their health and
prosperity.   What a difference between them and some
human beings, who, " if they lived in a glass bottle, would
insist on keeping the cork in! "

Impure air, one would think, is bad enough ; but all its
inherent vileness is stimulated to still greater activity by air-
tight, or rather *lung-tight* stoves, which can economize fuel
only by squandering health and endangering life.   Not only
our private houses, but all our places of public assemblage,
are either unimproved with any means of ventilation, or to

a great extent, supplied with those so deficient, that they only

> "Keep the word of promise to our ear,
> To break it to our hope."

Men may, to a certain extent, resist the injurious influences of foul air; as their employments usually compel them to live more out of doors: but alas, alas! for the poor women! In the very land where they are treated with such merited deference and respect, often no provision is made to furnish them with that first element of health, cheerfulness, and beauty, heaven's pure, fresh air.

Observing Hives.

374. For nearly a century, hives have been in use containing only one comb, inclosed on both sides by glass. These hives are darkened by shutters, and, when opened, the queen is as much exposed to observation as the other bees. Mr. Langstroth has discovered that, with proper precautions, colonies can be made to work in observing-hives, even when exposed continually to the full light of day; so that observations may be made at all times, without interrupting by any *sudden* admission of light, the ordinary operations of the bees. In such hives, many intelligent persons from various States in the Union have seen the queen-bee depositing her eggs in the cells, while surrounded by an affectionate circle of her devoted children. They have also witnessed with astonishment and delight, all the mysterious steps in the process of raising queens from eggs, which with the ordinary development would have produced only the common bees. Often for more than three months, there has not been a day in our Apiary, in which some colonies were not engaged in rearing new queens to supply the place of those taken from them; and we have had the pleasure of

exhibiting these facts to bee-keepers, who never before felt
willing to credit them.

·375. An Apiarist may use the box hives a whole
life-time, and, unless he gains his information from other
sources, may yet remain ignorant of some of the most important principles in the physiology of the honey-bee;
while any intelligent cultivator may, with an observing-hive
and the use of movable-frames, in a single season, verify
for himself the discoveries which have been made only by
the accumulated toil of many observers, for more than two
thousand years.

"An opportunity of beholding the proceedings of the queen, in
hives of the old form, is so very rarely afforded, that many Apiarists have passed their lives without enjoying it; and Réaumur
himself, even with the assistance of a glass-hive, acknowledges
that it was many years before he had that pleasure."—(Bevan.)

Swammerdam, who wrote his wonderful treatise on bees,
before the invention of observing-hives, was obliged to tear
hives to pieces in making his investigations! When we see
what important results these great geniuses obtained, with
means so imperfect, if compared with the facilities which
the veriest tyro now possesses, it ought to teach us a becoming lesson of humility.

The sentiments of the following extract from Swammerdam, ought to be engraven upon the hearts of all engaged
in investigating the works of God:

"I would not have any one think that I say this from a love of
fault-finding"—he had been criticising some incorrect drawings
and descriptions—"my sole design is to have the true face and
disposition of Nature exposed to sight. I wish that others may
pass the like censure, when due, on my works; for I doubt not
that I have made many mistakes, although I can, from the heart,
say, that I have not, in this treatise designed to mislead."

376. This hive is a simplified form, but Mr. D. F. Savage suggests a still more simple one, by making the top so

narrow as not to conceal any of the bees, and leaving off
the shutters entirely, to replace them with a dark cloth
thrown over the hive. But this cloth can be used only when
the hive is established inside the house. Its main advantages are to do away with the noise and jar of opening
the shutters.

Fig. 80.

OBSERVING-HIVE.

(From Alley's ''Handy-Book.'')

a, stand; B, CC, movable glass frame; F, moulding under which the
top of the shutter H slips, to darken the hive, if needed; F, movable top,
held in place by hooks. The comb of brood and bees is put in, by removing the top and one side.

377. A parlor observing-hive of this form may be conveniently placed in any room in the house; the alighting-
board being outside, and the whole arrangement such that
the bees may be inspected at all hours, day, or night, without the slightest risk of their stinging. Two such hives
may be placed before one window, and put up or taken
down in a few minutes, without cutting or defacing the woodwork of the house.

An observing-hive will prove an unfailing source of pleasure and instruction; and those who live in crowded cities,

may enjoy it to the full, even if condemned to the penance of what the poet has so feelingly described as an "endless meal of brick." The nimble wings of the agile gatherers will quickly waft them above and beyond "the smoky chimney-pots;" and they will bear back to their city homes the balmy spoils of many a rustic flower, "blushing unseen," in simple loveliness. Might not their pleasant murmurings awaken in some the memory of long-forgotten joys, when the happy country child listened to their soothing music, while intently watching them in the old homestead-garden, or roved with them amid pastures and hill-sides, to gather the flowers still rejoicing in their "meadow-sweet breath," or whispering of the precious perfumes of their forest home?

> " To me more dear, congenial to my heart,
> One native charm than all the gloss of art ;
> Spontaneous joys, where nature has its play,
> The soul adopts and owns their first-born sway ;
> Lightly they frolic o'er the vacant mind,
> Unenvied, unmolested, unconfined,
> But the long pomp, the midnight masquerade,
> With all the freaks of wanton wealth array'd,
> In these, ere triflers half their wish obtain,
> The toilsome pleasure sickens into pain ;
> And e'en while fashion's brightest arts decoy,
> The heart distrusting asks, if this be joy."
>
> GOLDSMITH.

CHAPTER V.

HANDLING BEES.

The Honey-bee Capable of Being Tamed.

378. If the bee had not such a formidable weapon both of offense and defense, many who now fear it might easily be induced to enter upon its cultivation. As the present system of management takes the greatest possible liberties with this insect, it is important to show how all necessary operations may be performed without serious risk of exciting its anger.

Many persons are unable to suppress their astonishment, when they see an Apiarist, with the help of a little smoke, opening hive after hive, removing the combs covered with bees, and shaking them off in front of the hives; forming new swarms, exhibiting the queen, transferring the bees with all their stores to another hive; and in short, dealing with them as if they were as harmless as flies. We have sometimes been asked, whether the hives we were opening had not been subjected to a long course of training; when they contained swarms which had been brought only the day before to our Apiary.

We shall, in this chapter, show that any one favorably situated may enjoy the pleasure and profit of a pursuit which has been appropriately styled, "the poetry of rural economy," without being made too familiar with a sharp little weapon, which speedily converts all the poetry into sorry prose.

It must be manifest to every reflecting mind, that the Creator intended the bee, as truly as the horse or the cow, for the comfort of man. In the early ages of the world, and indeed until quite modern times, honey was almost the only natural sweet; and the promise of "a land flowing with milk and honey" had once a significance which it is difficult for us fully to realize. The honey-bee, therefore, was created not merely to store up its delicious nectar for its own use, but with certain propensities, without which man could no more subject it to his control, than he could make a useful beast of burden of a lion or a tiger.

379. One of the peculiarities which constitutes the foundation of the present system of management, and indeed of the possibility of domesticating at all so irascible an insect, has never to our knowledge been clearly stated as a great and controlling principle by any one before Mr. Langstroth. It may be thus expressed:

A honey-bee when heavily laden with honey never volunteers an attack, but acts solely on the defensive. *

This law of the honeyed tribe is so universal, that a stone might as soon be expected to rise into the air, without any propelling power, as a bee *well filled* with honey to offer to sting, unless crushed or injured by some direct assault. The man who first attempted to hive a swarm **(428)** of bees, must have been agreeably surprised at the ease with which he was able to accomplish the feat; for it is wisely ordered that bees, when intending to swarm, should fill their honey-bags to their utmost capacity. They are thus so peaceful that they can easily be secured by man, besides having materials for commencing operations immediately in their new habitation, and being in no danger of starving, if several stormy days should follow their emigration.

380. While swarming, bees issue from their hives in the most peaceable mood imaginable; and unless abused allow themselves to be treated with the greatest familiarity. The

---

* This statement has been contradicted by a high authority, but we persist in affirming it, and will adduce several proofs in different passages.

hiving of them might always be conducted without risk, if there were not, *occasionally*, some improvident or unfortunate ones, who, coming forth without a sufficient amount of the soothing supply, are filled instead with the bitterest hate against any one daring to meddle with them. Such thriftless radicals are always to be dreaded, for they must vent their spleen on something, even though they perish in the act. (**84.**)

If a whole colony, on sallying forth, possessed such a ferocious spirit, no one could hive them unless clad in a coat of mail, bee-proof; and not even then, until all the windows of his house were closed, his domestic animals bestowed in some place of safety, and sentinels posted at suitable stations, to warn all comers to keep at a safe distance. In short, if the propensity to be exceedingly good-natured after a hearty meal, had not been given to the bee, it could never have been domesticated, and our honey would still be procured from the clefts of rocks or the hollows of trees. Probably the good nature resulting from a hearty meal is not the only cause of the above fact. There is another physiological fact connected with it (**85**). When her stomach is empty, a bee can curve her abdomen easily to sting. If her honey-sack is full, the rings of the abdomen are distended, and she finds more difficulty in taking the proper position for stinging.

381. A second peculiarity, in the nature of bees, gives an almost unlimited control over them, and may be expressed as follows:

Bees, when frightened, usually begin to fill themselves with honey from their combs.

If the Apiarist only succeeds in frightening his little subjects, he can make them as peaceable as though they were incapable of stinging. By the use of a little smoke, the largest and most fiery colony may be brought into complete subjection. As soon as the smoke is blown among them,

they retreat from before it, raising a subdued or terrified
note; and, seeming to imagine that their honey is to be
taken from them, they cram their honey-bags to their utmost
capacity. They act either as if aware that only what they
can lodge in this inside pocket is safe, or, as if expecting to
be driven away from their stores, they are determined to
start with a full supply of provisions for the way. The
same result may be obtained by shutting them up in their
hive and drumming upon it for a short time, but this latter
process is only successful with some races of bees easily
frightened, like the black bees (**559**).

382. The bellows-smokers, in present use, for smoking
bees and controlling them, are as far superior to the old
method of blowing smoke on them with the mouth from a

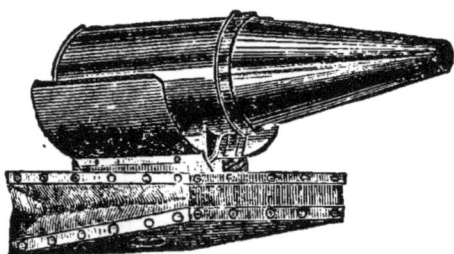

Fig. 81.   Fig. 82.
BINGHAM BEE-SMOKER.   MUTH BEE-SMOKER.

piece of punk or rotten wood, or a bunch of rags, as the
movable-frame hive is superior to the box hive of old. The
writer of this, who kept bees in large numbers in several
Apiaries before the introduction of the practical bellows-
smoker, has many a time felt dizzy from the fatigue of blowing smoke on the bees.

Bellows-smokers were used in Europe long ago, but they
were not practical, as they could not be used with one
hand.

Quinby, one of the veterans of progressive Apiculture, invented the first bellows-smoker that had the bellows on the side of the fire-box, that could stand up and draw like a chimney, and that could practically be held with one hand. Bingham afterwards greatly improved on this smoker. Since then, others have made different styles, all based on Quinby's or on Bingham's ideas.

The Improved Quinby-Bingham smokers have been imitated all over the world, especially in England and France, and we are sorry to say, some of these imitations have been sold as personal inventions, without any credit being given to the real inventors.

A bee-smoker is indispensable to any Apiarist, and should be properly filled, when used, with dry wood, lighted at the bottom by a few hot coals. With a good smoker any kind of wood may be used. When the bees are located in an orchard, dead limbs of apple-trees, are handiest and will make good smoke. Shavings, leaves, rags, can also be used, if no wood is at hand. By setting the smoker upright, when not held in the hand, so as to create a good draft, and refilling it from time to time, a good smoke can be kept up from morning till night, if necessary.

383. Some Apiarists of England have tried several liquids, for rubbing on the hands, to pacify the bees. Most of these liquids are hydro-carbonous fluids, or volatile oils of plants, such as wintergreen, turpentine, bergamot, cloves, thyme, etc. Mr. Grimshaw, after divers trials, invented a compound of several of these oils, to which he seems to have added ether and chloroform, if our sense of smell does not mislead us. He calls it *Apifuge.*

Several Apiarists praise this drug, while others say that their bees did not mind it, and sting them as usual; and some complain of blisters on their hands after its use. (*British Bee-Journal.*)

Mr. Cowan presented us with a vial of Apifuge, but,

after trying, we cannot see much advantage to be derived
from its use.

384. Mr. Raynor advises the use of a carbolized sheet,
to frighten bees:

"Make a solution of 3 oz. carbolic acid in a quart of water,
and preserve for use. Mix 1½ oz. of this solution with 1½ oz. of
glycerine; put the mixture in a quart of water, shake well before
using; steep in the mixture a piece of calico, or cheese cloth,
sufficiently large to cover the top of the hive, wring out dry and
spread over the hive as soon as the quilt is removed.

"You may use the same to drive the bees out of the sections.
Keep the bottles well corked for future use."—(Rev. G. Raynor,
in the *British Bee-Journal.*)

The same liquid may be forced among the bees through
an atomizer. As it evaporates it leaves no bad smell behind.

385. A neighbor of ours, who is a magnetist, told our
foreman-Apiarist that bees could be pacified by simply lay-

Fig. 83.
VEIL ABOVE THE HAT.

ing one's hands above the combs while the cloth is carefully removed. We have seen bees withdraw from the frames inside the hive, under this laying on of hands; but we are not sure that such magnetism, if there be magnetism in it, is sufficient to prevent the bees from stinging.

386. A bee-veil, although objectionable to some beekeepers, who prefer to handle their bees barefaced, is really a necessity in a large Apiary. Timid persons feel safer in using it, and even the boldest bee-keepers recognize the necessity of wearing one, when colonies become aroused by accident. The best veils are sewed to outer-edge of the rim of a straw-hat; with a rubber at their lower extremity, to fasten around the neck. The veil can be slipped on and off in a twinkling, if necessity requires; when not in use, it is simply folded into the crown of the hat, where it is always at hand.

We keep a number of these veil hats in our bee-house, for the accommodation of visitors, who wish to look through the wonders of the bee-hive, without fear of stings.

Some veils are made removable, with a rubber at each end; the upper one being slipped over the crown of the hat. This veil can be taken off at will, and carried in the pocket.

Fig. 84.

VEIL SEWED AROUND RIM OF HAT.

In his "Success In Bee-Culture," Mr. Heddon says: "A bee-veil should never be any color but black, as all other shades are more or less difficult to see through clearly," and we fully agree with him. White veils are most especially objectionable. Green is the best color after black.

387. The hands may be protected by india-rubber gloves, such as are now in common use. These gloves, while impenetrable to the sting of a bee, do not materially interfere with the operations of the Apiarist. As soon, however, as he acquires confidence and skill, he will much prefer to use nothing but the bee-hat, even at the expense of an occasional sting on his hands.

An English Apiarist advises persons using gloves to cut the tips of the fingers so as to handle the frames more dexterously, and to wash their fingers with some kind of Apifuge.

Stings on the hands usually cause but little suffering or swelling, while stings on the face are quite painful; and the grotesque appearance which the swelling often gives to the human face, makes it much more desirable to protect the head than the hands.

If the hands are wet with honey, they will seldom be stung.

388. All woolen clothes are more objectionable to bees than linen or cotton, for wool resembles the hair of animals, being made of it, while linen or cotton resembles the twigs and leaves of plants, being made of vegetable fibre. Butler says:

" They use their stings against such things as have outwardly some offensive excrement, such as hair or feathers, the touch whereof provoketh them to sting. If they alight upon the hair of the head or beard, they will sting if they can reach the skin. When they are angry their aim is most commonly at the face, but the bare hand that is not hairy, they will seldom sting, unless they be much offended."—("Feminine Monarchy," 1609.)

389. In handling bees, it is not always necessary to compel them to fill themselves with honey. With the quiet Italians (**551**), a few puffs of smoke, at the entrance, when opening the hive, and occasionally on the combs, if they show any disposition to anger, are quite sufficient to keep them down. Some of our best Apiarists often open their hives and handle the bees without smoke. It takes practice, patience and firmness.

While the timid, if unprotected, are almost sure to be stung, there is something in the fearless movements of a skillful operator, that seems to render a colony submissive to his will.

390. Some races, however, like the Cyprian (**559**), cannot be controlled without a cloud of smoke, but they promptly retreat before the overpowering argument of a good smoker.

391. Bees can be handled at all times; but they are quietest in the middle of the day. At such a time, the old bees, which are the crossest in the colony, are out in the field. In cold, cloudy, or stormy weather, they are most irritable, especially if there is a scarcity of honey, as the lurking robbers excite the bees. Old bees that come home *loaded*, are not cross, while those going out *empty*. are easily angered. During a plentiful honey flow, when the hives are crowded for room, the bees are nearly *all full* of honey, and the colonies can then be handled without smoke (**379**).

By our methods you can superintend a large Apiary, performing every operation necessary for pleasure or profit, without as much risk of being stung, as must frequently be incurred in attempting to manage a single hive in the old way.

392. Let all your motions about your hives be gentle and slow; never crush or injure the bees; acquaint yourself fully with the principles of management detailed in this treatise, and you will find that you have little more reason to dread the sting of a bee, than the horns of a favorite cow, or the heels of your faithful horse.

13

Cotton, quoting from Butler, who, in these remarks, follows mainly Columella, says:

393. "Listen to the words of an old writer: — 'If thou wilt have the favour of thy bees, that they sting thee not, thou must avoid such things as offend them: thou must not be unchaste or uncleanly; for impurity and sluttiness (themselves being most chaste and neat) they utterly abhor; thou must not come among them smelling of sweat, or having a stinking breath, caused either through eating of leeks, onions, garlick, and the like, or by any other means, the noisomeness whereof is corrected by a cup of beer; thou must not be given to surfeiting or drunkenness; thou must not come puffing or blowing unto them, neither hastily stir among them, nor resolutely defend thyself when they seem to threaten thee; but softly moving thy hand before thy face, gently put them by; and lastly, thou must be no stranger unto them. In a word, thou must be chaste. cleanly, sweet, sober, quiet, and familiar; so will they love thee, and know thee from all others. When nothing hath angered them, one may safely walk along by them; but if he stand still before them in the heat of the day, it is a marvel but one or other spying him, will have a cast at him.'*

"Above all, never blow † on them; they will try to sting directly, if you do.

"If you want to catch any of the bees, make a bold sweep at them with your hand; and if you catch them without pressing them, they will not sting. I have so caught three or four at a time. If you want to do anything to a single bee, catch him 'as if you loved him,' between your finger and thumb, where the tail joins on to the body, and he cannot hurt you."

When gorged with honey, they may be taken up by handfuls, and suffered to run over the face, and may even have their glossy backs gently smoothed as they rest on our persons; and all the feats of the celebrated Wildman may be

---

* Many persons imagine themselves to be quite safe, if they stand at a considerable distance from the hives; whereas, cross bees delight to attack those whose more distant position makes them a surer mark to their long-sighted vision, than persons who are close to their hives.

† While bees resent the *warm* breath exhaled *slowly* from the lungs, we have ascertained, that they will run from a blast of cold air blown upon them by the mouth of the operator, almost as quickly as from smoke. Before employing smoke Mr. Langstroth often used a pair of bellows.

safely imitated by experts, who, by securing the queen, can make the bees hang in large festoons from their chin, without incurring any risk of being taken by the beard.

> " Such was the spell, which round a Wildman's arm,
> Twin'd in dark wreaths the fascinated swarm ;
> Bright o'er his breast the glittering legions led,
> Or with a living garland bound his head.
> His dextrous hand, with firm yet hurtless hold,
> Could seize the chief, known by her scales of gold,
> Prune 'mid the wondering train her filmy wing,
> Or o'er her folds the silken fetter fling."

394. *The ignorance of most bee-keepers of the almost unlimited control which may be peaceably acquired over bees, has ever been regarded by the author of this treatise as the greatest obstacle to the speedy introduction of movable-frame hives.* Such ignorance has led to the invention of costly and complicated hives, all the ingenuity and expense lavished upon which, are known, by the better informed, to be as unnecessary as a costly machine for lifting up bread and butter, and gently pushing it into the mouth and down the throat of an active and healthy child.

We have before us a small pamphlet, published in London in 1851, describing the construction of the "Bar and Frame Hive" of W. A. Munn, Esq. The object of this invention is to elevate frames, one at a time, *into a case with glass sides*, so that they may be examined without risk of annoyance from the bees. Great ingenuity is exhibited by the inventor of this very costly and very complicated hive, who seems to imagine that smoke "must be injurious both to the bees and their brood."

395. In opening a hive, little danger may be feared from the bees that are exposed to the light, unless quick motions are made, as they are completely bewildered by their sudden exposure, and removal from the hive.

It is not merely the *sudden* admission of light, but its introduction from an *unexpected quarter*, that *for the time*, disarms the hostility of the bees. They appear, for a few moments, almost as much confounded as a man would be, if, without any warning, the roof and ceiling of his house should suddenly be torn from over his head. Before they recover from their amazement, they are saluted with a puff of smoke, which, by alarming them for the safety of their treasures, induces them to snatch whatever they can. In the working season, the bees near the *top* are gorged with honey : and those coming from *below* are met in their threatening ascent, by a small amount of harmless smoke, which excites their fears, but leaves no unpleasant smell behind. *No genuine lover of bees ought* ever to use the sickening fumes of tobacco.

396. Heddon says ("Success in Bee-Culture," page 18) : "I know of but one instance where the use of smoke can do harm, and that is in smoking the guards of a colony that is in danger of being robbed." (**664.**) To this important statement, we would add, that too much smoke to a colony already subdued, will drive them from their combs, and often cause them to get in the way of the Apiarist.

But the greatest care should be taken to repress by smoke, the *first* manifestations of anger ; for, as bees communicate their sensations to each other with almost magic celerity, while a whole colony will quickly catch the pleased or subdued notes uttered by a few, it will often be roused to fury by the angry note of a single bee. When once they are thoroughly excited, it will be found very difficult to subdue them, and the unfortunate operator, if inexperienced, will often abandon the attempt in despair.

It cannot be too deeply impressed upon the beginner, that nothing irritates bees more than breathing upon them, or jarring their combs. Every motion should be deliberate, and no attempt whatever made to strike at them. If inclined to be cross, they will often resent even a quick *pointing* at them with the finger, by darting upon it, and leaving their stings behind.

397. The first thing to be done, after having opened a hive and removed the cloth (**352**), is to remove the division-board (**349**) from the inside of the hive—to give room for handling the frames,—with the help of a common wood chisel. Then the frames which have been glued (**236**) fast to the rabbets by the bees, must be very gently pried loose; this may be done without any serious jar, and without wounding or enraging a single bee. They may be all loosened for removal in less than a single minute.

If there is no division-board (**349**) in the hive, the Apiarist should *gently* push the third frame from either end of the hive, a little nearer to the fourth frame; and then the second as near as he can to the third, to get ample room to lift out the end one, without crushing its comb, or injuring any of the bees. To remove it, he should take hold of its two shoulders which rest upon the rabbets, and carefully lift it, so as to crush no bees by letting it touch the sides of the hive, or the next frame. If it is desired to remove any particular frame, room must be gained by moving, in the same way, the adjoining ones on each side. As bees usually build their combs slightly waving, it will be found impossible to remove a frame safely, without making room for it in this way. If the combs are built on foundation (**674**), however, they will be much easier to remove, as they are then perfectly straight. In handling heavy frames in hot weather, be careful not to incline them from their *perpendicular*, or the combs will be liable to break from their own weight, and fall out of the frames.

If more combs are to be examined, after lifting out the outside frame, set it carefully on end, near the hive, when the second one may be easily moved towards the vacant space, and lifted out. After examination, put it in the place of the one first removed; in the same way, examine the

third, and put it in the place of the second, and so proceed
until all have been examined. If a division-board is used,
it will not be necessary to set any of the frames down outside of the hive, as the removal of this board will leave one
vacant space in the hive.

If the frames, as they are removed, are put into an empty
hive, or a comb-bucket, they may be protected from the
cold, and from robber-bees.

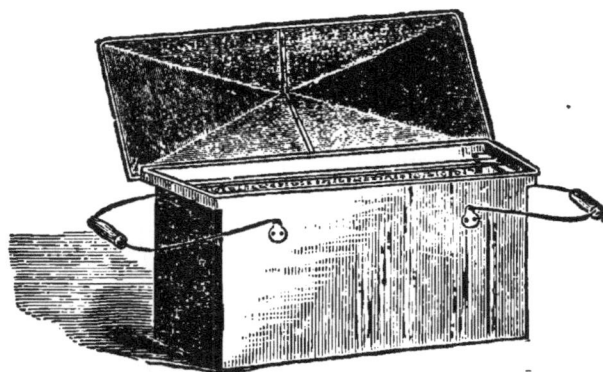

Fig. 85.

COMB-BUCKET.

The inexperienced operator, who sees that the bees have
built small pieces of comb, or bridges (**237**), between
the outside of the frames and the sides of the hive, or
slightly fastened together some parts of their combs, may
imagine that the frames cannot be removed at all.   Such
slight attachments, however, offer no practical difficulty to
their removal.*   The great point to be gained, is to secure

---

* If sufficient room for storing surplus honey is not given to a strong colony,
in its anxiety to amass as much as possible, it will fill the smallest accessible
places. If the bees build comb between the tops of the frames and the under
side of the upper story, it can be easily cut off, and used for wax. If this
shallow chamber were not used, they would fasten the upper story to the
frames so tightly, that it would be very difficult to remove it; and every time

a single comb on each frame; and this is effected by the use of the triangular comb-guides, or better, by comb-foundation (**674**).

If bees were disposed to fly away from their combs, as soon as they are taken out, instead of adhering to them with such remarkable tenacity, it would be far more difficult to manage them; but even if their combs, when removed, are all arranged in a continued line, the bees, and most especially the Italian bees, instead of leaving them, will stoutly defend them against the thieving propensities of other bees.

398. In *returning* the frames, care must be taken not to crush the bees between them and the rabbets on which they rest; they should be put in so *slowly*, that a bee, on feeling the slightest pressure, may have a chance to creep from under them before it is hurt.

The frames should be returned, as far as possible, in the same position, as they were found, with the brood in the forward part of the hive, and the honey in the back, for bees always live and breed in *front of their stores*, to more easily defend their treasures against intruders.

In shutting up the hive, the surplus story, if any is there, should be carefully *slid* on, so that any bees which are in the way may be pushed before it, instead of being crushed. A beginner will find it to his advantage to practice—using an empty hive—the directions for opening and shutting hives, and lifting out the frames, until confident that he fully understands them. If any bees are where they would be imprisoned by closing the upper cover, it should be propped up a little, until they have flown to the entrance of the hive, or, they may be brushed away gently.

it was taken off, they would glue it still faster, so that, at last, it would be well nigh impossible, in getting it off, not to start the frames so as to crush the bees between the combs.

MISMANAGEMENT OF BEES.

399. When a colony of bees is unskillfully dealt with, they will "compass about" their assailant with savage ferocity; and woe be to him, if they can creep up his clothes, or find a single unprotected spot on his person.

Not the slightest attempt should be made to act on the offensive; for, if a single one is struck at, others will avenge the insult; and if resistance is continued, hundreds, and at last, thousands, will join them. The assailed party should quickly retreat to the protection of a building, or, if none is near, should hide in a clump of bushes, and lie perfectly still, with his head covered, until the bees leave him. When no bushes are at hand, they will generally give over the attack, if he lies still on the grass, with his face to the ground. A practical Apiarist, sheltered with a veil and armed with a well lighted smoker, will not retreat much before the most ferocious swarm of bees.

Those who are alarmed if a bee enters the house, or approaches them in the garden or fields, are ignorant of the important fact, that *a bee, at a distance from its hive, never volunteers an attack.* Even if assaulted, they seek only to escape, and never sting, unless they are hurt.

If they were as easily provoked away from home, as when called to defend those sacred precincts, a tithe of the merry gambols, in which our domestic animals indulge, would speedily bring about them a swarm of infuriated enemies; we should be no longer safe in our quiet rambles among the green fields; and no jocund mower could whet or swing his peaceful scythe, unless clad in a dress impervious to their stings. The bee, instead of being the friend of man, would, like savage wild beasts, provoke his utmost efforts for its extermination.

Let none, however, take encouragement from the contrast between the conduct of bees at home and abroad, to reserve all their pleasant ways for other places than the domestic roof; for, towards the members of its own family the bee is all kindness and devotion; and while, among human beings, a *mother* is often treated by her own children with disrespect or neglect, among bees she is always waited upon with reverence and affection.

400. Huber has demonstrated, that bees have an exceedingly acute sense of smell, and that unpleasant odors quickly excite their anger.[*] Long before his time, Butler said, "Their smelling is excellent, whereby, when they fly aloft into the air, they will quickly perceive anything under them that they like, even though it be covered." They have, therefore, a special dislike to those whose habits are not neat,[†] and who bear about them a perfume not in the least resembling

"Sabean odors
From the spicy shores of Araby the blest."

A horse, when assailed by them, is often killed; as, instead of running away, like most other animals, it will plunge and kick until it falls overpowered. The Apiary should be fenced in, to prevent horses and cattle from molesting the hives. We have known of a horse, which happening to be loose in a bee-yard, was attacked by a few bees. In trying to defend himself against them by kicking and rolling he upset one hive and then another, till tens of thousands of bees assailed him, and the poor animal was

[*] Strong perfumes, however pleasant to us, are disagreeable to bees; and Aristotle observes, that they will sting those scented with them. We have known persons ignorant of this fact to be severely treated by bees.

[†] Some persons, however cleanly, are assaulted by bees as soon as they approach their hives. It is related of a distinguished Apiarist that, after a severe attack of fever, he was never able to be on good terms with his bees. That they can readily perceive the slightest differences in smell, is apparent from the fact that any number of bees, fed from a common vessel, will be gentle towards each other, while they will assail the first strange bee that alights on the feeder.

stung to death, before his owner could come to the rescue.
We were informed by an eye-witness, that although the carcass remained unburied two days, neither dogs, crows,
buzzards, nor any of the usual scavengers of decaying flesh,
attempted to feed upon it, so great was the amount of poison
(**79**) instillèd into it by the revengeful bees.

401. The sting of a bee (**78**) upon some persons, produces very painful, and even dangerous effects. We have
often noticed that, while those whose systems are not sensitive to the venom, are rarely molested by bees, they seem
to take a malicious pleasure in stinging those upon whom
their poison produces the most virulent effect. Something
in the secretions of such persons may both provoke the
attack and render its consequences more severe.

The smell of their own poison (**87**) produces a very irri-
tating effect upon bees. A small portion of it offered to them
on a stick, will excite their anger.

"If you are stung," says old Butler, "or any one in the company—yea, though a bee hath stricken but your clothes, especially in hot weather—you were best be packing as fast as you
can, for the other bees, smelling the rank flavor of the poison,
will come about you as thick as hail."

Remedies for the Sting of a Bee.

402. If only a few of the host of cures, so zealously
advocated, could be made effectual, there would be little
reason to dread being stung.

The first thing to be done after being stung, is to pull—
or rather push—the sting out of the wound *as quickly as
possible.* When torn from the bee, the poison-bag, and all
the muscles which control the sting, accompany it; and it
penetrates deeper and deeper into the flesh, injecting continually more and more poison into the wound. If extracted
at once, it will very rarely produce any serious consequen-

ces; but, in extracting it, it should not be taken between the fingers. In so doing, most of the poison will be pressed into the wound. It must be rubbed or scraped off by a quick motion of the finger-nail, so as to prevent any more of the poison of the sack from getting into the flesh. After the sting is removed, the utmost care should be taken not to irritate the wound by the *slightest rubbing*. However intense the smarting, and the disposition to apply friction to the wound, *it should never be done*, for the moment that the blood is put into violent circulation, the poison is quickly diffused over a large part of the system, and severe pain and swelling may ensue. On the same principle, by severe friction, the bite of a mosquito, even after the lapse of several days, may be made to swell again. As most of the popular remedies are *rubbed in*, they are worse than nothing.

When the operator is perspiring abundantly, the stings are less painful, as some of the poison exudes with the sweat.

If the mouth is applied to the wound, unpleasant consequences may follow; for, while the poison of snakes, affecting only the circulating system, may be *swallowed* with impunity, the poison of the bee acts with great power on the organs of digestion. Distressing headaches are often produced by it, as any one, who has been stung, or has tasted the poison, very well knows.

403. In our own experience, we have found *cold water* to be the best remedy for a bee-sting. The poison is quickly dissolved in it; and the coldness of the water has also a powerful tendency to check inflammation.

The leaves of plantain, crushed and applied to the wound, are a very good substitute, when water cannot at once be procured. Bevan recommends the use of spirits of hartshorn, and says that, in cases of severe stinging, its internal use is also beneficial. In very serious cases, the ammonia may be taken, in quantities of from five to twenty drops,—for an adult, less for a child,—in hot tea, with beneficial results. It causes an increased perspiration, and neutralizes the effects of the poison. ("Commentaires Thérapeutiques," Gubler, Paris, 1874.)

404. It may be some comfort to novices to know that the poison will produce less and less effect upon their system. Old bee-keepers, like Mithridates, appear almost to thrive upon poison itself. When we first became interested in bees, a sting was quite a formidable thing, the pain being often very intense, and the wound swelling so as sometimes to obstruct our sight. At present, the pain is usually slight, and, if the sting is quickly extracted, no unpleasant consequences ensue, even if no remedies are used. Huish speaks of seeing the bald head of Bonner, a celebrated practical Apiarist, covered with stings, which seemed to produce upon him no unpleasant effects. The Rev. Mr. Kleine advises beginners to allow themselves to be stung frequently, assuring them that, in two seasons, their system will become accustomed to the poison!

An old English Apiarist advises a person who has been stung, to catch another bee as speedily as possible, and make it sting on the same spot. Even an enthusiastic disciple of Huber might hesitate to venture on such a singular homœopathic remedy; but, as this Apiarist had stated, what we had verified in our own experience, that the oftener a person is stung the less he suffers from the venom, the writer determined to make trial of his prescription. Allowing a sting to remain until it had discharged all of its poison, he compelled another bee to insert its sting, as nearly as possible, in the same spot. He used no remedies of any kind, and had the satisfaction, in his zeal for new discoveries, of suffering more from the pain and swelling than for years before.

That the bee-keeper becomes inoculated with the poison

of the bee, and usually becomes proof against it, is no more to be doubted than the fact that vaccination is a preservative against small-pox. The recent discoveries of Pasteur, for the cure of hydrophobia, are another evidence of the efficiency of inoculation.

BEES AS MEANS OF DEFENSE.

405. "A small corsair, equipped with forty or fifty men, and having on board some bees, purposely taken from a neighboring island, and confined in earthen hives (**275**), was pursued by a Turkish galley. As the latter boarded her, the sailors threw the hives from the masts down into the galley. The earthen hives broke into fragments and the bees dispersed all over the boat. The Turks who had looked on the small corsair with contempt, as an easy prey, did not expect so singular an attack. Finding themselves defenseless against the stings, they were so frightened, that the men of the corsair, who had provided themselves with masks and gloves, took possession of the galley, almost without resistance."

"Amurat, Emperor of Turkey, having besieged Alba, and made a breach in the walls, found the breach defended by bees, whose hives had been brought on the ruins. The Janissaries, the bravest militia of the Ottoman empire, refused to clear the obstacle."—(Della Rocca, 1790,)

CHAPTER VI.

NATURAL SWARMING.

406. In the Spring, as soon as the combs of a hive, well filled, can no longer accommodate its teeming population, the bees prepare for emigration, or in other words, for departing with their queen, by building a number of royal-cells (**104**). These cells are begun about the time that the drones make their appearance in the open air; and when the young queens arrive at maturity, the males are usually very numerous (**186**).

The swarming of bees is one of the most beautiful sights in the whole compass of rural economy. Although those who use movable-comb hives prefer the artificial multiplication of colonies, it being more profitable. all Apiarists delight in the pleasing excitement of natural swarming.

> " Up mounts the chief, and to the cheated eye
> Ten thousand shuttles dart along the sky:
> As swift through æther rise the rushing swarms,
> Gay dancing to the beam their sun-bright forms;
> And each thin form, still ling'ring on the sight,
> Trails. as it shoots. a line of silver light.
> High pois'd on buoyant wing, the thoughtful queen,
> In gaze attentive, views the varied scene.
> And soon her far-fetch'd ken discerns below
> The light laburnum lift her polish'd brow,
> Wave her green leafy ringlets o'er the glade,
> And seem to beckon to her friendly shade.
> Swift as the falcon's sweep, the monarch bends
> Her flight abrupt: the following host descends.
> Round the fine twig, like cluster'd grapes, they close
> In thickening wreaths, and court a short repose."
>   EVANS.

407. Bees sometimes abandon their hives very early in Spring, or even late in Summer or Fall (**264**).   Although exhibiting the appearance of natural swarming, they leave,

not because the population is so crowded that they wish to form new colonies, but because it is either so small, or the hive so destitute of supplies, that they are driven to desper-

ation. Seeming to have a presentiment that they must
perish if they stay, instead of awaiting the sure approach
of famine, they sally out to see if they cannot better their
condition. Such desertions should not be mistaken for
natural swarming.

408. The time, when new swarms may be expected,
depends, of course, upon the climate, the forwardness of the
season, and the strength of the colonies. In our Northern
and Middle States, they seldom issue before the latter part
of May ; and June may there be considered as the great
swarming month. In Brownsville, Texas, on the lower Rio
Grande, bees often swarm quite early in March.

Swarming does not always take place in Spring, although
this is the usual time for it. Swarms are likely to issue in
any locality, whenever the hive is crowded for room, or
nearly so, during a good and prolonged honey-harvest. In
warm latitudes, it lasts for several months, owing to a continuous flow of honey. Wherever there are two distinct
honey crops (**705**), there are also two swarming seasons,
especially along the low lands or river bottoms, where
Fall pasturage is abundant. Swarms, hived during the fore-
part of either of these honey seasons, are always the best ;
having a few weeks of honey crop before them, they have
ample time to build comb (**198**), and fill it with honey and
brood ; while swarms which are cast during the latter part
of either the clover or the Fall harvest, coming as they do,
just before a dearth of honey, are unable to build comb and
raise brood, and easily perish, if left to themselves. Thus,
a swarm harvested in August, in this latitude, at the opening of the Fall crop, stands better chances than one harvested in July, at the close of the clover and basswood
crop.

First or Primary Swarm

409. The first swarm is almost invariably led off by the old queen, unless she has died from accident or disease, when it is accompanied by one of the young ones reared to supply her loss. There are no signs from which the Apiarist can predict the certain issue of a *first* swarm. For years, we spent much time in the vain attempt to discover some *infallible* indications of first swarming; until facts convinced us that there can be no such indications.

410. If the weather is unpleasant, or the blossoms yield an insufficient supply of honey, bees often change their minds, and refuse to swarm at all. If, in the swarming season, but few bees leave a strong hive, on a clear, calm, and warm day, when other colonies are busily at work, we may look with great confidence for a swarm, unless the weather prove suddenly unfavorable.

If the weather is very sultry, a swarm will sometimes issue as early as seven o'clock in the morning; but from ten, A. M., to two, P. M., is the usual time; and the majority of swarms come off when the sun is within an hour of the meridian. Occasionally, a swarm ventures out as late as five, P. M.; but an old queen is seldom guilty of such an indiscretion.

411. We have repeatedly witnessed in our observing-hives (**374**) the whole process of swarming. On the day fixed for departure, the queen is very restless, and instead of depositing her eggs in the cells, roams over the combs, and communicates her agitation to the whole colony. The emigrating bees usually fill themselves with honey, just before their departure; but in one instance, we saw them lay in their supplies more than two hours before they left. A short time before the swarm rises, a few bees may generally be seen sporting in the air, with their heads turned

14

always to the hive; and they occasionally fly in and out, as though impatient for the important event to take place. At length, a violent agitation commences in the hive; the bees appear almost frantic, whirling around in circles continually enlarging, like those made by a stone thrown into still water, until, at last, the whole hive is in a state of the greatest ferment, and the bees, rushing impetuously to the entrance, pour forth in one steady stream. Not a bee looks behind, but each pushes straight ahead, as though flying "for dear life," or urged on by some invisible power, in its headlong career.

412. Often, the queen does not come out until many have left; and she is sometimes so heavy, from the number of eggs in her ovaries, that she falls to the ground, incapable of rising with her colony into the air (**40**). The bees soon miss her, and a very interesting scene may now be witnessed. Diligent search is at once made for their lost mother; the swarm scattering in all directions, so that the leaves of the adjoining trees and bushes are often covered almost as quickly with anxious explorers, as with drops of rain after a copious shower. If she cannot be found, they commonly return to the old hive, in from five to fifteen minutes.

413. The ringing of bells and beating of kettles and frying-pans to cause swarms to settle, is probably not a whit more efficacious, than the hideous noises of some savage tribes, who, imagining that the sun, in an eclipse, has been swallowed by an enormous dragon, resort to such means to compel his snakeship to disgorge their favorite luminary.

Many who have never practiced "tanging," have never had a swarm leave without settling. Still, as one of the "country sounds," and as a relic of the olden-times, even the most matter-of-fact bee-man can readily excuse the enthusiasm of that pleasant writer in the *London Quarterly Review*, who discourses as follows :

"Some fine, warm morning in May or June, the whole atmosphere seems alive with thousands of bees, whirling and buzzing, passing and repassing, wheeling about in rapid circles, like a group of maddened bacchanals. Out runs the good housewife, with frying-pan and key—the orthodox instruments for *ringing*—and never ceases her rough music, till the bees have settled. This custom, as old as the birth of Jupiter, is one of the most pleasing and exciting of the countryman's life; and there is an old colored print of bee-ringing still occasionally met with on the walls of a country-inn, that has charms for us, and makes us think of bright, sunny weather in the dreariest November day. Whether, as Aristotle says, it affects them through pleasure or fear, or whether, indeed, they hear it at all, is still as uncertain as that philosopher left it; but we can wish no better luck to every bee-master that neglects the tradition, than that he may lose every swarm for which he omits to raise this time-honored concert."

414. The queen sometimes alights first, and sometimes joins the cluster after it has begun to form. The bees do not usually settle, unless she is with them; and when they do, and then disperse, it is frequently the case that, after first rising with them, she has fallen, from weakness, into some spot where she is unnoticed by the bees.

Perceiving a hive in the act of swarming, the writer on two occasions, contracted the entrance, to secure the queen when she should make her appearance. In each case, at least one-third of the bees came out before she joined them. As soon as the swarm ceased searching for her, and were returning to the parent-hive, he placed her, with her wings clipped, on a limb of a small evergreen tree, when she crawled to the very top of the limb, as if for the express purpose of making herself as conspicuous as possible. The few bees, that first noticed her, instead of alighting, darted rapidly to their companions; in a few seconds, the whole colony was apprised of her presence, and flying in a dense cloud, began quietly to cluster around her. Bees, when on the wing, intercommunicate with such surprising rapidity, that telegraphic signals are scarcely more instantaneous.

415. That bees send out *scouts* to seek a suitable abode, admits of no serious question. Swarms have been traced directly to their new home. in an air-line flight, from the place where they clustered after alighting. Now this precision of flight to an unknown home, would plainly be impossible, if some of their number, by previous explorations, were not competent to act as guides to the rest. The sight of bees for distant objects is so wonderfully acute. that, after rising to a sufficient elevation, they can see, at the distance of several miles, any prominent objects in the vicinity of their intended abode. **(13-14.)**

Whether bees send out scouts *before* or *after* swarming, may admit of more question. but these scouts are usually absent for an hour or more. after the alighting of the swarm.

It is probable that most of the scouts are sent during the alighting ; otherwise how could they know where the swarm alighted, so as to come back to it?

The necessity for scouts or explorers seems to be unquestionable, unless we admit that bees have the faculty of flying in an "*air line,*" to a hollow tree, which they have never seen, and which may be the only one among thousands where they can find a suitable abode.

These views are confirmed by the repeated instances in which a few bees have been noticed inquisitively prying into a hole in a hollow tree, or the cornice of a building, and have, before long, been followed by a whole colony.

About fifty yards from our home Apiary, there was a large hollow oak tree, which we called "The Squirrel's Oak," because every season it sheltered a family of these pretty animals. One Summer we noticed for several days some bees flying, in and out of a hole. in one of its largest limbs. It seemed to us that they were cleaning the hollow, and we supposed that a swarm had taken possession of it. A change in the weather having taken place. the swarming preparations were discontinued, and we never again noticed any bees around the limb. The tree was cut down the following Winter, and no trace of comb was found in the hollow. It proved conclusively, that the bees we had seen, were scouts in search of a lodging.

416. The swarm sometimes remains until the next day, where bees have clustered in leaving the hive, and instances are not unfrequent of a more protracted delay.

If the weather is hot when they first cluster, and the sun shines directly upon them, they will often leave before they have found a suitable habitation. Sometimes the queen of emigrating bees, being heavy with eggs, and unaccustomed to fly, is compelled to alight, before she can reach their intended home. Queens, under such circumstances, are occasionally unwilling to take wing again, and the poor bees sometimes attempt to lay the foundations of their colony on fence-rails, hay-stacks, or other unsuitable places.

Mr. Wagner once knew a swarm of bees to lodge under the lowermost limb of an isolated oak-tree, in a corn-field. It was not discovered until the corn was harvested, in September. Those who found it, mistook it for a recent swarm, and in brushing it down to hive it, broke off three pieces of comb, each about eight inches square. Mr. Henry M. Zollickoffer, of Philadelphia, informed us that he knew a swarm to settle on a willow-tree in that city, in a lot owned by the Pennsylvania Hospital; it remained there for some time, and the boys pelted it with stones, to get possession of its comb and honey.

If the Apiary is located in the woods, and the bees are allowed to swarm, they may settle on high trees, and the bee-master, unless some special precautions are used, will lose much time in hiving his swarms.

417. Having noticed that swarming bees will almost always alight wherever they see others clustered, we found that they can be determined to some *selected* spot by an old black hat, or even a mullen-stalk, which, when col-

ored black, can hardly be distinguished, at a distance, from
a clustering swarm. A black woolen stocking or piece of
cloth, fastened to a shady limb, or to a pole, in plain sight
of the hives, and where the bees can be most conveniently
hived, would answer as good a purpose. Swarms are not
only attracted by the bee-like color of such objects, but are
more readily induced to alight upon them, if they furnish
something to which they can easily cling, the better to support their grape-like clusters.

Still better than the above, a frame of dry comb, as dark
as possible, will often attract the bees and cause them to
cluster. None of these devices however are infallible; hence
the advisability of locating an Apiary among low trees or
bushes, or in an orchard, if possible.

When no trees or bushes are to be found, and no settling
place has been provided, they will settle wherever the queen
may happen to alight, on a grape-vine, on weeds, on the
ground, on the corner of a building, etc.

418. It will inspire the inexperienced Apiarist with more
confidence, to remember that almost all the bees in a swarm,
are in a very peaceable mood, having filled themselves with
honey before leaving the parent-stock (**380**). Yet there
are, in nearly every swarm, a few bees that have either
joined from a neighboring hive, or have not filled their
honey-sack completely before leaving. These bees are liable
to get angry, when the swarm is harvested. So, if the Apiarist is timid, or suffers severely from the sting of a bee, he
should, by all means, furnish himself with the protection of
a bee-dress (**386**). The use of a smoker (**382**). is also
advisable, both in preventing the bees from stinging, and in
helping to drive them into the hive; but it must not be used
plentifully, as it might cause the bees to abscond, or to
return to the clustering spot.

419. *A new swarm should be hived as soon as the bees have
quietly clustered around their queen;* although there is no

necessity for the headlong haste practiced by some, which increases their liability to be stung. Those who show so little self-possession, must not be surprised, if they are stung by the bees of other hives ; which, instead of being gorged with honey, are on the alert, and very naturally mistake the object of such excited demonstrations. The fact that the bees have clustered, makes it almost certain, that, unless the weather is very hot, or they are exposed to the burning heat of the sun, *they will not leave for at least one or two hours.* All convenient dispatch, however, should be used in hiving a swarm, lest the scouts have time to return,—which will entice them to go,—or lest other colonies issue, and attempt to add themselves to it.

420. Should you give the scouts time to return, you would first see a few bees flying around the cluster. Slowly their number would increase, till the whole swarm took wing, and it would be almost useless to try to stop it, or to follow it. When a swarm thus takes flight, it knows no bounds. Hedges, fences, woods, walls, ditches, rivers, are barriers only to the breathless and disappointed owner. The only thing that we ever have known to stop a departing swarm is throwing water among them. Flashing the sun's rays on them by the use of a looking-glass is advised by some. We tried it, but did not succeed in a single instance.

421. As a matter of course, we suppose that the Apiarist has an empty hive in readiness, clean and cool. Bees, when they swarm, being unnaturally heated, often refuse to enter hives that have been standing in the sun, or at best are slow in taking possession of them. The temperature of the parent-stock, at the moment of swarming, rises very suddenly, and many bees are often so drenched with perspiration, that they cannot take wing to join the emigrating colony. To attempt to make swarming bees enter a heated hive in a blazing sun, is, therefore, as irrational as it would be to force a panting crowd of human beings into the suffocating atmosphere of a close garret. If the process of hiving cannot be conducted in the shade, the hive should be covered with a sheet or with leafy boughs.

422. In the movable-comb hive, every good piece of worker-comb), if large enough to be attached to a frame, should be used, both for its intrinsic value, and because bees are so pleased when they find such unexpected treasure in a hive, that they will seldom forsake it. A new swarm often takes possession of a deserted hive, well stored with comb; whilst, if dozens of empty ones stand in the Apiary, the bees very seldom enter them of their own accord. It once seemed to us that an instinct impelling them to do so, would have been much better for us than the present arrangement: but further reflection has shown us that, on the contrary, it would have been a fruitful source of disputes among neighboring bee-keepers; and that in this, as in so many other things, the instincts of the honey-bee have been devised with special reference to the welfare of man.

" The bee-keepers of Greece used to attract the swarms into their hives by rubbing the entrance and the inside of their empty hives with bees-wax and propolis. But such practice was often the cause of contests between neighbors, for their bees did not inquire about the ownership of the hive selected."—(Della Rocca, 1790.)

Drone-combs (**224**) *should never be put up in frames*, or the bees may follow the pattern, and build comb suitable only for breeding a horde of useless consumers.

423. Frames containing worker combs, from colonies that have died in the previous Winter are very good, if the comb is dry and clean. Combs of honey will do if the swarm is hived on a propitious day, otherwise they will attract robbers (**664**) and the presence of the latter will prevent the swarm from entering the hive. For this reason, combs containing honey should not be given to the swarm until the following evening.

But when a few combs only are given to a swarm, as the queen will not follow the builders (**229**), too much drone comb (**224**) will be built. Then, in hiving a swarm, the Apiarist had better dispense with giving any, unless he fills the hive (**234**).

424. In the absence of combs or comb-foundation, (**674**) the triangular comb-guide will greatly help to secure straight combs, in the frames, but it cannot be depended upon, in every case. Comb-foundation in full sheets is so far superior, and is now in such general use, that the triangular comb-guide (**319**) is discarded by most Apiarists. By the use of comb-foundation, crooked combs,—the bane of the Apiary—are no longer found, and every comb hangs in its frame, as straight as a board.

425. It has been held, of late, by some writers, that the use of empty combs, or comb-foundation, was detrimental, in hiving natural swarms, because the bees filled the combs given them, with honey, and left but little room for the queen to lay. This actually takes place in extraordinary seasons and locations, but in the greater number of instances, the empty combs help the colony greatly, and, in bad seasons, a hive-full of empty combs, furnished to a swarm, is equivalent to saving it from starvation, since the combs of a hive cost the bees almost as much honey as is necessary for them to winter on (**223**). Should they fill the combs nearly full of honey, this honey will be partly used up during the dearth which usually comes after the honey harvest, and will serve in rearing brood to strengthen the hive before Winter. Better be safe than sorry.

426. It is very important that the frames should hang true in the hive, and at the proper distance apart (**316**). If the hive has to be removed, they should be previously fastened in their places, by the use of small wire nails only partly driven, and removed later. The cloth (**352**) and mat (**353**) should be carefully placed over the frames, or the swarm would build and raise brood in the upper story, intended only for surplus honey.

427. When the hive is thus prepared and placed in a convenient position, the entrance should be opened as wide as possible. If it has a movable-bottom-board, it should be raised from it in front (344), and the entrance-blocks inserted under its edges, so as to leave a larger passage for the swarm, that the bees may get in as soon as possible; and a well-stretched sheet, or coarse cloth, should be securely fastened to the alighting-board, to keep them from becoming separated, or soiled by dirt; for, if separated, they are a long time in entering; and a bee covered with dust or dirt is very apt to perish. Bees are much obstructed in their travel, by any *corner*, or great inequality of surface; and if the sheet is not smoothly stretched, they are often so confused, that it takes them a long time to find the entrance to the hive.

428. If the bees have alighted on a small limb, which can be cut with sharp pruning-shears, without jarring the swarm, or damaging the value of the tree, they may be gently carried on it to the hiving-sheet, in front of their new home. If they seem at all reluctant to enter it, gently scoop up a few of them with a large spoon, or a leafy twig, or even with the fingers (72), and shake them close to its entrance. As they go in with fanning wings, they will raise a peculiar note, which communicates to their companions the joyful news that they have found a home; and in a short time, the whole swarm will enter, without injury to a single bee.

When bees are once shaken down on the sheet, they are quite unwilling to take wing again; for, being loaded with honey, they desire, like heavily-armed troops, to march slowly and sedately to their place of encampment.

429. When they alight on a high limb, which cannot be reached, or when the limb is too valuable to be sacrificed, the swarm can be hived by using a light box or swarm-sack,

at the end of a pole of proper length. This swarm sack (fig. 87) is made of strong muslin, about two feet deep, fastened around a wire hoop, about one foot in diameter, and is similar to a butterfly net. A piece of braid, is sewed at the bottom, inside and outside, to help in emptying it.

Fig. 87.
SWARM-SACK.

When the sack is placed under the swarm, the bees are suddenly shaken into it by a single tap on the limb. Hold the sack firmly, as the sudden weight will draw it down in a most unexpected manner. To prevent the bees from escaping, hold the handle perpendicularly, as this will close the opening of the bag instantly.

430. In bringing it to the hive, and turning it inside out, by holding the braid with the fingers, some care must be exercised, as this unceremonious imprisoning of the bees is apt to cause some to be angry. A little smoke (**282**) should be used, or a few seconds should be allowed to elapse before they are gently liberated in front of the hive.

431. The sack is preferable to a box or a basket, as the latter do not close readily, and a number of the bees are apt to fly back to the clustering spot, before they are emptied in front of their intended abode.

If this happens, the process of hiving must be repeated, unless the queen has been secured, when they will quickly form a line of communication with those on the sheet. If the queen has not been secured, the bees will either refuse to enter the hive, or will speedily come out* and take wing,

---

* It is a mistake to suppose that a swarm will not enter a hive unless the queen is with them. If some start for it, the others will speedily follow, all seeming to take it for granted that the queen is somewhere among them. Even after they begin to disperse in search of her, they may often be induced to return, by pouring out a fresh lot of bees, which, by entering the hive with fanning wings, cause the others to believe that the queen is coming at last.

to join her again. This happens oftenest with after-swarms, whose young queens, instead of exhibiting the gravity of an old matron, are apt to be frisking in the air.

When the swarm is clustered so high that the sack cannot be raised to it on a pole, it may be carried up to the cluster, and the bee-keeper, after shaking the bees into it, may gently lower it, by a string, to an assistant below.

432. When a colony alights on the trunk of a tree, or on anything from which the bees cannot easily be gathered in a basket, or in the sack, fasten a leafy bough, or a comb over them, and with a little smoke, compel them to ascend it. If the place is inaccessible, they will enter a well-shaded basket, inverted, and elevated just above the clustered mass. We once hived a neighbor's swarm, which settled in a thicket, on the inaccessible body of a tree, by throwing water upon the bees, so as to compel them gradually to ascend the tree, and enter an elevated box. If proper alighting places are not furnished, the trouble of hiving a swarm will often be greater than its value.

433. If the swarm is noticed, when it begins to issue from the parent hive, the practical bee-keeper often harvests it without trouble, by catching the queen (**100**). Provided with a queen cage (**536**), he watches for her exit, and as she comes out, he seizes her and places her in the cage. He then removes the old hive, and places the new one, ready for the swarm, on its stand, with the caged queen on the platform. The swarm may alight, but as soon as the bees notice their loss, they will return, and will cluster around her; and the hiving of the swarm takes but a few minutes. In a circumstance of this kind, it is well to return the parent colony to its stand, after the swarm is hived, for, if entirely removed, it would lose all the bees that were in the field, when the swarm left, and would be too much weakened.

434. To prevent primary swarms from escaping, some

bee-keepers clip one of the wings of their queens previous to the swarming season.*

As an old queen leaves the hive only with a new swarm, the loss of her wings in no way interferes with her usefulness, or the attachment of the bees. If, in spite of her inability to fly, she is bent on emigrating, though she has a "will," she can find "no way," but helplessly falls to the ground, instead of gaily mounting into the air. If the bees find her, they cluster around her, and may be easily secured by the Apiarist; if she is not found, they return to the parent-stock, to await the maturity of the young queens.

This method will do, provided the Apiary ground is bare, so that the queen runs no risk of getting lost in the grass. We abandoned it, after having tried it, for several years, but we know of some owners of large Apiaries who are successful with it. We notice that Mr. Heddon, in his interesting work, "Success in Bee-Culture," is of our opinion on this subject.

435. Where a great many colonies are kept, several swarms may issue at the same time, and unite in a single cluster.

If two swarms cluster together, they may be advantageously kept together, if abundant room for storing surplus honey can be given them. Large quantities of honey are generally obtained from such colonies, if they issue early, and the season is favorable.

"When more than two swarms have clustered together, it is better to divide them. Let us suppose that three have united. After putting three hives near each other, so as to form a triangle, the sack (**429**) or box, in which the bees have been captured,

---

* Virgil speaks of clipping the wings of queens, to prevent them from escaping with a swarm. Mr. Langstroth had devised a way of doing this, so as to designate *the age of the queens:*—With a pair of scissors, let the wings, on one side, of a young queen be carefully cut off: when the hives are examined next year, let one of her two remaining wings be removed, and the last one the third year.

is shaken on a cloth just between the three. If most of the bees
seem to go into the same hive, this should be removed a little
farther. Great care should be exercised to find the queens, and
to direct one towards each hive. But if only one queen is seen,
it is better to cage (**536**) her till the greater part of the bees
have entered. Then, as soon as the bees of one of the hives
show signs of uneasiness, and seem ready to join the bees in the
others, release the queen, and direct her towards this queenless
hive and all will be well."—(Hamet, "Cours d'Apiculture,"
Paris, 1866.)

436. If two queens have entered the same hive, they
can often be found on its bottom-board, each in a ball
(**538**) of angry bees, strangers to them. Open the ball,
and give one of the queens to the queenless hive, if the bees
have not already deserted it. When queens have been
"balled" by mixed swarms, it is well to keep them caged,
in the hive, for a few hours, or till the bees have quieted.
The quantity of bees in each hive can be equalized, by
shaking a few from the strongest in front of the weakest
(**72**).

437. Dr. Scudamore, an English physician, who has
written a tract on the Formation of Artificial Swarms,
says that he once knew as "many as ten swarms go forth at
once, and settle and mingle together, forming, literally, a
monster meeting." There are instances recorded of a still
larger number having clustered together. A venerable
clergyman in Western Massachusetts, told us that in the
Apiary of one of his parishioners, five swarms once clustered together. As he had no hive which would hold them,
they were put into a large box, roughly nailed together.
When taken up in the Fall, it was evident that the five
swarms had lived together as independent colonies. Four
had begun their work, each near a corner of the box, and
the fifth in the middle; and there was a distinct interval
separating the works of the different colonies. In Cotton's "My Bee Book," is a cut illustrating a similar sepa-

ration of two colonies in one hive. By hiving, in a large
box, swarms which have settled together, and leaving them
undisturbed till the following morning, they would sometimes be found in separate clusters, and might easily be put
into different hives.

If the Apiarist fears that another swarm will issue, to
unite with the one he is hiving, he may cover the latter from
the sight of other swarms, with a sheet.

438. If, while hiving a swarm, he wishes to secure the
queen, the bees should be shaken from the hiving-basket, a
foot or more from the hive, when a quick eye will generally
see her as she passes over the sheet. If the bees are reluctant to go in, a few must be directed to the entrance, and
care be taken to brush them back, when they press forward
in such dense masses that the queen is likely to enter unob-
served. An experienced eye readily detects her peculiar
color and form (**100**).

It is interesting to witness how speedily a queen passes
into the hive, as soon as she recognizes the joyful note (**76**)
announcing that her colony has found a home. She quickly
follows in the direction of the moving mass, and her long
legs enable her easily to outstrip, in the race for possession,
all who attempt to follow her. Other bees linger around
the entrance, or fly into the air, or collect in listless knots
on the sheet; but a fertile mother, with an air of conscious
importance, marches straight forward, and looking neither
to the right hand nor to the left, glides into the hive with
the same dispatchful haste that characterizes a bee returning fully laden from the nectar-bearing fields.

439. Swarms sometimes come off when no suitable hives
are in readiness to receive them. In such an emergency,
hive them in any old box, cask, or measure, and place
them, with suitable protection against the sun, where their
new hive is to stand; when this is ready, they may, by a
quick, jerking motion, be easily shaken out before it, on a
hiving-sheet.

Persons unaccustomed to bees, may think that we speak about "scooping them up," and "shaking them out," almost as cooly as though giving directions to measure so many bushels of wheat; experience will soon convince them, that the ease with which they may be managed (**72**) is not at all exaggerated.

440. Bees which swarm early in the day will generally begin to range the fields in a few hours after they are hived. or even in a few minutes, if they have empty comb; and the fewest bees will be lost when the hive is removed to its permanent stand, as soon as the bees have entered it. If it is desirable, for any reason, to remove the hive before all the bees have gone in, the sheet, on which the bees are lying, may be so folded that the colony can be easily carried to their new stand, where the bees may enter at their leisure.

While the hive should be set so as to incline slightly from rear to front (**328**), to shed the rain. there ought not to be the least pitch from *side to side*, or it will prevent the frames from hanging plumb, and compel the bees to build crooked combs.

441. If several rainy days, or a dearth of honey, should occur immediately after the hiving of bees, it is well to feed (**606**) them a little to keep them from starving, till there is honey in the blossoms.

442. The Apiarist has already been informed of the importance of securing straight worker combs for his hives (**318**). To a stock-hive, such combs are like cash capital to a business man; and so long as they are fit for use, they should never be destroyed.

Mr. S. Wagner had a colony over 21 years old, whose young bees appeared to be as large as any others in his Apiary. Mr. J. F. Racine, an old settler of Wallen, Indiana, lost a colony in the Winter of 1884-5 which he had had ever since 1855, without changing the combs. He considered it one of the best in his Apiary.

Those who have plenty of good worker-comb, will unquestionably find it to their advantage to use it in the place of comb-foundation (**674**) or artificial guides. Those who use the guides (**319**), should examine a swarm two or three days after it is hived, when, by a little management, any irregularities in their combs may be easily corrected. Some combs may need a little compression, to bring them into their proper positions, and others may even require to be cut out, and fastened as guides in other frames; but no pains should be spared to see that they are all right, before the work has gone so far as to make it laborious to remedy any defects. If a swarm is small, it ought to be confined, by a movable partition (**349**), to such a space in the hive as it can occupy with comb—as well for its encouragement, as to economize its animal heat. Varro, who flourished before the Christian Era, says (Liber III, Cap. xviii), that bees become dispirited, when placed in hives that are too large.

PRIMARY SWARM WITH A YOUNG QUEEN.

443. We have already stated (**157**) that queens die of old age, when about four years old. If the preparations for queen rearing (**489**) are begun during the swarming season, from th's cause, or by her death through accident, or because she has been removed by the Apiarist, it very often happens that bees prevent the first hatched queen from destroying her riva's (**112**), and the result is that a swarm leaves the hive with her. These primary swarms with young queens, are cast as unexpectedly, and may be as strong as those that are accompanied by the old queen. They have that in common with secondary swarms, that they behave like them, both in their exit and afterwards.

15

SECONDARY OR AFTER–SWARMS.

444. Having described the method commonly pursued for hiving a new swarm, we return to the parent-colony from which they emigrated.

From the immense number which have abandoned it, we should naturally infer that it must be nearly depopulated. To those who limited the fertility of the queen to four hundred eggs a day, the rapid replenishing of a hive, after swarming, must have been inexplicable; but to those who have seen her lay from one to four thousand eggs a day, it is no mystery at all (**40**). Enough bees remain to carry on the domestic operations of the hive; and as the old queen departs only when there is a teeming population, and when thousands of young are daily hatching, and tens of thousands rapidly maturing, the hive, in a short time, is almost as populous as it was before swarming.

Those who suppose that the new colony consists wholly of young bees, forced to emigrate by the older ones, if they closely examine a new swarm, will find that while some have the ragged wings of age, others are so young as to be barely able to fly.

After the tumult of swarming is over, not a bee that did not participate in it, attempts to join the new colony, and not one that did, seeks to return. What determines some to go, and others to stay, we have no certain means of knowing. How wonderful must be the impression made upon an insect, to cause it in a few minutes so completely to lose its strong affection for the old home, that when established in a hive only a few feet distant, it pays not the slightest attention to its former abode!

445. It has already been stated that, if the weather is favorable, the old queen usually leaves near the time that the young queens are sealed over to be changed into nymphs. In about a week, one of them hatches; and the question must be decided whether or not, any more colonies shall be formed that season. If the hive is well filled with bees, and the season is in all respects promising, it is generally decided in the affirmative; although, under such circumstances, some very strong colonies refuse to swarm more than once.

If the bees of the parent-colony decide to prevent the first hatched queen from killing the others, a strong guard is kept over their cells, and as often as she approaches them with murderous intent, she is bitten, or given to understand by other most uncourtier-like demonstrations, that even a queen cannot, in all things, do just as she pleases.

446. About a week after first swarming, should the Apiarist place his ear against the hive, in the morning or evening, when the bees are still, if the queens are "piping," he will readily recognize their peculiar sounds (**115**). The young queens are all mature, at the latest, in sixteen days from the departure of the first swarm, even if it left as soon as the royal cells were begun.

The second swarm usually issues on the first or second day after piping is heard; though the bees sometimes delay coming out until the fifth day, in consequence of an unfavorable state of the weather. Occasionally, the weather is so very unfavorable, that they permit the oldest queen to kill the others, and refuse to swarm again. This is a rare occurrence, as young queens are not so particular about the weather as old ones, and sometimes venture out, not merely when it is cloudy, but when rain is falling. On this account, if a very close watch is not kept, they are often lost. As piping ordinarily commences about a week after first-swarming, the second swarm usually issues eight or nine days after the first; although it has been known to issue as early as the third, and as late as the seventeenth; but such cases are very rare.

447. It frequently happens, in the agitation of swarming, that the usual guard over the queen-cells is withdrawn, and several hatch at the same time, and accompany the colony; in which case the bees often alight in two or more separate clusters. In our observing-hives, we have repeatedly seen young queens thrust out their tongues from a hole in their cell, to be fed by the bees. If allowed to issue at will, they are pale and weak, like other young bees, and for some time unable to fly; but if confined the usual time, they come forth fully colored, and ready for all emergencies. We have seen them issue in this state, while the excitement caused by removing the combs from a hive, has driven the guard from their cells.

The following remarkable instance came under our observation, in Matamoras, Mexico: A second swarm deserting its abode the *second* day after being hived, settled upon a tree. On examining the abandoned hive, *five* young queens were found lying dead on its bottom-board. The swarm was returned, and, the next morning, two more dead queens were found. As the colony afterwards prospered, *eight* queens, at least, must have left the parent-colony in a single swarm!

Young queens, whose ovaries are not burdened with eggs, are much quicker on the wing than old ones, and frequently fly much farther from the parent-stock before they alight.

447 (*bis*). The bee-keepers of old, who were not acquainted with the habits of bees, noticing that primary-swarms were more populous than after-swarms, used to brimstone (**276**) the old colony which had swarmed, and its after-swarm, considering the first swarm as the best of the three; but this apparent superiority was often of short duration, for the first swarm is nearly always accompanied by the old queen. We know better now, since we consider the age of the queen as one of the qualities of a colony.

448. After-swarms are much more prone to abscond or

leave, after hiving, than primary-swarms. It is probably owing to the fact that the young queen has to go out for her bridal trip (**121**), and the bees sometimes leave with her. A comb of unsealed brood (**166**) given them will usually prevent this (**109**). An absconding swarm often leaves without settling.

449. After the departure of the second swarm, the oldest remaining queen leaves her cell; and if another swarm is to come forth, piping will still be heard; and so before the issue of each swarm after the first. It will sometimes be heard for a short time after the issue of the second swarm, even when the bees do not intend to swarm again. The third swarm usually leaves the hive on the second or third day after the second swarm, and the others, at intervals of about a day. We once had five swarms from one stock, in less than two weeks. In warm latitudes, more than twice this number of swarms have been known to issue, in one season, from a single stock.

After-swarms, or casts—these names are given to all swarms after the first—seriously reduce the strength of the parent-stock; since by the time they issue, nearly all the brood left by the old queen has hatched, and no more eggs can be laid until all swarming is over. If, after swarming, the weather suddenly becomes chilly, and the hive is thin, or the Apiarist continues the ventilation which was needed only for a crowded colony, the remaining bees being unable to maintain the requisite heat, great numbers of the brood may perish.

PREVENTION OF NATURAL SWARMING.

450. The prevention of natural swarming, in the present state of bee-keeping, is an important item, for several reasons.

1st, Bee-keeping has so spread in the last few years, that

many bee-keepers are possessors of as many colonies as
they desire to keep. Most Apiarists, especially farmers,
keep bees only for the honey, and as it is impossible to
produce both an increase of stock, and a large yield of
honey in average seasons, they prefer the production of
honey to that of swarms.

2nd, Another objection to natural swarming arises from
the disheartening fact, that bees are liable to swarm so
often, as to destroy the value of both the parent-stock,
and its after-swarms. Experienced bee-keepers obviate
this difficulty, by making one good colony out of two second
swarms, and returning to the parent-stock all swarms after
the second, and even this if the season is far advanced.
Such operations often consume more time than they are
worth.

3d, The bees may be located in a town, near a public thoroughfare where people pass constantly, and accidents may take place; or perhaps near the woods where
the swarm would cluster on such high limbs that it would
be difficult or impossible to hive them.

4th. It is very troublesome to have to watch the bees for
weeks, or to have them swarm at unexpected or unwelcome
times, when the family is away, or at dinner, or while the
owner is engaged with his business, for many bee-keepers
are also lawyers, doctors or merchants, occupied in daily
labors, which require a definite part of their time. The farmer may be interrupted in the business of hay-making, by
the cry that his bees are swarming; and by the time he has
hived them, perhaps a shower comes up, and his hay is injured more than the swarm is worth. Thus the keeping of
a few bees, instead of being a source of profit, may prove
an expensive luxury; while in a large Apiary, the embar-
rassments are often seriously increased. If, after a succession of days unfavorable for swarming, the weather becomes
pleasant, it often happens that several swarms rise at once,

aud cluster together; and not unfrequently, in the noise and confusion, other swarms fly off, and are lost. We have seen the bee-master, under such circumstances, so perplexed and exhausted as to be almost ready to wish he had never seen a bee.

451. Mr. J. F. Racine, of Wallen, Allen Co., Indiana, had 505 natural swarms from 165 colonies in the summer of 1883. Sixty-one swarms came out on the 3d of July. We will let him tell the story in his own way:

" In the morning, as soon as the watchword had been given for the first swarm, there was no rest. Primary, secondary, and after-swarms, all passed under the same limb of the same tree. The bees were no sooner shaken in a basket, and emptied in front of a hive, than there was another cluster gathered, in the same spot. Some swarms had no queen, while others had 3, 4, and even 5 of them. Some were young queens, some were old queens. When we could find a queen, we caged her (**536**) to preserve her from being balled (**538**). The sixty-one swarms were hived in 20 hives, and surplus cases were given them at once. A man, who had come with 5 hives to buy swarms, said that he had never seen the like, neither had I, although I have kept bees for 57 years. And the best of it is, I did not want any swarms at all that season. "

452. *5th.* It is admitted, by all progressive people, that man can achieve a great deal by artificial selection and cultivation of plants and animals. The same selection is advisable in the reproduction of the honey-bee, and an increase from selected colonies or selected races, cannot always be had by natural swarming. In this, artificial swarming is much better, and gives much more satisfactory results when ever an increase is desirable.

453. *6th.* The numerous swarms lost every year, is a strong argument against natural swarming.

An eminent Apiarist has estimated, that, taking into account all who keep bees, one-fourth of the best swarms are lost every season. While some bee-keepers seldom lose a swarm, the majority suffer serious losses by the flight of

their bees to the woods; and it is next to impossible, even
for the most careful, to prevent such occurrences, if their
bees are allowed to swarm.

Apiarists will then recognize that it is very important to
follow a method, which will nearly, if not altogether, prevent natural swarming. But in order to prevent it, we
must know the causes of it.

454. Natural swarming, so far, has been considered as a
natural impulse in bees. Yet, it can be prevented, for it is
always caused by uneasiness, as we will show in the next
paragraph, or by an abnormal condition of the colony (**465**).
It is caused:

1st. In the majority of instances, by the want of room in
the comb. By want of room, we do not mean want of
empty space in the hive, but want of empty comb for the
queen to deposit her eggs (**97**), or for the workers to deposit their honey. So long as bees have an abundance of
empty space below their main hive, they very seldom swarm;
but if it is on the *sides* of their hive, or *above* them, they
often swarm rather than take possession of it.

This happens, not only in the Southern latitudes, where
the swarming instinct is so powerful, but even in our Northern or Middle States. This fact is corroborated by Simmins, whose non-swarming system is based on the idea of
keeping "open space and unfinished combs at the front, or
adjoining the entrance." (Rottingdean, England, 1886.)
Persons who are unacquainted with the details of bee-keeping have no idea how suddenly the honey harvest comes,
and how rapidly the combs can be filled, when it once begins. Strong colonies which were almost destitute, just at
the opening of the crop, owing to the large amount of brood
they were raising, have been known to harvest twenty
pounds, and more, in one day. When bees are thus gathering large quantities of honey, and the combs are becoming crowded, so that the cells, from which the young bees

hatch, are filled with honey as fast as they are vacated, they feel the necessity of emigrating, especially as the constant hatching workers add daily to their large population. The building of additional combs, by a part of the bees, is sometimes insufficient to keep them from making preparations for swarming, as it does not give employment to all. The reader must remember that in a good colony, at this season, there are between 50,000 and 120,000 bees, according to the laying capacity of the queen and the size of the breeding room. There is also an additional increase over mortality of perhaps 2,000 bees daily. In spite of the admirable order of these wonderful little insects, there cannot help be more or less crowding, unless there is ample room in the combs.

455. If some of the bees decide that they are too crowded, queen-cells are raised (**104**) and the colony gets what Apiarists call the "*swarming fever.*" It is a very appropriate name indeed, since the so-called fever is cured only by swarming. In some extraordinary seasons, after this "swarming fever" has taken possession of their little brains, no amount of room given, even by dividing (**470**) will prevent them from executing their purpose, unless the weather and the honey crop become unfavorable. We have repeatedly, in such seasons, divided a colony into several nuclei (**520**) without avail, each nucleus swarming in spite of its weakness.

456. *2d.* The heat of the Summer sun, which alone would not cause them to swarm, hastens their preparations, when the bees are disposed to emigrate.

457. *3d.* The hatching of a great number of drones (**189**)—due to an excess of drone-comb (**224**) in the brood chamber, in which the queen has deposited eggs,—is also an incitation to the "swarming fever." These big, burly, noisy fellows help to make the already crowded comb quite uncomfortable. This is why a great many bee-keepers of the old school report having noticed that hives which raise the most drones cast the greatest number of swarms.

458. *4th.* An improperly ventilated hive (**336**), or surplus arrangement, strongly induces natural swarming. We have seen ignorant bee-keepers, owners of box-hives wonder why their bees swarmed and did not work in the surplus honey receptacle. The average box-hive (**274**) is made about twelve inches square inside, often with only a shallow entrance a couple of inches long, at the bottom, and the surplus arrangement is reached by the bees, through an auger-hole an inch or two in diameter. In order to ventilate the honey receptacle, the bees have to form a line (**61**) from the outside of the hive through the thickly covered combs to this hole, and force in air enough to enable them to breathe and live there. How can we wonder that they refuse to work in such a place, especially when the hive is exposed to the heat of the sun in a June afternoon?

Under such circumstances, hordes of useless consumers often blacken, for months, the outside of the hives, to the great loss of their disappointed owners.

459. *1st.* It results from the above that *the principal condition for the prevention of natural swarming is, a sufficient amount of empty comb*, and this empty comb must be given in an easily accessible place near and above, or in front of the brood.

The giving of comb foundation (**674**) instead of empty combs, will be sufficient if the crop is not flowing too fast. But in a very good season, if the harvesting workers bring the honey faster than the young bees can stretch the foundation into comb, it will not be sufficient.

460. If the breeding story is full of comb, and the surplus arrangement is placed above with a wooden division or honey board (**352**) between, the bees will often consider the latter as too remote from their breeding room, especially if the holes which connect the two are few, and ventilation cannot be readily given from one apartment to another.

461. *The giving of combs in a place of easy access, must be attended to, just before the crop begins, or the bees may make preparations which would render all later enlargements of the hive completely useless, as far as prevention of swarming is concerned.* The breeding room must be large enough to accommodate the most prolific queen (**155**).

462. *2nd. The hive must be located where the sun will not strike it directly* in the hottest hours of the day. It can easily be sheltered artificially with a roof, if there is no shrubbery around it (**369**).

463. *3d. The drone-comb must be carefully removed*, in Spring, as far as possible, *and replaced by worker-comb* (**675**). It is impossible to remove every cell of drone-comb, but a few drones will not hurt. It is the excess, the breeding of thousands of drones which is objectionable, and an incentive to swarming. The removal of drone-comb is highly advisable for other reasons (**512**).

464. *4th.* The hive should be thoroughly ventilated, so that the bees will find themselves comfortable in it.

465. This system, which gives the smallest possible number of swarms, and the largest possible amount of surplus-honey, was inaugurated by us, years ago, and has been adopted on both continents. Mr. Cowan, the worthy editor of the *British Bee-Journal*, says of it, page 148, April 1886, "Hives managed in this way, will give the maximum of honey with the least amount of labor."

If the above directions are followed, the natural swarms will not exceed three to five per cent. These swarms will be very large—Mr. DeLayens once had a swarm weighing 11½ lbs—and after-swarms will be scarce. The few hives that swarm are those which, having old queens, attempt to replace them during the swarming season (**499**), or those whose queens die while the crop is abundant.

In the first case, one or more young queens being raised in the hive, it often happens that the old queen tries to destroy them; the bees prevent her (**112**), and swarming is the result. The same reason *may* cause swarming in a strong colony, in which a queen has been introduced (**533**) by the Apiarist, during a good yield of honey. Perhaps the bees accept her "*under protest*," and soon begin raising queen-cells (**104**) to replace her, but the abundant honey harvest causes them to change their preparations, and they swarm with this introduced queen. A hive which has been made queenless during the honey crop, may swarm for the same reasons (**443**), as soon as the young queens are old enough.

466. The prevention of natural swarming, when comb-honey is raised in sections (**721**), is not so successful, because the Apiarist cannot furnish his bees with empty combs. But very good results can be obtained, by following as nearly as possible all the directions above given.

467. As the queen cannot get through an opening $\frac{5}{32}$ of an inch high—which will just pass a loaded worker, if the entrance to the hive be contracted to this dimension, she will not be able to leave with a swarm.

This is done with drone or queen-traps, perforated zinc, entrance-blocks, and other fixtures. (See Drone-trap, **191**).

This method of preventing swarming requires great accuracy of measurement, for a very trifling deviation from the dimensions given, will either shut out the loaded workers, or let out the queen. It should be used only to imprison old queens; for young ones, if confined to the hive, cannot be impregnated (**120**). These fixtures, if firmly fastened, will exclude mice from the hive in the Winter. When used to prevent all swarming, it will be necessary to adjust them a little after sunrise and before sunset, to take out, or allow the bees to carry out any drones that have died.

We have seen colonies kill their queen, and raise another, because she had thus been unable to follow the swarm, hence, these appliances will do only in small Apiaries, where bee-keepers can examine each colony daily; and even there, we would not advise their constant use.

Mr. Langstroth had formerly devised a non-swarmer block, with a metallic slide, to prevent the escape of the queen. This was abandoned, because it annoyed the bees and interfered with ventilation (**333**), as all such arrangements do.

Fig. 88.

NON-SWARMER BLOCK.

It is shown attached on the hive in Fig. 56.

468. After-swarms have been prevented from issuing, by a method invented by Jas. Heddon, who is one of the noted and successful Apiarists of Michigan. The Heddon method consists in placing the first swarm, side by side with the parent hive, and one week after the issue of the swarm, or just previous to the expected departure of the second swarm, removing the parent hive to a new location, thus giving all its old bees to the first swarm. This is virtually preventing a natural issue by a forced issue, but making the first swarm strong, at the expense of the mother colony. The sole objection to this method is that it does away only with the annoyance of catching the swarm, and leaves the parent colony much weakened.

CHAPTER VII.

ARTIFICIAL SWARMING.

469. Every practical bee-keeper is aware of the uncertainty of natural swarming (**406**). Under no circumstances, can it be confidently relied on. While some colonies swarm repeatedly, others, apparently as strong in numbers, and rich in stores, refuse to swarm, even in seasons in all respects highly propitious. Such colonies, on examination, will often be found to have taken no steps for raising young queens. Besides, it frequently happens that, when all the preparatio..s have been made for swarming, the weather proves so inclement that the young queens approach maturity before the old ones can leave, and are all destroyed. Under such circumstances, swarming, for that season, is almost certain to be prevented. The young queens are also sometimes destroyed, because of some sudden, and perhaps only temporary, suspension of the honey-harvest; for bees seldom colonize, even if all their preparations are completed, unless the blossoms are yielding an abundant supply of honey.

The numerous perplexities pertaining to natural swarming, have, for ages, directed the attention of cultivators to the importance of devising some more reliable method for increasing the number of their colonies.

Dr. Scudamore quotes Columella as giving directions for making artificial swarms. Although he taught how to furnish a queen to a destitute colony, and how to transfer brood-comb, with maturing bees, from a strong stock to a weak one, he does not appear to have formed entirely new colonies by any artificial process. His treatise on bee-keeping shows not only that he was well acquainted with previous writers on the subject, but that he was also a successful practical Apiarist. Its precepts, with but few exceptions, are truly admirable, and prove that in his time bee-keeping, with the masses, must have been far in advance of what it was fifty years ago.

We have spoken of the bar-hive, (**282**) as at least two hundred years old. From "A Journey into Greece, by George Wheeler, Esq.," made in 1675-6, it appears that it was, at that time, in common use there, and, probably, even then an old invention; he described its uses in forming artificial swarms, and removing spare honey. As the new swarms were made by dividing the combs between two hives, and no mention is made of giving the queenless one a royal cell, those old observers were probably acquainted with the fact that they could rear one from the worker-brood. Huber says:—"Monticelli, a Neapolitan Professor, claims that the plan of artificial swarming was borrowed from Favignana, and that the practice is so ancient that even the Latin names are preserved by the inhabitants in their procedure."

470. Huber, after his splendid discoveries in the physiology of the bee, felt the need of some way of multiplying colonies, more reliable than that of natural swarming. He recommends forming artificial swarms, by dividing one of the hives, and adding six empty frames to each half.

"Dividing-hives," (**278–279**) of various kinds, have been used in this country. The principle seems to have all the elements of success; but it was ascertained, that, however modified, such hives are all practically worthless for purposes of artificial increase.

It is one of the laws of the hive, that *bees which have no mature queen, seldom build any cells except such as are designed merely for storing honey, and are too large for the rearing of workers* (**228**).

471. Messrs. Langstroth and Dzierzon were the first observers who had noticed the bearing of this remarkable fact on artificial increase. It may, at first, seem unaccountable that bees should build only comb unfit for breeding, when their young queen will so soon require worker-cells for her eggs; but it must be borne in mind, that at such times they are in an "*abnormal*" condition. In a state of nature, they seldom swarm until their hive is full of comb; or if they do, their numbers are so reduced, that they are rarely able to resume comb-building, until the young queen has hatched.

The determination of bees having no mature queen, to build comb designed only for storing honey, and unfit for rearing workers, shows very clearly the folly of attempting to multiply colonies by dividing-hives, unless the greater part of the bees are given to the queen, and the greater part of the combs to the queenless half.

When the queenless part proceeds to supply her loss, if it has bees enough to build new comb, it will build such as is designed only for storing honey. The next year, if this hive is divided, one-half will contain nearly all the brood, while the other, having most of its combs fit only for storing honey, or raising drones, will be a complete failure.

So uniformly do bees with an unhatched queen build coarse, or drone-comb, that often a glance at the combs of a new colony, will show either that it is queenless, or that, having been so, it has just reared a new queen (**229**).

472. Some Apiarists have attempted to multiply their colonies, by removing, when thousands of its inmates are ranging the fields, a strong stock to a new stand, and setting in its place an empty hive, with a frame of brood-comb, suitable for raising a queen. This method is still worse than the one just described. One half of the dividing-hive was filled with breeding comb, while this empty hive having next to none, all that is built before the queen hatches, will be of a size unsuitable for rearing workers. The queenless part of the divided hives might also have contained a young queen almost mature, so that the building of large combs would have quickly ceased; for it is not always necessary that a queen should have commenced laying eggs to induce her colony to build worker-cells; we have known a strong swarm with a virgin queen, to build beautiful worker-comb, before a single egg was deposited in the cells.

When a new colony is formed by dividing the old hive, the queenless part has thousands of cells filled with brood and eggs, and young bees will be hatching for at least three weeks : by this time, the young queen will ordinarily be laying eggs, so that there will be an interval of not more than three weeks, during which the colony will receive no accessions. But when a new swarm is formed, in the way above described, not an egg will be laid for nearly three weeks, and not a bee hatched for nearly six. During all this time, the colony will rapidly decrease,* and by the time the progeny of the young queen begins to mature, the new hive will have so few bees, that it would seldom be of any value, even if its combs were of the best construction (**182.**)

473. One *strong forced swarm*, can be obtained in any style of hive, including box-hives, by the driving process (**574** to **577**) as follows : When it is time to form artificial colonies, we mean a few days before swarming time, or as soon as the hives are about full of bees,—drum a strong stock—which call *A*—so as to secure *all* its bees.

They may be driven either into a forcing box, or into the upper story of a movable frame hive, and hived like a new swarm, when, if placed on their old stand, they will work as vigorously as a natural swarm. If they were driven, at first,

---

*Every observing bee-keeper has noticed how rapidly even a large swarm diminishes in number, for the first three weeks after it has been hived. So great is the mortality of bees during the height of the working-season, that often, in less than that time, it does not contain one half its original number.

16

into a hive which will suit the Apiarist, it may be returned to their old location, without disturbing the bees.

If any bees are abroad when this is done, they will join this new colony. Remove to a new stand in the Apiary a second strong stock—which call *B*—and put *A* in its place.

Thousands of the bees that belong to *B*, as they return from the fields, will enter *A*, which thus secures enough to develop the brood, and rear a new queen. In fact, this colony often becomes so strong, by the help of the field workers of *B*, as well as through its own constantly hatching bees, that there is some danger of its casting off a swarm when the first young queen hatches, unless again divided at that time.

474. It is quite amusing to observe the actions of the bees that return to their old stand, when their homes have been exchanged as above.

If the strange hive is like their own in size and outward appearance, they go in as though all was right, but soon rush out in violent agitation, imagining that by some unaccountable mistake, they have entered the wrong place. Taking wing to correct their blunder, they find, to their increasing surprise, that they had directed their flight to the proper spot ; again they enter, and again they tumble out, in bewildered crowds, until at length, if they find a queen or the means of raising one, they make up their minds that if the strange hive is not home, it looks like it, stands where it ought to be, and is, at all events, the only home they are likely to get. No doubt they often feel that a very hard bargain has been imposed upon them, but they are generally wise enough to make the best of it. They will be altogether too much disconcerted to quarrel with any bees that were left in the hive when it was forced, and these on their part give them a welcome reception, especially if they come in with a heavy load.

This method of artificial swarming will not weaken either

of the mother-colonies. If *B* had been first forced, and then removed, it would have been seriously injured ; but as it loses fewer bees than if it had swarmed, and retains its queen, it will soon become almost as powerful as before it was removed.

The Apiarist, by treating a natural swarm as he has been directed to treat a forced one, can secure an increase of one colony from two ; and of all the methods of conducting natural swarming, in regions where rapid increase is not profitable, this is the best, provided the colonies do not stand too close together, and the hives used in the process are somewhat similar in shape and color.

475. Whenever the bee-keeper learns how to handle the movable-frames safely he must dispense with the forcing-box, and make his swarms by lifting out the frames from the parent-stock, and shaking the bees from them, by a quick jerking motion, upon a sheet, directly in front of the new hive.

If the hive contains much fresh honey, which is usually very thin, the bees must be brushed off, for shaking them off would also shake out a large amount of nectar (**247**).

As soon as a comb is deprived of its bees, it should be returned to the parent-stock. If one or two combs containing brood, eggs, and stores, are given to the forced swarm, it will be much encouraged, and will need no feeding (**605**) if the weather should be unfavorable. In removing the frames, the bee-keeper should look for the queen, and give the comb she is on, to the forced swarm, without shaking off the bees. If he does not see her on the combs, he will seldom after a little practice, fail to notice her, as she is shaken on the sheet, and crawls towards the new hive. The queen is seldom left on a frame after it has been shaken so that most of the bees fall off (**439**).

476. The more combs with brood are taken from *A*, the less chance it will have to send forth a natural swarm with its first hatched queen.

If it is desirable to make a large number of swarms, and the parent colony is strong in hatching bees, only a few of the combs need be shaken in front of the new hive containing the queen, and the parent colony, with the adhering young bees, may be set in a new place.

By this method, one swarm is made from each of the hives set apart for increase, and although the colonies thus divided are not so strong as when one swarm is made from two hives; yet, in ordinary localities and seasons, they become strong enough for all purposes, long before the season is over, especially if young queens are introduced (**533**) in the colonies made queenless, and comb-foundation is used in full sheets in the frames (**674**).

477. If the mother-colony has not been supplied with a fertile queen, it cannot for a long time part with another swarm, without being seriously weakened.

Second-swarming, as is well known, often very much injures the parent-stock, although its queens are rapidly maturing; but the forced mother-stock may have to start them almost from the egg. By giving it a fertile (**533**) queen, and retaining enough adhering bees to develop the brood, another swarm may be taken away in ten or twelve days in a good season, and the mother-stock left in a far better condition than if it had parted with two natural swarms. In favorable seasons and localities, this process may be repeated two or three times, at intervals of ten days, and if no combs are removed, the mother-stock will still be well supplied with brood and mature bees. Indeed, the judicious removal of bees, at proper* intervals, often leaves it, at the close of the Summer, better supplied than non-swarm-

---

*If a stock of bees, in a hive of moderate size, is examined, at the height of the honey-harvest, nearly all the cells will often be found full of brood, honey, or bee-bread. The great laying of the queen is over—not as some imagine, because her fertility has decreased, but simply for want of room for more brood. A queen in such a colony, or in a hive having few bees, often appears almost as slender as one still unfertile; but if she has plenty of bees and empty comb given her, her proportions will soon become very much enlarged.

ing stocks with maturing brood; the latter having—in the expressive language of an old writer—" waxed over fat."

We have had stocks which, after parting with four swarms in the way above described, have stored their hives with Fall honey, besides yielding a surplus in boxes.

This method of artificial increase, which resembles natural swarming, in not taking away the combs of the mother-stock, is not only superior to it, in leaving a fertile queen, but obviates almost entirely all risk of after-swarming; for the forced swarm, containing the old queen, seldom attempts to send forth a new colony, and the parent hive, in which the young queen is placed, is too destitute of field-workers to swarm soon. The young queen herself is equally content— except in very warm climates, or in extraordinary seasons —to stay where she is put. Even if the old queen is allowed to remain in the mother-stock, she will seldom leave, if sufficient room is given for storing surplus honey; and it makes no difference—as far as liability of swarming is concerned—where the young one is put.

478. Artificial increase may be also made, by simply giving several frames of hatching bees to a nucleus (**520**) containing a fertile queen, and placing the colony thus built up on the stand of a strong hive, removing the latter to a new location.

If, from some cause, the parent-colony could not be moved, the forced swarm might be made to adhere to a new location as follows: Secure their queen, when the bees are shaken out of the hive; and when they show that they miss her, confine them to their hive, until their agitation has reached its height. Then open the hive, and as the bees begin to take wing, present their queen to them. When they have clustered around her, *they may be treated like a natural swarm.* To do this with every forced swarm would take too much time; but it would answer well when the forced swarm is to be moved, a short distance.

479. If no queens have been raised previously (**514**), by making a few forced swarms, from select colonies (**513**), nine days before the time in which the most are to be made, there will be an abundance of sealed queens, almost mature, so that every parent-stock may have one. If the forced swarms were made a short time before natural swarming would have taken place, some of the parent-colonies will contain a number of maturing queens, which may be removed, a few days before hatching, and given to such as have started none. But it is far better to rear the queens first, as they can be bred from choice stock (**513**).

480. A nucleus (**520**) may be built up after its queen has commenced laying, by helping it with a comb of brood and young bees, from a full colony, adding, at proper intervals, a third, and a fourth, until they are strong enough to take care of themselves. This mode of increase is laborious, and requires skill and judgment; for, the bee-keeper should be very careful never to give a weak colony more brood than its bees can cover, remembering that, should the temperature become colder, the brood might be chilled and perish.

As a number of nuclei are to be simultaneously strengthened, the Apiarist cannot complete his artificial processes by a single operation, and must always be on hand, or incur the risk of ending the season with a number of starving colonies. For these and other reasons, we much prefer the other methods, above given, dispensing with so much opening of hives and handling of combs. If, however, any of the new colonies are weak enough to need it, they must be helped to combs from stronger ones.

481. *Whatever method of artificial increase is pursued by the Apiarist, he should never reduce the strength of his mother-stocks, so as seriously to cripple the reproductive power of their queens.* This principle should be to him as "the law of the Medes and Persians, which altereth not;" for, while a queen, with an abundance of worker-comb and bees, may, in a single season, become the parent of a number of prosperous families, if her colony, at the beginning of the swarming season, is divided into three or four parts, not one of them will ordinarily acquire stores enough to survive the Winter.

The practical bee-keeper should remember that no drone-comb is built when the queen is with the builders (**224**), and that the less increase he takes, from the colonies on which he relies for surplus-honey, the better.

482. With the movable-frame hive, and the improved system, the Apiarist, by raising his queens or queen-cells (**514**) previously (*and this is very important*) can take the increase that he wishes to make, *from colonies that would have produced little, if any, surplus, and preserve his best colonies for honey production.* Let it not be understood by this, that we advise taking the increase from weak colonies. In every Apiary, there are some colonies, which, though of fair strength, do not become populous in time to harvest more than their supply. Such colonies can furnish good swarms, with but little help, owing to the fact that the greater number of their bees raised during the harvest, instead of before it, are too young to go to the field (**162**).

If our method is followed, the colonies, which have been kept for honey production, can furnish help, if necessary, towards the end of the season, for those of the artificial swarms that need it.

To the prudent Apiarist, they are as a reserve body of select troops to the skillful general, a timely help, in an emergency.

Remember that populous colonies, that are raising queen-cells, during the early part of a good honey harvest, are strongly inclined to swarm when the young queens hatch. (**465.**)

483. *The colonies that are raising young queens, either* *from worker-brood or from queen-cells given them, must be well supplied with honey, must have enough young bees to keep the brood warm and to take care of it, and no comb-building to do* (**224**).

One artificial swarm made at the opening of the honey harvest, when the hive is full of brood, is better than two swarms made at its close.

When new colonies are made by purchasing queens (**594**) with *bees by the pound* (**599**), shipped from a distance (**587**), they should be hived on as many combs of brood, taken from other hives, as they can well cover. If full frames of foundation (**674**) are added, from time to time, strong colonies may be built out of them, quite readily.

If the colonies are gathering much honey, when artificial swarms are made, but little smoke (**382**) will be needed in the operations. The frequent use of smoke makes the queen leave the combs, for greater security. This often causes great delay in the formation of artificial swarms by removing the frames, and in operations where it is desirable to catch the queen, or to examine her upon the comb.

484. *Artificial operations of all kinds are most successful when bee-forage is abundant;* when it is scarce, they are quite precarious, even if the colonies are well supplied with food.

When bees are not busy in honey-gathering, they have leisure to ascertain the condition of weak colonies, which are almost certain to be robbed, if they are incautiously opened. When forage is scarce, the Apiarist who does not guard against robbing (**664**) will seriously impair the value of his colonies, and entail upon himself much useless and vexatious labor. *Beware of demoralizing bees, by tempting them to rob one another.*

485. *During a good honey flow, bees from different hives may be mixed without quarrelling,* owing to their more peaceable disposition, when full of honey, hence all manipulations become much easier. But at other times, great caution is requisite not only in giving a hive a strange queen, but in all attempts *to mix bees belonging to different colonies.* Bees having a fertile queen will often quarrel with those having an unimpregnated one.

Members of different colonies (30) recognize their hive-companions by the sense of smell, and if there should be a thousand hives in the Apiary, any one will readily detect a strange bee; just as each mother in a large flock of sheep is able, by the same sense, in the darkest night, to distinguish her own lamb from all the others. Colonies might always be safely mingled, by sprinkling them with sugar-water, scented with peppermint or any other strong odor, which would make them all smell alike.

Bees also recognize strangers by their *actions*, even when they have the same scent; for a *frightened bee curls herself up with a cowed look*, which unmistakably proclaims that she is conscious of being an intruder. If, therefore, the bees of one colony are left on their *own stand*, and the others are suddenly introduced, in a time of scarcity, the latter, even when both colonies have the same smell, are often so frightened that they are discovered to be strangers, and are instantly killed. If, however, *both* colonies are removed to a *new stand*, and shaken out together on a sheet, they will peaceably mingle, when scented alike. We find substantially the same thing recommended, in 1778, by Thomas Wildman (page 230 of the 3d edition of his valuable work on Bees), who says, that bees will " unite while in fear and distress, without fighting, as they would be apt to do, if strange bees were added to a hive in possession of its honey."

486. The forcing of a swarm ought not to be attempted when the weather is cool, nor after dark. Bees are always much more irascible when their hives are disturbed after it is dark, and as they cannot see where to fly, they will alight on the person of the bee-keeper, who is almost sure to be stung. It is seldom that night work is attempted upon bees, without making the operator repent his folly.

487. We would strongly dissuade any but the most experienced Apiarists, from attempting, at the furthest, to do more than double their colonies in one year. It would take another book to furnish directions for rapid multiplication, sufficiently full and explicit for the inexperienced; and even then, most who should undertake it, would be sure, at first, to fail. With ten strong colonies of bees, in movable-comb hives, in one propitious season, we could so increase them, in a favorable location, as to have, on the approach of Winter, one hundred good colonies; but we should expect to purchase queens, foundation, and perhaps hundreds of pounds of honey, devoting much of our time to their management, and bringing to the work the experience of many years, and the judgment acquired by numerous lamentable failures.

In one season, being called from home after our colonies had been greatly multiplied, the honey harvest was suddenly cut short by a drought, and we found, on our return, that most of our stocks were ruined by starvation.

The time, care, skill, and food required in our uncertain climate for the rapid increase of colonies, are so great, that not one bee-keeper in a hundred* can make it *profitable;* while most who attempt it, will be almost sure, at the close of the season, to find themselves in possession of colonies which have been *managed to death.*

A *certain* rather than a *rapid* multiplication of colonies, is most needed. A single colony, doubling every year, would, in ten years, increase to 1,024 colonies, and in twenty

* Many a person who reads this will probably imagine that he is the one in a hundred.

years to over a million!*　At this rate, our whole country might, in a few years, be over-stocked with bees; and even an increase of one-third, annually, would soon give us enough.

488. All the methods of increase above given, and several others of less importance, were described by Mr. Langstroth years ago.　He never hesitated to sacrifice several colonies, in order to ascertain a single fact; and it would require a large volume, to detail his various experiments on the single subject of artificial swarming.　The practical bee-keeper, however, should never lose sight of the important distinction between an Apiary managed principally for purposes of observation and discovery, and one conducted exclusively with reference to pecuniary profit.†　Any bee-keeper can easily experiment with movable-frame

---

* The following calculation of *possible* profits from bee-culture, taken from "Sydserff's Treatise on Bees," published in England, in 1792, is a perfect gem of its kind:

"Suppose a swarm of bees at the first to cost 10s. 6d., and neither them nor the swarms to be taken, but to do well, and swarm once every year"—bees must be naughty, indeed, if they dare to do otherwise!—"what will be the product for fourteen years, and what the profit, if each hive is sold at 10s. 6d.?

Years.	Hives.	Profits.		
		£	s.	d
1	1	0	0	0
2	2	1	1	0
3	4	2	2	0
4	8	4	4	0
**	**	*	*	*
14	8192	4300	16	0

"N.B.—Deduct 10s. 6d., what the first hive cost, and the remainder will be clear profit; supposing the second swarms to pay for hives, labor, etc." The modesty with which this writer, who seems to have had as much faith in his bees as in the doctrine that "figures cannot lie," closes his calculation at the end of fourteen years, is truly refreshing. No bee-keeper, on such a royal road to wealth, could ever find it in his heart to stop under twenty-one years, by which time, probably, he would be willing to close his bee-business, by selling it for over two and three-quarter millions of dollars! The attention of all venders of humbug bee-hives, is respectfully invited to this antique specimen of the art of puffing.

† Professor Siebold says, that Berlepsch told him, that some of his hives "had been very much prejudiced by the various scientific experiments."

hives; but he should do it, at first, only on a small scale.
and if pecuniary profit is his object, should follow our directions, until he is *sure* that he has discovered others which
are better.   These cautions are given to prevent serious
losses in using hives which, by facilitating all manner of
experiments, may tempt the inexperienced into rash and
unprofitable courses.   Beginners, especially, should follow
the directions here given as closely as possible ; for, although
they may doubtless be modified and improved, it can only
be done by those experienced in managing bees.

Let us not be understood as wishing to intimate that perfection has been so nearly attained, that no more important
discoveries remain to be made.   On the contrary, we believe that apiculture is a growing science.   Those who
have time and means should experiment on a large scale
with the movable-comb hives ; and we hope that every intelligent bee-keeper who uses them, will experiment, at least,
on a small scale.   In this way, we may hope that those
points in the natural history of the bee still involved in
doubt, will, ere long, be satisfactorily explained.

There is a large class of bee-keepers—not " bee-masters"
—who desire a hive which will give them, however ignorant
or careless, a large yield of honey from their bees.   They
are easily captivated by the shallowest devices, and spend
their money and destroy their bees, to fill the purses of un-
principled men.   There never will be a " royal road " to
profitable bee-keeping.   Like all other branches of rural
economy, it demands care and experience ; and those who
are conscious of a strong disposition to procrastinate and
neglect, will do well to let bees alone, unless they hope, by
the study of their systematic industry, to reform evil habits
which are well nigh incurable.

CHAPTER VIII.

QUEEN REARING.

489. We have shown (**109**) that when a colony is deprived of its queen, the bees soon raise another, if they have worker eggs or young larvæ.

In general, they select, first, some of the oldest among those whose milky " pap " has not yet been changed for coarser food (**107**). Such a selection is wise, for the older the larva is, the sooner the colony will recover a queen.

490. But some Apiarists fear that the bees will secure poorer queens, if they use larvæ, for they suppose that the food given to these during the first three days, may be different from the food given to the queen-larvæ, although it looks the same, and for this reason, they prefer to raise their queens, from the egg.

491. A learned bee-keeper, of Switzerland, Mr. De Planta, has made comparative chemical experiments, on the milky food which is first given to the larvæ of drones, queens, and workers, and has ascertained that this food is composed of the same substances for all, albumen, fat, sugar, and water, and that the only difference is in the proportions of these substances. Yet he concludes that these variations are but accessory, and not premeditated by the bees.

We think that these conclusions are right, for Mr. De Planta, to get a sufficient quantity of this food, had to take it from different hives, and at different seasons of the year ; and as this milky food is apparently the product of glands, (**64**), as is the milk of our cows, the proportions of substances in the " milk " of bees, may vary, as they do in the milk of cows, which contains more or less caseine, fat, sugar, or water, according to the race, the age, and the food eaten.

492. Other bee-keepers suppose that the newly-hatched larvæ, intended by the bees to be raised as queens, are more plentifully fed from the first, than worker-larvæ. But we have always noticed, that, except during a scarcity, the latter have as much of this pap as they can eat, during the first three days, since they float on the milky food (**166**). The wise bee-keeper can ward against the rearing of poor queens, by feeding his bees abundantly, if necessary, a few days in advance, and during the queen-breeding.

493. Lastly, some bee-keepers think that bees sometimes use larvæ more than three days old, and which consequently, have already received coarser food. One of our leaders in bee-culture, Mr. Doolittle, writes that one of his colonies must have used a larva four and one-half days old, since this colony hatched a queen in eight and one-half days, instead of about ten, as usually (**110**). (Cook's Guide, pages 70 and 72). But we cannot admit that the nurses were guilty of such blunder, especially since they would have had the trouble of replacing with better food, the coarse pap already given. Most likely, some already constructed queen-cell had passed unnoticed. Every one of us, old bee-keepers, has made similar errors. (See "Deceptive Queen-Cells (**519**)."

494. The worker-larvæ are fed with milky food for three days, and with coarse food for the three following days. Not only does this coarse food change their organism, but it retards their growth, since the queens are mature in sixteen days, from the time that the egg is laid (**197**), while the workers do not hatch before twenty-one days, on average. Thus the three days of coarse food have prolonged the growth five days, or in other words, each day of coarse feeding has delayed the maturity forty hours. Therefore, if we suppose that bees could, and would use, larvæ four and one-half days old, queens thus produced would hatch two and one-half days later than those raised from larvæ three days old. They would consequently hatch in eleven and one-half days instead of ten as usual.

495. If some Apiarists have noticed that their best queens were reared during the swarming fever (**455**), it is because the colonies are then in the best conditions to produce healthy queens. They have pollen and honey in abundance ; as they are numerous, they keep the combs very warm ; and, in addition, they have a large number of young bees, or nurses, to take care of the larvæ (**164.**)

496. The following accidental experiment has proved to us that most of the old workers are unable to act as nurses. Years ago, one of our neighbors moved three colonies of bees about half a mile, in the Summer, without taking proper precautions; we were informed the next day, that quite a number of the oldest bees had returned, and had clustered under an old table. We brought a hive there, with a comb containing eggs and young larvæ. They took possession of it, but neglected to raise a queen, and soon dwindled away.

497. By placing the colonies, intended to raise queens, in the same condition as to food, heat, and nursing, as during the swarming fever (**455**), we will raise as good queens as are then raised. If, to these conditions, we add the selection of brood, from our best queens (**315**), we will greatly improve the quality of our stock.

For over twenty years, we have used all the precautions described above, and, although our queens have never been reared from the egg, they are very prolific and long-lived. Using hives with ten or eleven large Quinby-frames (**340**), we are enabled to ascertain, beyond doubt, the prolificness of our queens. Our preventing swarming (**459**) enables us also to reckon their longevity.

498. The interposition of the Apiarist, in queen-rearing, may be necessary :

1st. To supply the loss of a queen in a colony that has not the means of raising another **(109)**.

2d. To breed a superior race of bees **(550)**, or improve the present stock **(315)**.

3d. To provide for the artificial increase of colonies. **(469.)**

We will study the rearing of queens, in view of these requirements ; but as each queen-breeder has his pet method, we will give only the main outlines, leaving our readers to their own choice, according to their judgment and circumstances.

Loss of the Queen.

499. That the Queen-Bee is often lost, and that her colony will be ruined unless such a calamity is seasonably remedied, ought to be familiar facts to every bee-keeper.

Queens sometimes die of disease, or old age, when there is no brood to supply their loss. Few, however, perish under such circumstances ; for, either the bees build royal cells, aware of their approaching end, or they die so suddenly as to leave young brood behind them. Queens are not only much longer-lived **(157)** than the workers, but are usually the last to perish in any fatal casualty. As many die of old age, if their death does not occur under favorable circumstances, it would cause, yearly, the loss of a very large number of colonies. As they seldom die when their strength is not severely taxed in breeding, drones are usually on hand to impregnate their successors.

500. Young queens are sometimes born with wings so imperfect that they cannot fly ; and they are often so injured in their contests with each other, or by the rude treatment they receive when driven from the royal-cells, that they cannot leave the hive for impregnation **(123)**.

501. We have yet, however, to describe under what circumstances the majority of hives become queenless. *More queens, whose loss cannot be supplied by the bees, perish when they leave the hive to meet the drones, than in all other ways.* After the departure of the first swarm, the mother-stock and all the after-swarms have young queens which must leave the hive for impregnation; *their larger size and slower flight* make them a more tempting prey to birds, while others are dashed, by sudden gusts of wind, against some hard object, or blown into the water: for, with all their queenly dignity, they are not exempt from mishaps common to the humblest of their race.

502. *In spite of their caution to mark the position and appearance of their habitation, the young queens frequently make a fatal mistake, and are destroyed, when attempting to enter the wrong hive.*

This accounts for the fact that ignorant bee-keepers, with forlorn and rickety hives, no two of which look just alike, are sometimes more successful than those whose hives are of the best construction. The former—unless their hives are excessively crowded—lose but few queens, while the latter lose them in almost exact proportion to the taste and skill which induced them to make their hives of *uniform size, shape* and *color* (**356**).

503. We first learned the full extent of the danger of crowded Apiaries, in the Summer of 1854. To protect our hives against extremes of heat and cold, they were ranged, side by side, over a trench, so that, through ventilators in their bottom-boards, they might receive, in Summer, a cooler, and in Winter, a much warmer air, than the external atmosphere. By this arrangement—which failed entirely to answer its design—many of our colonies became queenless, and we soon ascertained under what circumstances young queens are ordinarily lost.

From the great uniformity of the hives in size, shape,

17

color, and height, it was next to impossible for a young
queen to be sure of returning to her hive. The difficulty
was increased, from the fact that the ground before the
trench was free from bushes or trees, and no hive—except
the two end ones, which did not lose their queens—could
have its location remembered, from its relative position to
some external object. Most of the hives thus placed, which
had young queens, became queenless, although supplied
with other queens, again and again; and many, even of the
workers, were constantly entering hives adjoining their
own.

504. If a traveler should be carried, in a dark night, to
a hotel in a strange city, and on rising in the morning,
should find the streets filled with buildings precisely like it,
he would be able to return to his proper place, only by previously ascertaining its number, or by counting the houses
between it and the corner. Such a numbering faculty,
however, was not given to the queen-bee; for who, in a
state of nature, ever saw a dozen or more hollow trees or
other places frequented by bees, standing close together,
precisely alike in size, shape, and color, with their entrances all facing the same way, and at exactly the same height
from the ground?

On describing to a friend our observations on the loss of
queens, he told us that in the management of his hens, he
had fallen into a somewhat similar mistake. To economize
room, and to give easier access to his setting hens, he had
partitioned a long box into a dozen or more separate apartments. The hens, in returning to their nests, were deceived
by the similarity of the entrances, so that often one box
contained two or three unamiable aspirants for the honors
of maternity, while others were entirely forsaken. Many
eggs were broken, more were addled, and hardly enough
hatched to establish one mother as the happy mistress of a
flourishing family. Had he left his hens to their own in-

stincts, they would have scattered their nests, and gladdened his eyes with a numerous offspring.

Every bee-keeper, whose hives are so arranged that the young queens are liable to make mistakes, must count upon heavy losses. If he puts a number of hives, under circumstances similar to those described, upon a bench, or the shelves of a bee-house, he can never keep their number good without constant renewal.

505. The bees are sometimes so excessively agitated when their queen leaves for impregnation (**120**), that they exhibit all the appearance of swarming. They seem to have an instinctive perception of the dangers which await her, and we have known them to gather around her and confine her, as though they could not bear to have her leave. If a queen is lost on her wedding excursion, the bees of an old colony will gradually decline; those of an after-swarm, will either unite with another hive, or dwindle away (**182**).

506. It would be interesting, could we learn how bees become informed of the loss of their queen. When she is taken from them under circumstances that excite the whole colony, we can easily see how they find it out; for, as a tender mother, in time of danger, is all anxiety for her helpless children, so bees, when alarmed, always seek first to assure themselves of the safety of their queen. If, however, the queen is very carefully removed, several hours may elapse before they realize their loss. How do they first become aware of it? Perhaps some dutiful bee, anxious to embrace her mother, makes diligent search for her through the hive. The intelligence that she cannot be found being noised abroad, the whole family is speedily alarmed. At such times, instead of calmly conversing, by touching each other's antennæ, they may be seen violently striking them together, and by the most impassioned demonstrations manifesting their agony and despair (**181**).

We once removed the queen of a small colony, the bees of which took wing and filled the air, in search of her. Although she was returned in a few minutes, royal-cells were found two days later. The queen was unhurt, and the cells untenanted. Was this work begun by some that did not believe the others, when assured that she was safe? or from the apprehension that she might be removed again?

507. As soon as the bees begin to fly briskly in the Spring, a colony which does not industriously gather pollen, *or accept of flour (**267**), is almost certain to have no queen, or one that is not fertile—unless it is on the eve of perishing from starvation.

A colony is sure to be queenless, if, after taking its first Spring-flight, the bees, by roaming, in an enquiring manner in and out of the hive show that some great calamity has befallen them. Those that come from the fields, instead of entering the hive with that dispatchful haste so characteristic of a bee returning, well loaded, to a prosperous home, usually linger about the entrance with an idle and dissatisfied appearance, and the colony is restless, late in the day, when others are quiet. Their home, like that of a man who is cursed in his domestic relations, is a melancholy place, and they enter it only with reluctant and slow-moving steps.

508. And here, if permitted to address a word of friendly advice, we would say to every wife—Do all that you can to make your husband's home a place of attraction. When absent from it, let his heart glow at the thought of returning to its dear enjoyments; as he approaches it, let his countenance involuntarily assume a more cheerful expres-

---

*" Mr. Randolph Peters, of Philadelphia, had a colony which he was satisfied was queenless, as the bees did not carry in pollen for 28 days. I put a queen into the hive, he holding a watch in his hand, and in 3½ minutes from the time she was introduced, a bee was seen to enter with pollen on her legs! We both observed the entrance for some time, and saw many bees carry in pollen."— P. J. Mahan.

sion, while his joy-quickened steps proclaim that he feels that there is no place like the cheerful home where his chosen wife and companion presides as its happy and honored Queen.* If your home is not full of dear delights, try all the virtue of winning words and smiles, and the cheerful discharge of household duties, and exhaust the utmost possible efficacy of love, and faith, and prayer, before those words of fearful agony,

> "Anywhere, anywhere
> Out of the world!"

are extorted from your despairing lips, as you realize that there is no home for you, until you have passed into that habitation not fashioned by human hands, or inhabited by human hearts.

509. The neglect of a colony to expel drones (**192**), when they are destroyed in other hives, is always a suspicious sign, and generally an indication either that it has no queen, or else a drone-laying one (**134**), or drone-laying workers (**176**). A colony, in these circumstances, will not even destroy the drones of other hives, which may come to it, until a healthy queen has been raised in the hive, and is fertilized (**133**), and laying worker-eggs.

510. In opening a queenless hive, the plaintive hum of the bees (**76**), the listless and intermittent vibrating of their wings, and the total lack of eggs, or young worker brood, tell their condition.

A comb, with hatching bees,† should be given to it from

---

*" The tenth and last species of women were made out of a bee; and happy is the man who gets such a one for his wife. She is full of virtue and prudence, and is the best wife that Jupiter can bestow."—SPECTATOR, No. 209.

† That class of bee-keepers who suppose that all such operations are the "new fangled" inventions of modern times, will be surprised to learn that Columella, 1800 years ago, recommended strengthening feeble colonies, by *cutting* out combs from stronger ones, containing workers "just gnawing out of their cells."

a stronger colony, together with another comb, of eggs and larvæ, from the best colony in the Apiary; and the number of its combs should be reduced to suit the size of the cluster.

A better way yet to supply the loss, is to give the colony a queen-cell (**104**) or a young queen raised in the manner to be now described.

Rearing Improved Races.

511. We will see (**550**) that some races of bees are superior to others. Even in the same Apiary, some colonies are better than others, in prolificness, honey-gathering, endurance, gentleness, etc. It is very important to improve the Apiary by rearing queens from the best breeds, for the increase of colonies, as well as to replace the inferior ones.

To this end, the bee-keeper should select two or more of the best colonies in his Apiary, one for the production of drones, the others for the production of queens. Italian (**551**) bees are universally preferred; and as they are now almost as easily found as common bees, and are very cheap, we advise the novice to begin with at least two queens of this race.

A slight mixture of Cyprian or Syrian (**559**) blood is good, provided the issue be gentle and peaceable. Hybrids of common bees and Italians are generally inferior, both in quality and disposition.

512. In selecting a colony for drone production, the color and size of the drones should not be considered so much, as the prolificness of its queen, and the qualities of its workers, unless you wish to breed for beauty, in preference to honey-production.

Place two drone-combs (**224**) in the center of the brood-chamber of this colony, as soon as it has recuperated from its winter losses. If the colony is kept well supplied with honey, enough drones will be raised to impregnate all the queens in the neighborhood; otherwise, they might destroy these early drones after having raised them.

If our directions on the removal of drone-comb (**675**) are followed, but few drones will be raised outside of those colonies specially intended for drone-breeding. As soon as they begin to hatch, we may make preparations for queen-rearing, the best time being at the opening of fruit-blossoms. Some queen-breeders begin earlier, but early breeding gives much trouble and little pay, and our advice to Northern Apiarists, who want early queens, is to buy them from some reliable Southern Apiarist, as they can be raised earlier in the South, much more cheaply than in the North.

513. In an Apiary composed of several colonies, there are always some comparatively weak ones, either because their queens are old, or because they are not prolific. Such queens are of very little value, and should be replaced. Select one of these colonies—not the poorest, unless it is populous enough to raise good queens. Kill its queen, and exchange its brood-combs, after having brushed the bees off, for a less number of combs, containing eggs and larvæ, from your best queen. It may be well to feed the colonies containing the select queens beforehand, so as to incite the laying of eggs (**154**) and nursing of the brood.

514 If you desire to raise queens from eggs, (**490**), or larvæ just hatching, prepare for it, by giving your select colony some frames of dry comb, or comb foundation, (**674**) a few days ahead, for the queen to lay in. In this case, only those combs that contain eggs should be given to the queenless colony. It is always better to give but a small number of brood-combs to the colony intended for queen-raising, and to reduce its space with the division-board (**349**); as they can best keep it warm, in this manner, and raise better queens.

515. The largest number of queen-cells (**104**) can be obtained by cutting holes into the combs under the cells containing young larvæ or eggs, and feeding the bees plentifully. Some Apiarists hold that, by leaving them without brood of any kind for a few hours, they will raise more cells afterwards.

516. Nine days after the furnishing of the brood to the queenless colony, count the number of queen-cells raised, remembering that one has to be left to the colony that raised them. On the same day, make swarms, (**475**) or *nuclei*, (**522**) or destroy worthless queens (**155**) which you desire to replace next day.

517. The next day, with a sharp pen-knife, carefully remove a piece of comb, an inch or more square, that contains a queen-cell (Fig. 89), and in one of the brood combs of the hive to which this cell is to be given, cut a place just large enough to receive and hold it in a natural position. (Fig. 90.)

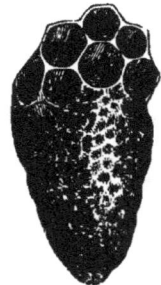

Fig. 89.
QUEEN-CELL,
REMOVED.

Each queenless stock can thus be supplied with a queen,

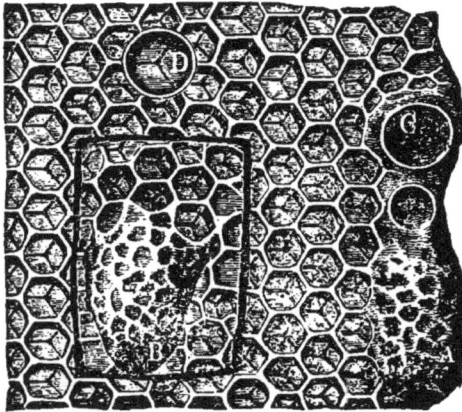

Fig. 90.
(From Gravenhorst.)
CUTTING OUT AND INSERT-
ING QUEEN-CELLS.
A, Unsealed cell. *B*, Insert-
ed cell. *C*, Unfinished cell.
D, Deceptive cell just be
gun.

ready to hatch, from the best breeding mother.

Unless *very* great care is used in transferring a royal cell, its inmates will be destroyed, as her body, until she is nearly mature, is so exceedingly soft, that a slight compression of her cell—especially near the base, where there is no cocoon—generally proves fatal. For this reason, it is best to defer removing them, until they are within three or four days of hatching. A queen-cell, nearly mature, may be known by its having the wax removed from the lid, by the bees, so as to give it a *brown* appearance.

518. If the weather is warm, and the hive, to which a queen-cell is given, is very populous, the cell may be introduced by simply inserting it in its natural position between two combs of brood. *It is very important to have the queen-cell in or near the brood, or the bees might neglect it.*

Sometimes, the bees so crowd their royal cells together (fig. 91) that it is difficult to remove one without fatally

(Fig. 91.)
CLUSTER OF CELLS.
(From Alley's Handy Book.)

injuring another, as, when a cell is cut into, the destruction and removal of the larva usually follows. Mr. Alley, by his method, given further on (**528**), found a remedy for this. If many queens are to be raised, it is well to have a new supply of cells started every week o. even oftener.

519. A day or two after introducing the queen-cells, the Apiarist can ascertain, by examination, whether they have been accepted. If they have not been accepted, the cells will be found torn open, on the side (fig. 92), instead of on the end, and the colonies will have begun queen-cells of their own brood. These queen-cells must be destroyed and replaced by other, from the next supply. In removing them, the greatest care should be taken not to pass the deceptive queen-cells, if any are there (fig. 90), which, although less apparent, would disappoint the end in view.

520. When queens are raised ahead of time for artificial increase, Italianizing, or for sale, it is more profitable to use *nuclei* instead of full colonies to hatch these queens. The word nuclei (plural of *nucleus*), from the Latin *nucleus* a nut, a kernel, was first applied by Mr. Langstroth to diminutive colonies of bees. This term is now universally adopted on both continents.

521. When we were raising queens for sale, we had contrived a divisible frame (fig. 93) to make these *nuclei* of combs taken from

(Fig 92)
QUEEN-CELLS.

a, hatched cell: *b*. sealed cell; *c*, rudimentary cell; *d*, cell torn by the bees.

full colonies. Our combs could be thus separated in two, and used in smaller hives, and in the Fall, these same combs were returned to the full colonies. Two small frames are more advantageous than one large frame, as they give more compactness to the cluster. Besides, these small colonies can be built up easily afterwards by coupling the frames, and uniting the combs of 3 or 4 nuclei into one large hive.

It is not necessary to have many of these frames in an Apiary, as a few are sufficient to make a number of nuclei, if they are placed in the centre of full colonies early in Spring.

(Fig. 93.)
DIVISIBLE FRAME.

Two frames thus made from one standard Langstroth frame measure about 8½ by 8½ inches each, a very convenient size for nucleus frames.

In the Fall, a number of nuclei may be united, in a full sized hive, on their own combs, by this method.

522. To make a nucleus, take from a colony, as late in the afternoon as there is light enough to do it, a comb containing worker-eggs, and bees just gnawing out of their cells, and put it, with the mature bees that are on it, into an empty hive. If there are not bees enough adhering to it, to prevent the brood from being chilled during the night, more must be shaken into the hive from other combs. If the transfer is made so late in the day that the bees are not disposed to leave the hive, enough may have hatched, by morning, to supply the place of those which will return to the parent stock.

523. In every case. when a swarm has left its hive for another quarter, each bee, as she sallies out, flies with her head turned towards it, that by marking the surrounding objects, she may find her way back.   If, however, the bees did not emigrate of their *own free will*, most of them appearing to forget, or not knowing, that their location has been changed, return to their familiar spot; for it would seem that,

> " A ' bee removed ' against her will,
> Is of the same opinion still."

Should the Apiarist, ignorant of this fact, place the nucleus on a new stand without providing it with a sufficient number of young bees, it would lose so many of the bees which ought to be retained in it, that most of its unsealed brood would perish from neglect.

If the comb used in forcing such a *nucleus* was removed at a time of day when the bees would be likely to return to the parent stock, they should be confined to the hive, until it is too late for them to leave ; and if the number of bees, just emerging from their cells, is not large, the entrance to the hive should be closed, until about an hour before sunset of the next day but one.   The hive containing this small colony, should be properly ventilated, and shaded—if thin—from the intense heat of the sun ; it should always be well supplied with honey.   The space unoccupied in the hive should be separated from the nucleus by a division board (**349**).

524. Beginners must remember that it is bet er to have these small nuclei strong with bees; but, in giving them young bees, care should be taken not to give them the queen. If a nucleus is made at mid-day, nearly all the bees given to it will be young bees, as the old bees are then in the field.

The best manner to add young bees from strange colonies to weak nuclei, is to shake or brush them, on the apron board in front of the entrance, as is done in swarming (**428**).

525. Hives, or nuclei in which queen-cells are to be introduced, should be aware of their queenless condition before a queen-cell is given them. Hence the necessity of preparing them 24 hours previous.

526. A vigilant eye should be kept upon every colony that has not an impregnated queen; and when its queen is about a week old it should be examined, and if she has become fertile, she will usually be found supplying one of the central combs with eggs. If neither queen nor eggs can be found, and there are no certain indications that she is lost, the hive should be examined a few days later, for some queens are longer in becoming impregnated than others, and it is often difficult to find an unimpregnated one, on account of her adroit way of hiding among the bees.

As soon as the young queen lays, she may be introduced to a queenless colony, or sold, and if queen-cells are kept on hand, another one can be given to the nucleus the next day. Thus, nuclei may be made to raise two queens or more in a month.

527. If the queens are to be multiplied rapidly, the nuclei must never be allowed to become too much reduced in numbers, or to be destitute of brood or honey. With these precautions, the oftener their queen is taken from them, the more intent they will usually become in supplying her loss.

There is one trait in the character of bees which is worthy of profound respect. Such is their indomitable energy and perseverance, that under circumstances apparently hopeless, they labor to the utmost to retrieve their losses, and sustain the sinking State. So long as they have a queen, or any prospect of raising one, they struggle vigor-

ously against impending ruin, and never give up until their
condition is absolutely desperate.   We once knew a colony
of bees not large enough to cover a piece of comb four inches
square, to attempt to raise a queen.   For two whole weeks,
they adhered to their forlorn hope ; until at last, when they
had dwindled to less than one-half their original number,
their new queen emerged, but with wings so imperfect that
she could not fly.   Crippled as she was, they treated her
with almost as much respect as though she were fertile.   In

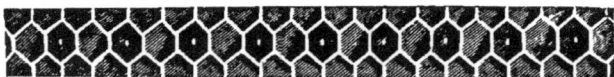

Fig. 92b. (From Alley.)

the course of a week more, scarce a dozen workers remained
in the hive, and a few days later, the queen was gone, and
only a few disconsolate wretches were left on the comb.

528. Mr. Alley, who raises queens by the thousand, has
published his method of queen-rearing.   His queens are all
raised in very small nuclei which he calls *miniature hives.*
From a light-colored worker-comb filled with hatching eggs,
he cuts strips with a sharp knife, as in fig. 92*b.*

"After the comb has been cut up, lay the pieces flat upon a board
or table, and cut the cells on one side down to within one fourth
of an inch of the foundation or septum, as seen in fig. 93*b* which
represents the comb ready to place in position for cell building.   While engaged in this work, keep a lighted lamp near

Fig. 93b. (From Alley.)

at hand, with which to heat the knife, or the cells will be
badly jammed   *   *   *   *

The strips of comb being ready, we simply destroy each alternate larva or egg, (fig. 92*b*. In order to do this, take the strips carefully in the left hand, and insert the end of a common lucifer match into each alternate cell, pressing it gently on the bottom of the cell, and then twirling it rapidly between the thumb and fingers. This gives plenty of room for large cells to be built without interfering with those adjoining, and permits of their being separated without injury to neighboring cells."—" Bee–keepers' Handy Book," Wenham, 1885.

This strip, Mr. Alley fastens under a trimmed comb cut slightly convex. by dipping the cells, which have been left full length, into a mixture of two parts rosin and one of

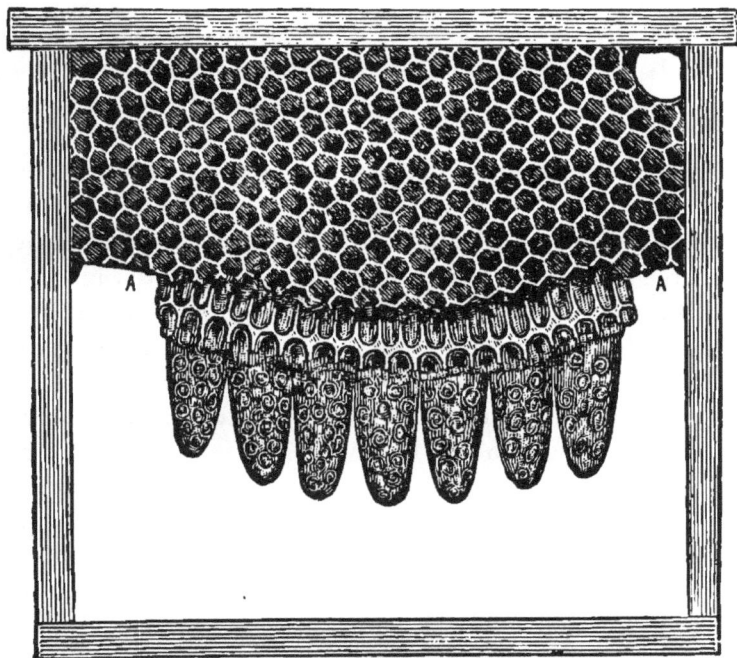

Fig. 94. (From Alley.)

bees-wax, taking care not to over-heat this mixture, as the heat might destroy the eggs (fig. 94). The comb thus prepared is given to a *miniature* colony, which has been queenless and without brood for ten hours, Mr. Alley having noticed that the eggs may be destroyed if given to a colony just made queenless.

This method is probably the most expeditious and the cheapest that can be followed, for raising a large number of queens; but we would hardly advise Apiarists to use as small nuclei as Mr. Alley does (5 combs, 4½ inches square). The stronger the colony in which a queen is raised, the better the queen.

529. As it happens very often, that more queen-cells are raised than are needed immediately, and as the bees usually destroy all after the first one has hatched, Apiarists have devised *queen-nurseries* to preserve the supernumerary cells until needed. It is not safe to leave the queen-cells under the control of the bees after ten days, as a queen may hatch at any time.

There are several ways to make queen-nurseries. Messrs. Root, Hayhurst, Heddon and Hutchinson, warm their nurseries with lamps, while the nurseries used by Messrs. Alley, Demaree and others, are placed in well populated hives.

530. The lamp-nursery is a doubled-walled tin box,* of the right size to receive the breeding frames. The space between the walls and the bottom is filled with water, and a kerosene lamp is lighted under it, with the flame about one foot from the bottom of the box. The temperature of this lamp-nursery is regulated by raising or lowering the flame, and is kept between 90° and 100°. The combs containing the sealed queen-cells are placed in this box, and if the brood in the combs is all of the same age, every queen will hatch, at least, five days before any of the workers. These queen-cells have to be examined every few hours, for the first queens hatched would destroy the others.

The Alley queen-nursery is composed of a number of small

---

* Mr. Hayhurst, of Kansas City, who is one of the most successful Western queen breeders, uses a galvanized iron nursery, packed in a chaff case.

cages, covered with wire cloth on each side and inserted in a frame. Each cage has two holes at the top, one for a sponge saturated with honey, the other to receive the queen-cell. The frame is inserted in a strong colony, not necessarily queenless, since these young queens are caged, and have feed at hand when they hatch.

The hatching of queens in nurseries properly belongs to the trade of the queen-breeder. The honey producer, who raises queens for himself only, does not need fresh queens every day. Besides, the introducing of these young virgin queens to nuclei, previous to impregnation, is quite difficult and uncertain. (**541**.)

531. Before we pass to the subject of introducing queens, we cannot refrain from noticing the rapid progress of the business of queen rearing in the last 20 years. The introduction of brighter races has greatly increased the spreading of Apiarian science, and many facts which, years ago, were known only to the few, now belong to the public domain.

532. In breeding the new races, let the novice remember that the qualities he should seek to improve are, first, prolificness and honey production ; second, peaceableness ; third, beauty.

Since their introduction into this country, the Italians have been bred too much for color, at the expense of their other qualities. We have seen queens, that had been so inbred for color, that their mating with a black drone hardly showed the hybridization of their progeny.

This in-and-in breeding, for color, has even produced white-eyed drones, stone blind, a degeneracy which would tend to the extinction of the race.

18

Introducing Impregnated Queens.

533. *Great caution is needed in giving to bees a stranger queen.* Huber thus described the way in which a new queen is usually received by a colony:

"If another queen is introduced into the hive within *twelve* hours after the removal of the reigning one, they surround, seize, and keep her a very long time captive, in an impenetrable cluster, and she commonly dies either from hunger or want of air. If eighteen hours elapse before the substitution of a stranger-queen, she is treated, at first, in the same way, but the bees leave her sooner, nor is the surrounding cluster so close; they gradually disperse, and the queen is at last liberated; she moves languidly, and sometimes expires in a few minutes. Some, however, escape in good health, and afterwards reign in the hive."

The manner in which strange queens are treated by the bees, when they are queenless, depends mainly on the state of the honey harvest.

534. But in order to meet with uniform success, the following conditions must be fulfilled:

The bees must be absolutely queenless. Sometimes a colony contains two (**117**) queens, and the Apiarist after removing one may imagine that he can introduce a stranger, safely. Many queens are thus killed.

535. As bees recognize one another by the scent, the new queen should be placed so as to get the odor of the hive, before being released among them. This can be effected readily by sprinkling the bees and the new queen with sweetened water scented with peppermint, and liberating her at once. But as this method generally causes some robbing (**664**) in times of scarcity, it is not always to be relied upon.

536. Our method consists in placing the queen in a small flat cage, made of wire cloth, between two combs, in the most populous part of the hive, near the brood and the honey, and keeping her there from 24 to 48 hours. These queen-cages were first used in Germany for introducing queens.

537. In catching a queen, she should be gently taken with the fingers, from among the bees, and if none are crushed, there is no risk of being stung. The queen herself will not sting, even if roughly handled.

If she is allowed to fly, she may be lost, by attempting to enter a strange hive.

To introduce her into the cage, she should be allowed to *climb up* into it. *It is a fact well known to queen breeders that a bee or a queen cannot be easily induced to enter a cage or a box turned downward.* The meshes of the wire cloth should not be closer than 12 to the inch, that the bees may feed the queen readily through them. This is important, for we have lost two queens successively in a cage with closer meshes.

The bees will cultivate an acquaintance with the imprisoned mother, by thrusting their antennæ through the openings, and will be as quiet as though the queen had her liberty. Such a cage will be very convenient for any temporary confinement of a queen.

538. It is necessary, when the queen is released, that the bees be in good spirits, neither frightened, nor angered, and there should be no robbers about, as they might take her for an intruder, and *ball* her.

This technical word is used to describe the peculiar way in which bees surround a queen whom they want to kill. The cluster that encloses her, is in the form of a ball, sometimes as large as one's fist, and so compact that it cannot readily be scattered. She may be rescued by throwing the ball into a basin of water. We have known bees to ball their own mother in such circumstances, for queens are of a timid disposition and easily frightened. When we release a strange queen, we put a small slice of comb honey, or honey cappings, in place of the stopper of the cage, and close the hive. It takes from 15 to 20 minutes for the bees to eat through, and by that time all is quiet, so the queen walks leisurely out of her cage, and is safe.

539. If the colony, in which a queen is to be introduced, is destitute, the bees should be abundantly fed on the preceding night (**605**). After she has been released, it is well to leave the colony alone for two or three days.

As a fertile queen can lay several thousand eggs a day, it is not strange that she should quickly become exhausted, if taken from the bees. "*Ex nihilo nihil fit*"—from nothing, nothing comes—and the arduous duties of maternity compel her to be an enormous eater. After an absence from the bees of only fifteen minutes, she will solicit honey, when returned; and if kept away for an hour or upwards, she must either be fed by the Apiarist, or have bees to supply her wants.

Mr. Simmins has taken advantage of this appetite, and of the propensity of bees to feed the queens, in introducing them directly, after keeping them without bees and food, for about 30 minutes. At dusk he lifts a corner of the cloth (**352**) of the hive in which he wants to introduce the queen, drives the bees away with a little smoke (**382**), and permits the queen to run between the combs. Then he waits 48 hours before visiting the hive. Several bee-keepers report having succeeded with this method. On account of this propensity of bees to feed queens, any number of fertile ones may be kept in a hive already containing a fertile queen, if they are placed in cages between the combs, near the honey and the brood.

540. Some Apiarists use chloroform, ether, puff-balls, or other ingredients, to stupefy the bees of mutinous colonies who persist in refusing to accept a strange queen and who show it by angrily surrounding the cage in which she is confined.

The Rev. John Thorley, in his "*Female Monarchy*," published at London, in 1744, appears to have first introduced the practice of stupefying bees by the narcotic fumes of the "puff ball" (*Fungus pulverulentus*), dried till it will hold fire like tinder. The bees soon drop motionless from their comb, and recover again after a short exposure to the air. This method was once much practiced in France (L'Apiculteur, page 17, Paris, 1856) but is very dangerous, as too large a dose of anæsthetics will cause death instead of sleep.

INTRODUCTION OF VIRGIN QUEENS.

541. The difference in looks between a virgin queen and an impregnated one is striking, and an expert will distinguish them at a glance. The virgin queen is slender, her abdomen is small, her motions quick, she runs about and almost flies over the combs, when trying to hide from the light. In fact, she has nothing of the matronly dignity of a mother.

Bees, in possession of a fertile queen, are quite reluctant to accept an unimpregnated one in her stead; indeed, it requires much experience to be able to give a virgin queen to a colony, and yet be sure of securing for her a good reception.

Mr. Langstroth was the first to ascertain, years ago, that the best time to introduce her, is just after her birth, as soon as she can crawl readily. If introduced too soon, the bees may drag her out, as they would any imperfect worker. Most queen-breeders liberate them on the comb, or at the entrance of a queenless nucleus. Mr. H. D. Cutting, of Clinton, Mich., recommends daubing the young queen with honey, as she comes out of her cell, and liberating her among the bees, without touching her with the fingers.

Nearly all breeders acknowledge that the introduction of
virgin queens to full colonies is an uncertain business, and
that they can be introduced safely only to small nuclei that
have been queenless some time.   In this, we fully agree.

Mr. G. W. Demaree, of Christianburg, Ky., is quite successful in the introduction of virgin queens several days old,
by much the same process as that given by us for the introduction of fertile queens.

We would advise novices to abstain from introducing virgin queens, until they become expert in the business of
queen rearing; the introduction of unhatched queen-cells
being much more easily performed, and more uniformly
successful.

542. In introducing queens or queen-cells to full colonies during the swarming season, it happens very often that
the bees also raise queen-cells of their own brood, and
swarm with the queen given them (**465**).   In view of this,
the Apiarist should watch, for a few days, the colony to
which a new queen has been introduced.

543. In hunting for a queen. it is necessary to remember that *she is on the brood combs* unless frightened away.
If the bees are not greatly disturbed, an Italian queen may
be found within five minutes after opening the hive.

A queen of common bees, or of hybrids. is more difficult
to find. as her bees often rush about the hive as soon as it is
opened.   If she cannot be found on the combs, and the hive
is populous, it is best to shake all the frames on a sheet, in
fro t of an empty box, and secure them in a closed hive, out
of the reach of robbers. until the search is over, when everything may be returned to its proper place.

544. After a queen is taken from a cage, the bees will
run in and out of it for a long time, thus proving that they
recognize her peculiar scent.   It is this odor which causes
them to run inquiringly over our hands, after we have caught

a queen, and over any spot where she alighted when her swarm came forth.

This scent of the queen was probably known in Aristotle's time, who says: " When the bees swarm, if the king (queen) is lost, we are told that they all search for him, and follow him with their sagacious smell, until they find him. " Wildman says: "The scent of her body is so attractive to them, that the slightest touch of her, along any place, or substance, will attract the bees to it, and induce them to pursue any path she takes. "

The intelligent bee-keeper has now realized, not only how queens may be raised or replaced, by the use of the movable-frame hive, but how any operation, which in other hives is performed with difficulty, if at all, is in this rendered easy and certain. No hive, however, can make the ignorant or negligent very successful, even if they live in a region where the climate is so propitious, and the honey resources so abundant, that the bees will prosper in spite of misman-agement or neglect.

CHAPTER IX.

RACES OF BEES.

545. The honey-bee is not indigenous to America. Thomas Jefferson, in his "Notes on Virginia," says:

"The honey-bee is not a native of our country. Marcgrave indeed, mentions a species of honey-bee in Brazil. But this has no sting, and is therefore different from the one we have, which resembles perfectly that of Europe. The Indians concur with us in the tradition that it was brought from Europe ; but when and by whom, we know not. The bees have generally extended themselves into the country, a little in advance of the white settlers. The Indians therefore call them, the white man's fly."

" When John Eliot translated the Scriptures into the language of the Aborigines of North America, no words were found expressive of the terms wax and honey." (A. B. J. July 1866.)

Longfellow, in his "Song of Hiawatha," in describing the advent of the European to the New World, makes his Indian warrior say of the bee and the white clover:—

> " Wheresoe'er they move, before them
> Swarms the stinging fly, the Ahmo,
> Swarms the bee, the honey-maker;
> Wheresoe'er they tread, beneath them
> Springs a flower unknown among us,
> Springs the White Man's Foot in blossom."

546. According to the quotations of the A. B. J., common bees were imported into Florida, by the Spaniards previous to 1763, for they were first noticed in West Florida in that year. They appeared in Kentucky in 1780, in New York in 1793. and West of the Mississippi in 1797.

547. "It is surprising in what countless swarms the bees have overspread the far West within but a moderate number of years.

The Indians consider them the harbingers of the white man, as the buffalo is of the red man, and say that, in proportion as the bee advances, the Indian and the buffalo retire ..... They have been the heralds of civilization, steadily preceding it as it advances from the Atlantic borders; and some of the ancient settlers of the West pretend to give the very year when the honey-bee first crossed the Mississippi. At present it swarms in myriads in the noble groves and forests that skirt and intersect the prairies, and extend along the alluvial bottoms of the rivers. It seems to me as if these beautiful regions answer literally to the description of the land of promise—'a land flowing with milk and honey;' for the rich pasturage of the prairies is calculated to sustain herds of cattle as countless as the sands upon the sea-shore, while the flowers with which they are enamelled render them a very paradise for the nectar-seeking bee."—WASHINGTON IRVING, " Tour on the Prairies," Chap. IX. (1832).

Many Apiarists contend that newly-settled countries are most favorable to the bee ; and an old German adage runs thus ;—

> " Bells' ding dong,
> And choral song,
> Deter the bee
> From industry :
> But hoot of owl,
> And ' wolf's long howl,'
> Incite to moil
> And steady toil."

It is evident that the bees spread Westward very rapidly, and to this day, many old bee-men can be found, who positively assert that a swarm never goes Eastward, even after it is proven to them that they usually go to the *nearest timber.*

548. Bees, like all other insects, are divided scientifically into genera, species, and varieties.

Aristotle speaks of three different varieties of the honey-bee, as well known in his time. The *best variety* he describes as " μιχρά, ϛρογγυλή χαι ποιχιλή "—that is, small, and round in size and shape, and variegated in color.

Virgil (Georgica, lib. IV., 98) speaks of two kinds as
flourishing in his time; the better of the two he thus describes:

> " Elucent aliæ, et fulgore coruscant,
> Ardentes auro, et paribus lita corpora guttis.
> Hæc potior soboles; hinc cœli tempore certo
> Dulcia mella premes."

" *The others glitter, and their variegated bodies shine like
drops of sprinkling gold. This better breed! Thanks to
them, if the weather of the sky is certain, you will have honey
combs to press.*"

This better variety, it will be seen, he characterizes as
spotted or variegated, and of a beautiful golden color.

549. The first bee introduced into America, was the
common bee of Europe, Western Asia, and Western Africa,
Apis mellifica, usually designated under the name of black,
or gray bee. Both names are appropriate, since the race
varies in shade, according to localities. In the greater part
of Africa, as well as in the European provinces of Turkey,
the common bees are dark, nearly black. In other places,
their color is grayish. They vary in size, as well. According to some French writers, the bees of Holland are small,
and denominated " *la petite Hollandaise* " (the little Hollander); on the other hand, the Carniolan* bees are quite
large. We have never seen queens as large as some Car-
niolans which we imported some ten years ago. But, in spite
of the prolificness and general good reputation of this race,
we did not attempt to propagate it, owing to the difficulty
of detecting their mating with the common bees, since they
are almost alike in color.

550. Besides the common bee, there are a great many
varieties. The best known are: *1st*, the *Ligurian*, *Apis
Ligustica*, so named by Spinola, because he found it first, in

---

* Carniolan is a province of Austria, near the Adriatic, but on the East slope
of the mountains.

the part of Italy called Liguria. The Rev. E. W. Gilman, of Bangor, Maine, directed the writer's attention to Spinola's "*Insectorum Liguriæ species novæ aut rariores*," from which it appears, that Spinola accurately described all the peculiarities of this bee, which he found in Piedmont, in 1805. He fully identified it with the bee described by Aristotle.

2d. The *apis fasciata* (banded bee). This bee, related to the Italian, or Ligurian, which has yellow bands also, is found in Egypt, in Arabia, along both sides of the Red Sea, in Syria, and in Cyprus.

3d. We shall mention also the large *Apis dorsata* of Southern Asia, and the *melipones* of Brazil and Mexico.

551. The Italian bee, *Apis Ligustica*, spoken of by Aristotle and Virgil as the best kind, still exists distinct and pure from the common kind, after the lapse of more than two thousand years.

The great superiority of this race, over any other race known, is now universally acknowledged; for it has victoriously stood the test of practical bee-keepers, side by side with the common bee. The ultimate superseding of the common bee by the Italian in this country is but a matter of time.

552. The following facts are evident:

1st. The Italian bees are less sensitive to cold than the common kind. *2d.* Their queens are more prolific. *3d.* They defend their hives better against insects. Moths (**802**) are hardly ever found in their combs, while they are occasionally found in the combs of even the strongest colonies of common bees. Their great vigilance is due to the mildness of the climate of Italy, whose Winters never destroy the moth. Having to defend themselves against a more numerous enemy, they are more watchful than the bees of colder regions. *4th.* They are less apt to sting. Not only are they less apt, but scarcely are they inclined to sting, though they will do so if intentionally annoyed, or irritated, or improperly treated.

Spinola speaks of the more peaceable disposition or this bee ; and Columella, 1800 years ago, had noticed the same peculiarity, describing it as "*mitior moribus*," (milder in habits). When once irritated, however, they become very cross.

5th. They are more industrious. Of this fact, all the results go to confirm Dzierzon's statements, and satisfy us of the superiority of this kind *in every point of view.*
6th. They are more disposed to rob than common bees, and more courageous and active in self-defense. They strive on all hands to force their way into colonies of common bees ; but when strange bees attack their hives, they fight with great fierceness, and with an incredible adroitness.

Spinola speaks of these bees as "*velociores motu*"— quicker in their motions than the common bees.

They however sooner grow tired of hunting, where nothing can be gained ; and if all the plunder is put out of their reach, they will give up the attempt at robbing (**664**) more promptly than common bees.

7th. Aside from their peaceableness, they are more easily handled than the common bees, as they cling to their combs and do not rush about, or cluster here and there, or fall to the ground, as the common bees do.

It is hardly necessary to add. that this species of the honey-bee, so much more productive than the common kind, is of very great value in all sections of our country. Its superior docility makes it worthy of high regard, even if in other respects it had no peculiar merits. Its introduction into this country, has helped to constitute the new era in bee-keeping, and has imparted much interest to its pursuit. It is one of the causes which have enabled America to surpass the world in the production of honey.

553. Their appearance can be described as follows:

"The first three abdominal rings (fig. 95) of the worker bee are trrnsparent, and vary from a dark straw or golden color to the deep yellow of ochre. These rings have a narrow dark edge or border, so that the yellow, which is sometimes called *leather color*, constitutes the ground, and is seemingly barred over by these black edges. This is most distinctly perceptible when a brood-comb, on which bees are densely crowded, is taken out of a hive, or when a bee is put on a window. When the bee is full of honey these rings extend and slide out of one another, and the yellow bands show to better advantage, especially if the honey eaten is of a light color. On the contrary, during a dearth of honey, the rings are drawn up, or telescoped in one another, and the bee hardly looks like the same insect. This peculiarity has annoyed many bee-keepers, who imagined their beautiful bees had suddenly become hybrids.

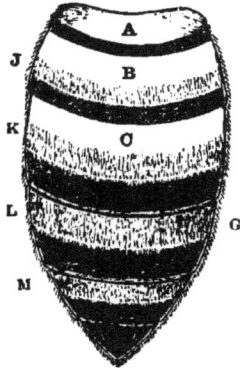

Fig. 95.

ABDOMEN OF THE ITALIAN BEE.

From A. I. Root.

In doubtful cases, as the purity of Italian bees is very important, it is well to follow the advice of A. I. Root: ‧‧If you are undecided in regard to your bees' purity, get some of the bees and feed them all the honey they can take; now put them on a window, and if the band C (fig. 95) is not plainly visible, call them *hybrids*.'' (‧‧A. B. C. page 145).

554. Aside from this test, their tenacity and quietness on the comb, while handled (**378**), are infallible signs of purity. We have repeatedly carried a frame of brood covered with pure Italian bees, from a hive to the house, and passed the comb from hand to hand among visitors, some of whom were ladies, without a single bee dropping off, or attempting to sting.

555. The drones (**185**) and the queens are very irregular in markings, some being of a very bright yellow color, others almost as dark as drones or queens of common bees.

" It is a remarkable fact that an Italian queen, impregnated by a common drone, and a common queen impregnated by an Italian drone, do not produce workers of a uniform intermediate cast, or hybrids ; but some of the workers bred from the eggs of each queen will be purely of the Italian, and others as purely of the common race, only a few of them, indeed, being apparently hybrids. Berlepsch also had several mismated queens, which at first produced Italian workers exclusively, and afterwards common workers as exclusively. Some such queens produced fully three-fourths Italian workers; others, common workers in the same proportion. Nay, he states that he had one beautiful orange-yellow mismated Italian queen which did not produce a single Italian worker, but only common workers, perhaps a shade lighter in color. The *drones*, however, produced by a mismated *Italian* queen are uniformly of the Italian race, and this fact, besides demonstrating the truth of Dzierzon's theory,(**133**) renders the preservation and perpetuation of the Italian race, in its purity, entirely feasible in any country where they may be introduced."—S. WAGNER.

556. The Italian bees from different parts of Italy are of different shades, but otherwise, preserve about the same characteristics all over the peninsula. But how can they keep pure, since there are common bees in Europe? A glance at the map will answer the question. Italy is surrounded on all sides by water or snow-covered mountains, which offer an insuperable barrier to any insects. This is further evidenced by the fact that the bees of the canton of Tessin (Italian Switzerland) are Italians, being on the South side of the Alps. while those of the canton of Uri (German Switzerland), on the other side of the mountains and only a few miles off, are common bees.*

557. The importation of Italian bees to another country was first attempted by Capt. Baldenstein.

---

* The idea that select Italian bees raised in America, **may be** *purer* **than any** Italians ever imported. has been gravely discussed by some persons.

" Being stationed in Italy, during part of the Napoleonic wars, he noticed that the bees, in the Lombardo-Venitian district of Valtelin, and on the borders of Lake Como, differed in color from the common kind, and seemed to be more industrious. At the close of the war, he retired from the army, and returned to his ancestral castle, on the Rhætian Alps, in Switzerland ; and to occupy his leisure, had recourse to bee-culture, which had been his favorite hobby in earlier years. While studying the natural history, habits, and instincts of these insects, he remembered what he had observed in Italy, and resolved to procure a colony from that country. Accordingly, he sent two men thither, who purchased one, and carried it over the mountains, to his residence, in September, 1843.

" His observations and inferences impelled Dzierzon—who had previously ascertained that the cells of the Italian and common bees were of the same size—to make an effort to procure the Italian bee; and, by the aid of the Austrian Agricultural Society at Vienna,* he succeeded in obtaining, late in February, 1853, a colony from Mira, near Venice."—S. WAGNER.

558. An attempt was made in 1856, by Mr. Wagner, to import them into America ; but, unfortunately, the colonies perished on the voyage. The first living Italian bees landed on this continent were imported in the Fall of 1859 by Mr. Wagner and Mr. Richard Colvin, of Baltimore, from Dzierzon's Apiary. Mr. P. G. Mahan, of Philadelphia, brought over at the same time a few colonies. In the Spring of 1860, Mr. S. B. Parsons, of Flushing, L. I., imported a number of colonies from Italy. Mr. William G. Rose, of New York, in 1861, imported also from Italy. Mr. Colvin made a number of importations from Dzierzon's Apiary ; and

*Some of the Governments of Europe have long ago taken great interest in disseminating among their people a knowledge of Dzierzon's system of Bee-Culture. Prussia furnishes monthly a number of persons from different parts of the Kingdom with the means of acquiring a practical knowledge of this system; while the Bavarian Government has prescribed instruction in Dzierzon's theory and practice of bee-culture, as a part of the regular course of studies in its teachers' Seminaries. We are glad to see that the United States is beginning to recognize the importance of bee-culture, and that an Apiarian department has been inaugurated under the control of the Agricultural Department at Washington.

in the Fall of 1863 and 1864 Mr. Langstroth also imported queens from the same Apiary, but the first large successful importations were made by Adam Grimm of Wisconsin, in 1867, from the Apiary of Prof. Mona of Bellinzona, and by us in 1874, from the Apiary of Signor Giuseppe Fiorini of Monselice, Italy. Since then, Mr. A. I. Root, and others, have succeeded well nearly every season.

This valuable variety of the honey-bee is now extensively disseminated in North America.

For directions on breeding and shipping Italian bees, see the chapters on Queen Raising (**497**) and Shipping Bees (**587**).

559. The Egyptian bees (Apis fasciata) are smaller and brighter than the Italian bee. The hairs of their body are more whitish, and their motions are quick and fly-like. Their prolificness is great, but their ill-disposition has caused many who have tried them to abandon them.

The Cyprian bees (a sub-race of Apis fasciata) were imported from Cyprus to Europe in 1872, and they were so much praised that, in 1880, two enterprising American Apiarists, Messrs. D. A. Jones and Frank Benton made a trip to Cyprus and the Holy Land, and brought bees from both countries to America.

The Cyprian bees resemble the Italian bees. The main difference between them, in appearance, is a bright yellow shield on the thorax of the Cyprians not to be seen in the Italians, and the yellow rings of the former are brighter, of a *copper color*, especially under the abdomen. Their drones are beautiful.

Their behavior is like that of the Egyptians; quick and ready, they promptly assail those who dare handle them. Smoke astonishes but does not subdue them. At each puff of the smoker (**382**), they emit a sharp, trilling sound, not easily forgotten, resembling that of "*meat in the frying pan,*" and as soon as the smoke disappears, they are again on the watch, ready to pounce on any enemy, whether man or beast, bee or moth. Their courage and great prolificness would make them a very desirable race, if they could be handled safely.

A *slight* mixture of this race with the Italian improves the latter wonderfully in color and working qualities.

560. The Holy Land or Syrian bees are almost similar in looks to the Egyptian, these two countries being contiguous. Those who have tried them do not agree as to their behavior ; some holding them to be very peaceable, others describing them as very cross. We have never tried them.

Among the different races of Eastern bees, the Caucasian are cited by Vogel, a German, as of such mild disposition, that it is hard to get them to sting. Yet it is said that these bees defend themselves well against robber bees.

According to Vogel, they resemble the Syrian bees, having also the shield of the Cyprians. It would seem that these bees exist in the temperate zone of Asia, from the shores of the Mediterranean to the Himalayas, for Dr. Dubini, in his book, writes that they were found at the foot of these mountains.

561. According to an article in the "*Scientific Review*" of England, although bees have been sent from this country and Europe, to Australia, there is an Australian native bee, which builds its nest on the Eucalyptus. These bees gather immense quantities of a kind of honey which, although very sweet, can be used as medicine, to replace the cod-liver oil, used with so much repugnance by consumptives.

562. *Apis dorsata*, the largest bee known, lives in the jungles of India. Mr. Benton attempted to import this bee at great expense and danger, but only succeeded in bringing one colony to Syria, where it died. Mr. Vogel tried also to bring some of them to Germany without success. At all events further attempts at importing or domesticating these bees would be so expensive, that private enter-

19

prise will be balked by the task. It behooves our government to take such matters in hand for the public good. Besides *Apis dorsata*, two other kinds exist in India, *Apis florea* and *Apis Indica*. The latter is cultivated by the natives with good results. Both are smaller than our common bee.

563. Another race of bees,* the Melipone, is found in Brazil and Mexico. More than twelve varieties of these have been described, all without stings.

Huber, in the beginning of this century, received a nest of them, but the bees died before reaching Geneva. Mr. Drory, while at Bordeaux, France, was more successful. One of his friends sent him a colony of Melipones, and he published in the "*Rucher du Sud-Ouest*" some very curious facts concerning them. The cells containing the stores of honey and pollen are not placed near those intended for brood, but higher in the hive; they are as large as pigeon eggs, and attached in clusters to the walls of the hive. The brood cells are placed horizontally in rows of several stories. The workers do not nurse the brood, but fill the cells with food, on which the queen lays. The cells are then closed till the young bees emerge from them.

A peculiarity of these bees is that the entrance to their home, which is very narrow, is usually watched by a single bee, acting as janitor, and withdrawing from the door to let the workers pass. They cannot stand the cold, and Mr. Drory could not save his, in spite of his care, in a location as mild as that of Bordeaux. Mr. T. F. Bingham of Abronia, Michigan, imported a nest of them, in the Spring of 1886, and lost them the same Fall. A part of their nest was exhibited by him at the Indianapolis Convention, in October 1868.

* These bees are scientifically classified as belonging to a different genus of Apidæ.

CHAPTER X.

The Apiary.

Location.

564. Any one can keep bees, successfully, if he has a liking for this pursuit and is not too timid to follow the directions given in this treatise. Even ladies can manage a large Apiary successfully, with but little help.

Almost any locality will yield a surplus of honey in average seasons. Mr. Chas. F. Muth of Cincinnati, with 22 colonies of bees, on the roof of his house, in the heart of this large city, harvested a surplus honey yield of 198 lbs. per colony in one season.

Mr. Muth informed us that this surplus was collected from white clover blossoms in 26 days.

565. But an intimate acquaintance with the honey resources of the country is highly important to those desirous of engaging largely in bee-culture. While, in some localities, bees will accumulate large stores, in others, only a mile or two distant, they may yield but a small profit.

"While Huber resided at Cour, and afterwards at Vevey, his bees suffered so much from scanty pasturage, that he could only preserve them by feeding, although stocks that were but two miles from him were, in each case, storing their hives abundantly." —BEVAN.

Those desirous of becoming specialists will find the subject of location and yield further treated in the chapter on Pasturage and Overstocking (**698**).

566. Inexperienced persons will seldom find it profitable to begin bee-keeping on a large scale. By using movable-frame (**286**) hives, they can rapidly increase their stock after they have acquired skill, and have ascertained, not simply that money can be made by keeping bees, *but that they can make it.*

While large profits can be realized by careful and experienced bee-keepers, those who are otherwise will be almost sure to find their outlay result only in vexatious losses. An Apiary neglected or mismanaged is worse than a farm overgrown with weeds or exhausted by ignorant tillage; for the land, by prudent management, may again be made fertile, but the bees, when once destroyed, are a total loss. Of all farm pursuits bee-culture requires the greatest skill, and it may well be called *a business of details.*

Fig. 96.

ORNAMENTAL GLASS HIVE; OLD STYLE; FRONT VIEW.

567. Wherever the Apiary is established, great pains should be taken to protect the bees against high winds. Their hives should be placed where they will not be annoyed by foot passengers or cattle, and should never be very near where horses must stand or pass. If managed on the swarming plan, it is very desirable that they should be in full sight of the rooms most occupied, or at least where the sound of their swarming (**406**) will be easily heard.

In the Northern and Middle States, the hives should have a South-Eastern, Southern, or South-Western exposure, to give the bees the benefit of the sun, when it will be most conducive to their welfare.

568. The plot occupied by the Apiary should be grassy, mowed frequently, and kept free from weeds.

Sand, gravel, saw-dust* or coal cinders, spread in front of the hive, will prevent the growing of grass in their (**382**) immediate vicinity, and be a great help to those overladen bees, that fall to the ground before reaching the entrance.

Hives are too often placed where many bees perish by falling into the dirt, or among the tall weeds and grass, where spiders and toads find their choice lurking-places.

A gentle slope southward will help to set the hives as they should be, slanting toward the entrance (**327**, **328**).

569. They should be placed on separate stands, entirely independent of one another, and, whenever practicable, room should be left for the Apiarist to pass around each hive. We prefer to place them in rows sixteen feet apart, with the hives about six feet apart in the rows. This isolates each hive completely, and, while handling one colony, the Apiarist is not in danger of being stung by the bees of another. The bees are also less likely to enter the wrong hives (**502**).

FIG. 97. ORNAMENTAL GLASS–HIVE. BACK VIEW; OLD STYLE.

Covered Apiaries.

570. Covered Apiaries, unless built at great expense, afford little or no protection against extreme heat or cold, and greatly increase the risk of losing the queens (**356**),

---

* Saw-dust is perhaps not very safe, owing to danger of fire from the smokers, in very dry weather.

and the young bees. The weak colonies are always the
losers, for their young bees, in returning from their first
trip (**173**), are attracted by the noise of other hives closely

(Fig. 98.)
HOUSE APIARY OF MR. JECKER IN SWITZERLAND.
From the *Revue Internationale.*

adjoining, and prove the truth of the French proverb "La
pierre va toujours au tas," (the stone always goes to the
heap).

When hives *must* stand too close together, they should be

of different colors. Even varying the color of the blocks will be of great usefulness.

John Mills, in a work published at London, in 1766, gives (p. 93) the following directions:—" Forget not to paint the mouths of your colonies with different colors, as red, white, blue, yellow, &c., in form of a half-moon, or square, that the bees may the better know their own homes."

Covered Apiaries are common in Germany and Italy; their only quality is that of being thief proof, when shut and locked. But such structures, especially when several stories high, cannot easily shelter top-opening hives.

571. Probably the most convenient covered Apiaries are simple sheds, facing South, and open in front during the Summer and warm days of Winter. House Apiaries, in which the hives are placed in several stories, facing every direction, are worse than nothing. Their only quality is to be ornamental and costly.

572. For ease of manipulation, out-door Apiaries are preferable.

In the Summer, no place is so congenial to bees as the shade of trees, if it is not too dense, or the branches so low as to interfere with their flight. As the weather becomes cool, they can, if necessary, be moved to any more desirable Winter location. If colonies are moved in the line of their flight, and *a short distance at a time*, no loss of bees will be incurred; but, if moved a few yards, *all at once*, many will be lost. A slanting board placed in front of the hive, so as to prevent the bees from flying in straight line from the entrance to the field, will incite them to mark the change of their position. By a *gradual process*, the hives in a small Apiary may, in the Fall, be brought into a narrow compass, so that they can be easily sheltered from the bleak Winter winds. In the Spring, they may be gradually returned to their old positions.

By removing the strongest colonies in an Apiary the

first day, and others not so strong the next, and continuing the process until all were removed, we have safely changed the location of an Apiary, when compelled to move bees in the working season. On the removal of the last hive, but few bees returned to the old spot. The change, as thus conducted, strengthened the weaker colonies, but we would advise bee-keepers to locate their hives in as permanent a position as possible, as this moving is *not practical*, especially with a large number of colonies. Those who do not winter their bees in the cellar, can easily protect them on their Summer stand. See chapter on Wintering (**619**).

If the hives have to be placed in an exposed location without shade, it is well to protect them with roofs (**369**). A roof will be found highly economical, as it not only sheds the rain, but wards off the heat of the sun.

Procuring Bees and Transferring.

573. The beginner will ordinarily find it best to stock his Apiary with swarms of the current year, thus avoiding, until he can prepare himself to meet them, the perplexities which often accompany either natural or artificial swarming. If new swarms are purchased, unless they are large and early, they may only prove a bill of expense. If old colonies are purchased, such only should be selected as are healthy and populous. If removed after the working season has begun, they should be brought from a distance of at least two miles (**13**).

If the bees are not all at home when the hive is to be removed, blow a little smoke into its entrance, to cause those within to fill themselves with honey, and to prevent them from leaving for the fields. Repeat this process from time to time, and in half an hour nearly all will have returned. If any are clustered on the outside, they may be driven within by smoke (**382**).

The best time to buy full colonies of bees, is Spring. A cool day may be selected, in which to move them, as the bees are not flying, none can be lost.   In the present thriving state of bee-keeping, colonies of pure Italian bees (**551**) in movable frame hives (**286**) can usually be bought at very reasonable figures.   If the Apiarist's means are very limited, black bees (**549**) in old style box-hives may prove the cheapest, if they can be found.   But they should be promptly transferred into more practical hives, and Italianized (**489**); these manipulations will help to give to the novice the practice which he lacks.   Italian bees and movable-frame hives are now a *sine quâ non* of success.

No colony should be purchased, unless it has brood in all stages, showing that it has a healthy queen.   For transporting bees, see (**587** and **603**).

Transferring Bees from Common to Movable-Frame Hives.

574. This process may be easily effected whenever the weather is warm enough for bees to fly.

It has sometimes been done in Winter, for purposes of experiment, by removing the bees into a warm room, but the best time for it, is when the bees have the least honey, at the beginning of the fruit bloom.   If it can be done on a warm day, when they are at work, there will be but little danger from robbers (**664**).

It is conducted as follows: Have in readiness a box—which we shall call the *forcing box*—whose diameter is about the same with that of the hive from which you intend to drive the swarm.   Smoke the hive, lift it from its bottom-board without the slightest jar, turn it over, and carefully carry it off about a rod, as bees, if disturbed, are much more inclined to be peaceable, when removed a short distance from their familiar stand.   If the hive is gently placed upside down on the ground, scarcely a bee will fly out, and there will be little danger of being stung. The timid and inexperienced should protect themselves with a bee-veil, and may blow more smoke among them, as soon as the hive is inverted. After placing it on the ground, the forcing-box must be put over it. If smooth inside, it should have slats fastened one-third of the distance from the top, to aid bees in clustering. Some Apiarists place the box slanting on the hive, so as to be able to see the bees climbing. This method, called open driving, is a little slower, but it may give the operator the chance of seeing the queen; when the driving can be considered as done.

575. As soon as the Apiarist has confined the bees, he should place an empty hive—which we call *the decoy-hive* —upon their old stand, which those returning from the fields may enter, instead of dispersing to other hives, to meet, perhaps, with a most ungracious reception. As a general rule, however, a bee with a load of honey or bee-bread, after the extent of her resources is ascertained, is pretty sure to be welcomed by any hive to which she may carry her treasure; while a poverty-stricken unfortunate that presumes to claim their hospitality is, usually, at once destroyed. The one meets with as flattering a reception as a wealthy gentleman proposing to take up his abode in a country village, while the other is as much an object of dislike as a poor man, who bids fair to become a public charge.

If there are in the Apiary several old colonies standing close together, it is desirable, in performing this operation, that the decoy-hive, and the forcing-box, should be of the same shape and even *color* with that of the parent-stock. If they are very unlike, and the returning bees attempt to enter a neighboring hive, because it resembles their old home, the adjoining hives should have sheets thrown over them, to hide them from the bees, until the operation is completed.

576. To return to our imprisoned bees: their hive should be beaten smartly with the palms of the hands, or two small rods, on the sides to which the combs are attached, so as to run no risk of loosening* them. These "rappings," although not of a very "spiritual" character, produce, nevertheless, a decided effect upon the bees. Their first impulse, if no smoke were used, would be to sally out, and wreak their vengeance on those who thus rudely assail their honied dome; but as soon as they inhale its fumes, and feel the terrible concussion of their once stable abode, a sudden fear, that they are to be driven from their treasures, takes possession of them. Determined to prepare for this unceremonious writ of ejection, by carrying off what they can, each bee begins to lay in a supply, and in about five minutes, all are filled to their utmost capacity. A prodigious humming is now heard, as they begin to mount into the upper box: and in about fifteen minutes from the time the rapping began—if it has been continued with but slight intermissions—the mass of bees, with their queen, will hang clustered in the forcing-box, like any natural swarm, and may, at the proper time, be readily shaken out on a sheet, in front of their intended hive.

Now put the forcing box on their old stand, and carry the parent-hive to some place where you cannot be annoyed by other bees.

577. It is important to make sure that the queen is removed, as she might be injured in the transfer of comb. Her presence among the driven bees can be ascertained in a few minutes, by the quietness of their behavior, or by the eggs which she drops on the bottom board, and which can easily be seen if a black cloth is spread under the forcing box **(155).**

* There is little danger of loosening the combs of an old colony. but the greatest caution is necessary when the combs of a hive are new. If, in inverting such a hive, the *broad sides* of the combs, instead of their *edges*, are inclined downwards, the heat, and weight of the bees. may loosen the combs. and ruin the colony.

If the queen is not with the bees, a few will come out and
run about, as if anxiously searching for something they
have lost.   The alarm is rapidly communicated to the whole
colony; the explorers are reinforced, the ventilators suspend their operations, and soon the air is filled with bees.
If they cannot find the queen, they return to their old stand,
and if no hive is there, will soon enter one of the adjoining
colonies.   If their queen is restored to them soon after they
miss her, those running out of the hive will make a half-cir-
cle, and return; the joyful news is quickly communicated
to those on the wing, who forthwith alight and enter the
hive; all appearance of agitated running about on the outside of the hive ceases, and ventilation, with its joyful hum,
is again resumed.*

If the queen has not left the old hive, it is safer to return
the bees and to resume the driving at another time.

578. To transfer the comb, have on hand tools for prying off a side of the hive; a large knife for cutting out the
combs; vessels for the honey; a table or board, on which
to lay the brood combs; and water for washing off, from
time to time, the honey which will stick to your hands.

Have also a number of pieces of wire, No. 16, cut a little
longer than the frame, and bent on the ends in this shape
⎣⎯⎯⎦ to be driven into the wood of the frame, and to hold
the combs in place.   Let a certain number of frames be in
readiness, with three or four of these wires fastened on one
side, and lay them on the table, *wire-side down.*   You must
also have your movable frame hive in readiness near the
table, with an extracting pan (**770**) under it, instead of a
bottom board, to receive what honey may drip.   All this
must be ready before disturbing the bees.

579. Having selected the *worker-combs,* carefully cut

---

* To witness these interesting proceedings, it is only necessary to catch the
queen and keep her until she is missed by her colony. For greater security,
she should be confined in a queen cage (**536**) during the experiment.

them rather large, so that they will just *crowd* into the frames, and retain their places in their natural position until the bees have time to fasten them.

Now tack as many wires over them as may be necessary to hold them securely, and hang them in the hive. *Drone combs should invariably be melted into wax.* If drone-brood (**168**) is found, it can be fed to young chickens, who are very fond of the larvæ. The bottom board should be put under the hive just before carrying it out.

When the hive is thus prepared, the bees may be put into it and confined, water being given to them, until they have time to make all secure against robbers (**664**).

If there is danger of robbers, it is preferable not to put the bees into the hive till late in the afternoon. They should be shaken in front of the new hive on a sheet (**427**) like a natural swarm.

When the weather is cool, the transfer should be made in a warm room, to prevent the brood from being fatally chilled. An expert Apiarist can complete the whole operation—from the driving of the bees to the returning of them to their new hive—in about an hour, and with the loss of very few bees, old or young.

580. When transferring in early Spring, it should be remembered that the worker-brood (**168**) is of great value; and not the least bit of it should be neglected or wasted unnecessarily. After a week, or more, according to the season, the hive may be opened and the fastening removed.

Dr. Kirtland thus spoke of the results of transferring some of his colonies to the movable-comb hives.

"I had three stocks transferred to an equal number of Mr. Langstroth's hives. The first had not swarmed in two years, and had long ceased to manifest any industry; the others had never swarmed. All the hives were filled with black and filthy comb, candied honey, concrete bee-bread, and an accumulation of the cocoons and larvæ of the moth. Within twenty-four hours, each colony became reconciled to its new tenement, and began

to labor with far greater activity than any of my old stocks. . . .
I have now no stronger colonies than these, which I considered
of little value till my acquaintance with this new hive."—Ohio
Farmer, Dec. 12, 1857.

Let not the novice, however, think that transferring bees
is a task that requires but little skill. *He who transfers successfully a large number of colonies may be called an expert in
handling bees.*

The process, as it has been conducted by careless Apiarists, has resulted in the wanton sacrifice of thousands of
colonies.

581. For the benefit of those who are timid in manipulations, we will give Mr. Jas. Heddon's method for transferring, (page 562 of "Gleanings" 1885). About swarming
time (**406**) Mr. Heddon drives the old queen and a majority of the bees into the forcing-box, he then removes the
old hive a few feet back, and places the new hive with
frames full of foundation (**674**) on its stand, and " runs
in " the forced swarm. It would be well to return a part
of the bees to the old hive, as its brood might be chilled if
the weather becomes cool.

Twenty-one days after the transfer of the bees, he drives
the old hive clean of all its bees, uniting them with the
former drive. As the worker brood of the old hive is all
hatched, there is nothing left in it but the combs and the
honey, which can be transferred at leisure in cool weather,
or, the honey may be extracted (**749**), and the comb melted
into wax (**858**).

Out–Apiaries.

582. When an Apiarist wishes to make bee-culture his
special occupation, he should expect to keep bees in more
than one location. If he owns more than 120 colonies, we
would advise his establishing an Out-Apiary. It is true

that there are many drawbacks to the cultivation of bees four or five miles off, but there are also some advantages. The crop sometimes fails in one locality, and is very good in another a short distance away. One Apiary may be in a hilly country, where white clover abounds, and another on low lands, where Fall blossoms never fail. It is well— according to a familiar proverb— not to "put all our eggs in one basket."

In many years' practice of keeping bees in five or six different Apiaries, occupying a range of country about twenty miles in width, we have found out that the crop will vary greatly in a few miles, owing to the different flora of the various localities, and more especially to the greater or less amount of rain-fall at the proper time. We have also learned that an Apiary placed near a large body of water (the Mississippi), will produce less honey than one a mile or two from it. owing to the smaller area of pasturage in reach of the bees.

583. In establishing an Out-Apiary on some farmer's land, the following must be taken into consideration: Select a farm on which a grove or an orchard is near the house, some distance from the road. The place ought to be, *at least*, three miles in a bee-line from your own bee-farm. It is not necessary that it should be more than four miles away.*

Locate your bees with some careful man. Do not trust a farmer who lets his fences fall, who leaves his mower in the yard over Winter, or puts his cows in his orchard. You will never rest easy, if you think that some of your hives may be upset any day by a vagrant cow.

Do not put your bees on land which is tenanted. Let

---

* Mr. J. M. Hambaugh, of Spring, Ill., harvested altogether different yields both in quality and quantity, from two Apiaries only two and a half miles apart. This agrees with our oft repeated experience in Apiaries three or four miles apart.

them be placed at some responsible farmer's own home, for
a tenant may leave on short notice, and you cannot remove
your bees at all seasons.

584. The terms usually made by us for a bee location
are as follows: The farmer furnishes us the Apiary ground,
one spare room during extracting, and a shed or a corner
in some empty room for our hives, combs, and fixtures. He
also furnishes board for the Apiarist and his help while at
work. In exchange, he gets one-fifth of the honey, and
seventy-five cents for every natural swarm he harvests. His
sole duties are, hiving swarms, and seeing that no accidents
happen to the Apiary. When bees are run for extracted
honey, the number of natural swarms is very limited (**454**).
We can always find more bee locations than we want. In
fact, we have never yet met a farmer who refused to take
bees on such terms.

We prefer giving the farmer a share of the crop, to giving him a stated sum for ground rent, etc., as some of our
leading bee-keepers do, because we thus give him an interest in our success, and he is more likely to pay attention to
our bees, and to produce crops that will yield some honey.
Association of interests means progress, peace, and harmony.

585. Six Apiaries, containing in all 600 colonies, are
probably the greatest number that one man can oversee.
In good localities, an Apiarist will find more profit from six
such Apiaries, than an intelligent farmer from half a Section
of land, and the outlay of money is less.

HONEY-HOUSE.

586. Few pursuits require so small an outlay for tools and
implements as practical bee culture. Outside of the cost of
hives, frames, sections, and honey packages, the total out-

lay need not amount to $50. Almost any spare room will do for a honey room.

' Yet when the Apiarist wishes to be at ease, we would advise him to build his honey-house in the middle of his Apiary. The windows and doors of this building must all be provided with wire cloth netting, to exclude bees, flies,

Fig. 99.

WINDOW-SCREEN.

etc. We here give an engraving of a simple method of placing the wire screen, so as to allow these insects to escape. The netting is nailed on the outside of the window projecting about six inches above. At the top three small slats are nailed between the frame and the netting, so as to leave a

20

space of ¼ of an inch between the wire cloth and the wall, at the top of the window. The bees and flies that have been brought in with the combs, or that have entered the room, at some time or other, fly against the wire cloth, and soon find the small fissure above, through which they escape ; but, in returning, they smell the honey through the wire cloth, and forgetting that they have escaped between the wire and the wall, they try in vain to pass through the wire cloth.

In the engraving, the window sashes have been removed, but their use in no way interferes with the screen, if the lower one is raised, or the upper one lowered, while there are bees in the room.

CHAPTER XI.

SHIPPING AND TRANSPORTING BEES.

587. In shipping colonies of bees by rail, it is not necessary to give them much ventilation, if they are sent during the cool weather of Spring. We have successfully shipped hundreds of colonies to all parts of the U. S., in early Spring, with no other ventilation than was afforded by the joints of a rough block nailed over the entrance of the hive. But, if the weather is warm, and the colony populous, plenty of air is needed. We usually replace the bottom board by a wire-cloth-frame protected by slats. The entrance should never be covered with wire-cloth, but should be entirely closed, for the old bees will worry themselves trying to get through it, and it will soon be clogged with dead bees. They should be given as much air as needed with the least possible amount of light.

When the colony is so populous, that draught through the hive cannot injure the brood, we nail a screen over the frames also, and shade it with a board nailed on slats, running across the ends of the hive. The closing of the portico alone, if there is one, with wire-cloth, is not practical, as a part of the swarm crowds into it and bars the ventilation.

588. The frames should, of course, be securely fastened in their places. For this purpose, Mr. Root uses sticks, or slats, of the depth of the hive, that fit between the frames and hold them.

New combs had better not be shipped at all. If there is plenty of fresh honey, we would advise the extracting of all that is unsealed, previous to shipment. When there is brood in every comb, and the weather is warm, it is safer to remove a part of the brood, and put frames of dry comb alternately with the frames of brood. The brood removed may be used to strengthen weak colonies.

As a rule, it is better to ship small lots by Express, but large lots may be sent in early Spring, by freight, if they are not to be more than a week on the way. We have sent bees safely, from Illinois to Utah, by freight.

589. In shipping bees, or colonies, it is important to place conspicuous cautionary cards or labels on the packages: Living Bees, Handle with Care, This side up, Keep out of the sun, etc.

The damage done by rough railroad handling, is the greatest item of loss, in the transportation of bees properly packed. If colonies are shipped in carloads, they should be so placed, that the combs will run lengthwise, and not from side to side, as in vehicles drawn by horses. Surplus racks or stories should be shipped separately.

590. Some Apiarists, among whom we will cite the firm of Flanagan and Illinski of Belleville, Ill., have practiced shipping bees by water routes to the Southern States in the Fall, for Winter, and returning them in Spring at the beginning of the honey harvest. If proper precautions are taken, this plan may be profitable, where low rates of transportation can be obtained, but much judgment must be exercised as to the time of returning them North. As the colonies become strong very early in the South, if they are brought back North before the warm weather, their brood may become chilled, and a tendency to the developement of foul-brood is encouraged.

591. Della Rocca, in his treatise on "Bee-culture in the Island of Syra," speaks of the Egyptian* method of keep-

---

* "Mr. Cotton saw a man in Germany who kept all his numerous stocks rich by changing their places as soon as the honey-season varied. 'Sometimes he sends them to the moors, sometimes to the meadows, sometimes to the forests, and sometimes to the hills. In France—and the same practice has existed in Egypt from the most ancient times—they often put hundreds of hives in a boat, which floats down the stream by night and stops by day.''—*London Quarterly Review*.

ing bees on boats, which were floated up and down the Nile to take advantage of the different crops of honey at different points.

It would even appear that the Greeks in the time of Columella transported their hives to Egypt by sea, "the season of blossoms being later than in Greece; for after the month of September there is no pasture in Achaia for bees, whilst in Egypt flowers are in full bloom even after that time, owing to the receding of the high waters of the Nile." He relates a laughable story about one of these floating Apiaries. One hive having been upset by accident on a boat, the enraged bees attacked the mariners unexpectedly, and forced them to jump into the river and swim to the shore, which likely, was not far distant, nor did they dare return, until they had provided themselves with a supply of smoke-producing ingredients.

592. There is a certain amount of fascinating romance connected with the idea of a floating Apiary, following the blossoms, on the waters of the great Mississippi, or of some of its tributaries. An attempt of this sort was made on a large scale, a few years ago, by a Chicago firm. It was a total failure, but we are inclined to think that the failure was due more to the lack of practical knowledge in bee-keeping, on the part of the managers, than to any other cause.

593. Transportation of bees from a location where blossoms are scarce to a good field, and returning them after the crop, is sometimes attended with fair success. Some Apiarists, located in places where the June crop alone can be depended upon, make it a practice to transport their hives to Fall pasturage every Summer. We, ourselves, have taken 120 hives of bees, about eighteen miles, to the Mississippi river bottoms, in August, 1880, when the drouth had destroyed all hopes of a Fall harvest on the hills. The high waters of the Mississippi, which had receded a few weeks before, had left those immense bottom lands covered with a luxuriant vegetation. The result fully answered our anticipations. Those lately starving colonies, yielded a bountiful surplus, while their sisters on the hills had to be fed for Winter. But the labor of transportation, the risk incurred, if the colonies are strong and heavy, and the difficulty of transporting old bee-hives, without danger of some bees escaping, make the habitual shipping of bees for pasturage hardly advisable.

Shipping Queens.

594. It was in the numerous and partially successful attempts, which we made before 1874, to import bees from Italy, that we became acquainted with the conditions necessary to the shipping of queens.

595. When they are to be confined a long time, the question of food is the most important. Many were the blunders made by the first shippers, who imagined that they required a large amount of food, and literally drowned them in honey. By repeated and costly experiments, we ascertained that the bees that arrived in the best condition were those that were fed on the *purest saccharine matter.* Those that suffered the most, were those that had the most watery (**249**), or the darkest, honey (**627**). Water (**271**), which some Italian shippers persisted in giving them, in spite of what we could say, was noxious; as the consumption of it, with the food, helped to load their abdomen with matter that could not be discharged (**73**), causing what is improperly called *dysentery* (**784**). *Water is needed only in brood rearing.*

596. Old bees, or rather, bees that have begun to work in the field, will stand a longer trip than young bees, as the latter consume more honey, and need to discharge their abdomen oftener.

The shipping boxes in which bees are usually sent from Italy, are about three inches deep, by three inches in width, and four inches in length, with two small frames of comb, one with thick sugar syrup, the other dry. From fifty to seventy-five bees are put with one queen in each box. Air holes are cut into the sides of the boxes, and these are fastened together in a pyramidal shape, with an outer covering of tin, to which is fastened the handle. Queens thus put up, have reached us after thirty-six days of confinement with very little loss, and it is in this way that the greatest number of imported queens are received.

The usual transit from Italy to New York, takes from ten to fourteen days. If the importer receives his bees, through a custom-house broker, they will not be delayed in the custom-house, but, if this precaution is neglected, the bees may be held at the custom-house for clearance, and the poor insects will die, martyrs to the protection (?) of the country's interests.

597. We might mention in connection with this, an oft-repeated incident, so touching and sweet, as to seem more like a romancer's fable, or a poetic idyl, than a mere fact. On receiving the boxes containing Italian queens, we noticed that frequently all the bees shipped with the queen had died, she being the only one alive in her prison. We afterward found out that the faithful little subjects had denied themselves nourishment, and starved to death, sacrificing themselves, that their queen might not be deprived of food.

MAILING QUEENS.

598. To Mr. Frank Benton is due the credit of first mailing queens safely across the ocean, but the mailing of them, with more or less success on the American continent, has been practiced for years. Messrs. J. H. Townley and H. Alley, appear to have been the first to succeed, as early as 1868.

The methods have been so far improved, that our friend Mr. Paul Viallon, a practical queen-breeder of Louisiana, sent us 150 queens in the season of 1885, by mail, with the loss of only three or four. The cages he used were the Peet

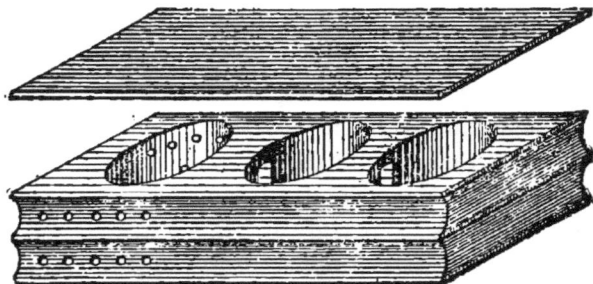

Fig. 100.
THE BENTON CAGE.
(From the "*Illustrierte Bienenzeitung.*")

cages. Yet the mails are so roughly handled generally, that we would not advise the sending of valuable queens in this way.

The food given is the Scholz candy (**613**) made of powdered sugar and honey kneaded together. A sufficient number of bees must be put with the queen to keep her warm, but not enough to crowd the cage—six to ten bees are sufficient, in Summer.

599. Of late years, at the suggestion of friend Root, the shipping of bees by the pound instead of in colonies, has been practiced, for the purpose of stocking Apiaries. Since the invention of comb foundation, a hive may be supplied with comb of the best quality, at comparatively small cost, and a choice queen, with a pound or two of bees, can build up a very fair colony, if purchased at the beginning of the clover harvest and properly cared for. They are shipped in wire-cloth cages (fig. 101) and fed with Scholz candy for the trip.

600. How many bees are there in a pound? This question has been propounded to us several times. *L'abbé Collin*, by careful experiments, found that in a normal condition it takes about 5,100 bees to weigh a pound; while in the swarm, when they are supplied with honey, it takes less than 4,300. Their weight will vary according to the quantity of honey they have absorbed.

601. Parties contemplating the breeding of bees and queens (**489**) for sale, will do well to locate themselves as far South as convenient for easy shipment, as it is by far more lucrative to raise them there than in the North. This is very easy to understand. In the South, the bees usually winter safely, and breed early, so that the colonies are strong, while those of the Northern latitudes are still confined in their hives, struggling against the rigors of Winter.

If an Apiarist purchases bees or queens at the proper time—Spring—to recruit his Winter loss, he will most likely buy them from some location South of him, as he can there obtain stronger colonies, and earlier queens, than in his own latitude.

602. On the other hand, as the honey of the Northern States is superior in quality to Southern honey, bee-culture for honey production can be made fully as profitable in the North, in spite of the difficulties of wintering (**619**).

Transporting Bees Short Distances.

603. The box-hives may be prepared for removal by inverting them and tacking a coarse towel or sack over them, or strips of lath may be laid over wire-cloth, and brads driven through them into the edges of the hive.

Confine the hive, so that it cannot be jolted, in a wagon with springs, and be sure, before starting, that it is *impossible* for a bee to get out. The inverted position of the hive will give the bees what air they need, and guard their combs from being loosened. It will be next to impossible, in warm weather, to move a hive which contains much new comb (**215**), or much fresh honey (**249**).

Indeed, we would strongly urge beginners not to transport bees in warm weather. Just before fruit-blossom is the best time to transport full colonies of bees. Some advise transporting them in Winter, on sleds, but after trial we condemned this method also. The joltings of a sleigh, though few, are hard, and will break combs; and disturbing bees in cold weather should always be discouraged. When hauling bees in warm weather, do not load or unload them while the horses are hitched to the wagon. We have seen serious accidents resulting from a hive dropping from a man's hands to the ground, causing the bees to escape, and to sting both the driver and the horses severely.

If a colony, in hot weather, is to be moved any distance in movable-frame hives, it will be advisable to fasten frames of wire-cloth, both to the top and bottom of the brood apartment, and to transport the bottom-board (**344**), cloth, mat, or surplus cap or cover (**355**), separately.

Glass hives ought never to be sent off for fear of accident. Hives with movable-frames should be arranged in such a position that the frames run from side to side, and not from front to rear, in the carriages.

603. (*bis.*) Upon arrival at the Apiary, if the weather is warm, you should at once set the hive in proper position, and release the bees. It is good policy to place a *shade board* in front of the entrance for a day or two. The object of this is to cause the old bees to notice that something is changed in their location, and to turn around and *mark the place*, instead of starting out as usual in a bee-line without looking behind.

604. New swarms may be brought home in any box which has ample ventilation. A tea-chest, with wire-cloth on the top, sides, and bottom-board, will be found very convenient. Of late years, Mr. A. I. Root, and others, have practiced the shipping of bees by the pound, with or without queens, to stock Apiaries. Their wire-cloth cages

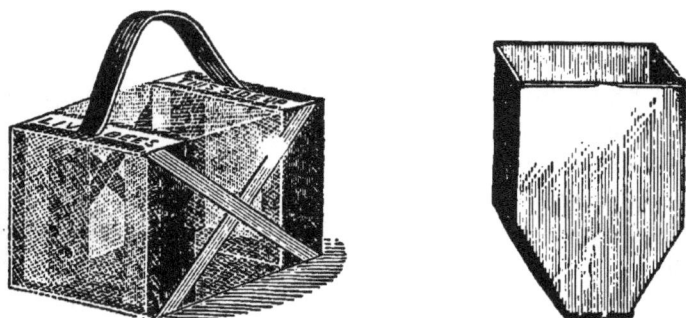

Fig. 101.

CAGE FOR ONE-HALF POUND   TIN FUNNEL FOR SHAKING THE
OF BEES.   BEES INTO THE CAGES.
(From Root's "A. B. C.")

or boxes for shipping bees, are just the thing for hauling natural swarms, if made large enough (fig. 101).

The bees may be shut up in the box as soon as they are hived. *New swarms require even more air than old colonies*, being full of honey and closely clustered together. They should be set in a cool place, and, if the weather is very sultry, should not be removed until night. Many swarms are suffocated by the neglect of these precautions. The bees may be easily shaken out from this temporary hive.

When movable-comb hives are sent away to receive a swarm, two strips of wood, with pieces nailed to them, to go between the frames and keep them apart, should be laid over the frames, or they may be tacked fast in their proper places.

The enamel-cloth (**352**) should be fastened on, by nailing strips all around over it.

For the further preparation of hives to receive swarms, see (**421**).

CHAPTER XII.

FEEDING BEES.

605. FEW things in practical bee-keeping are more important than the feeding of bees; yet none have been more grossly mismanaged or neglected. Since the sulphur-pit has been discarded, thousands of feeble colonies starve in the Winter, or early Spring; while often, when an unfavorable Summer is followed by a severe Winter, and late Spring, many persons lose most of their colonies and abandon bee-keeping in disgust.

In the Spring, the prudent bee-keeper will no more neglect to feed his destitute colonies, than to provide for his own table. At this season, being stimulated by the returning warmth, and being largely engaged in breeding, bees require a liberal supply of food, and many populous colonies perish, which might have been saved with but trifling trouble or expense.

> " If e'er dark Autumn, with untimely storm,
>   The honey'd harvest of the year deform ;
>   Or the chill blast from Eurus' mildew wing,
>   Blight the fair promise of returning Spring ;
>   Full many a hive, but late alert and gay,
>   Droops in the lap of all-inspiring May."   EVANS.

" If the Spring is not favorable to bees, they should be fed, because that is the season of their greatest expense in honey, for feeding their young. Having plenty at that time, enables them to yield early and strong swarms."—(Wildman.)

A bee-keeper, whose colonies are allowed to perish after the Spring has opened, is on a level with a farmer whose cattle are allowed to starve in their stalls ; while those who withhold from them the needed aid, in seasons when they cannot gather a supply, resemble the merchant who burns up his ships, if they have made an unfavorable voyage.

Columella gives minute instructions for feeding needy colonies, and notes approvingly the directions of Hyginus— whose writings are no longer extant — that this matter should be most carefully (" *diligentissime*") attended to.

Spring Feeding.

606. When bees first begin to fly in the Spring, it is well to feed them a *little*, as a small addition to their hoards encourages the production of brood. Great caution, however, should be used to prevent robbing. Feeding should always be attended to in the evening (**666**), and as soon as forage abounds, the feeding should be discontinued. If a colony is *over-fed*, the bees will fill their brood-combs, so as to interfere with the production of young, and thus the honey given to them is worse than thrown away.

The over-feeding of bees resembles, in its results, the noxious influences under which too many children of the rich are reared. Pampered and fed to the full, how often does their wealth prove only a legacy of withering curses, as, bankrupt in purse and character, they prematurely sink to dishonored graves.

Colonies, which have abundant stores, may be incited to breed, by simply bruising the cappings of a part of their honey. This causes them to feed their queen more plentifully, and more eggs are laid.

607. Bees may require feeding, even when there are many blossoms in the fields, before the beginning of the main harvest, if the weather is unfavorable to the honey flow. Large quantities of brood hatch daily, requiring much food, and a few days without honey sometimes endangers the life of colonies, on the eve of a plentiful harvest.

The best way to feed destitute colonies in Spring is to give them combs of honey, which have been saved from the previous season for this purpose. If such cannot be had, the food may be put into an empty comb, and placed where it can be easily reached by the bees.

Honey partially candied (**830**), may be given them, in small quantities, by pouring it over the top of the combs in which the bees are clustered. A bee deluged by sweets, when away from home, is a sorry spectacle; but what is thus given them does no harm, and they will lick each other clean, with as much satisfaction as a little child sucks its fingers while feasting on sugar candy.

If a colony has too few bees, its population must be replenished before it is fed. To build up small colonies by *feeding*, requires more care and judgment than any other process in bee-culture, and will rarely be required by those who have movable-frame hives. It can only succeed when everything is made subservient to the most rapid production of *brood*.

FALL FEEDING.

608. By the time the honey-harvest closes, all the colonies ought to be strong in numbers; and, in favorable seasons, their aggregate resources should be such that, when an equal division is made, there will be enough food for all. If some have more, and others less than they need, an equitable division may usually be effected in movable-frame hives. Such an agrarian procedure would soon overthrow human society; but bees thus helped, will not spend the next season in idleness; nor will those deprived of their surplus limit their gatherings to a bare competency.

After the first heavy frosts, when forage is over, all feeding required for wintering bees should be carefully attended to. If delayed to a later period, the bees may not have sufficient time to seal over their honey, which, by attracting moisture and souring, may expose them to dysentery.

609. Feeders of all descriptions are made and sold.*

In our opinion, the best feeder for liquid food is a simple fruit can or a jar. Mr. Root uses a can with perforated cover—we prefer the ordinary fruit can, because they are found in every house. After filling the can, we tie a cloth over the mouth of it, and invert it over a dish. The honey or syrup will leak through at first, but the atmospheric pressure soon stops its running, when the can may be carried to the hive in this position, and set immediately over

Fig. 102.
CAN FEEDER.

the cluster—without the dish—in the upper story or cap, a part of the enamel cloth being raised for this purpose.

The bees can then get their food, without being chilled even in cold weather, and they promptly store it away in the combs, for later use.

It is desirable to get through with Fall feeding as rapidly as possible,† as the bees are so excited by it that they consume more food than they otherwise would. In feeding a large amount for Winter supply, we have given as many as five quart-cans to one colony at one time. Wooden feeders in the shape of troughs, as made by Root, Shuck, and Heddon, have the advantage over the cans of not needing removal to be refilled, but they are not so well in reach of the cluster.

---

* Columella recommended wool, soaked in honey, for feeding bees. When the weather is not too cold, a saucer, bowl, trough, or vessel of any kind, filled with straw, makes a convenient feeder.

† Feeding colonies put late in the Fall into empty hives, unless combs can be given to them, will seldom pay expenses.

610. As honey is scarce in the seasons when Fall feeding has to be resorted to, we will give directions for making good syrup for Winter food: Dissolve twenty pounds of granulated sugar (use none but the best) in one gallon of boiling water, with the addition of five or six pounds of honey. Stir till well melted, and feed while lukewarm.

611. *Sugar candy*, for feeding bees, was first recommended by Mr. Weigel of Silesia. If the candy is laid on the frames just above the clustered bees, it will be accessible to them in the coldest weather. It may also be put between the combs, in an upright position, among the bees, or poured into combs before it is cold.

Fig. 103.
ROOT
FEEDER.

To make candy for bee-feed: add water to sugar, and boil slowly until the water is evaporated. Stir constantly so that it will not burn.

To know when it is done, dip your finger first into cold water and then into the syrup. If what adheres is brittle to the teeth, it is boiled enough. Pour it into shallow pans, a little greased, and, when cold, break it into pieces of a suitable size.

612. Before attempting to make candy for bee feed, the novice will do well to read the following advice from the witty pen of friend A. I. Root:

" If your candy is burned, no amount of boiling will make it hard, and your best way is to use it for cooking, or feeding the bees in Summer. Burnt sugar is death to them, if fed in cold weather. You can tell when it is burned by the smell, color and taste. If you do not boil it enough, it will be soft and sticky in warm weather, and will be liable to drip, when stored away. Perhaps you had better try a pound or two, at first, while you "get your hand in". Our first experiment was with 50 lbs. and it all got 'scorched' somehow . . . . . Before you commence, make up your mind, you will not get one drop of sugar or syrup on the floor or table. Keep your hands clean, and everything else clean, and let the women folks see that men have common sense; some of them at least. If you should forget yourself, and let the candy boil over on the stove, it would be very apt to get on the floor, and then you would be very likely to get "your foot in it", and before you got through, you might wish you had never heard of bees or candy either; and your wife, if she did not say so, might wish she had never heard of anything that brought a man into the kitchen. I have had a little experience in the line of feet sticking to the floor and snapping at every step you take, and with door knobs sticking to the fingers, but it was in the honey house." ("A. B. C." page 48.)

613. The Rev. Mr.Scholz, of Silesia, more than 30 years ago, recommended the following as a substitute for sugar-candy in feeding bees:

"Take one pint of honey and four pounds of pounded lump-sugar; heat the honey, without adding water, and mix it with the sugar, working it together to a stiff doughy mass. When thus thoroughly incorporated, cut it into slices, or form it into cakes or lumps, and wrap them in a piece of coarse linen and place them in the frames. Thin slices, enclosed in linen, may be pushed down between the combs. The plasticity of the mass enables the Apiarist to apply the food in any manner he may desire. The bees have less difficulty in appropriating this kind of food than where candy is used, and there is no waste."

This preparation has been used of late years with success, as food in mailing and shipping bees, under the name of "Good's candy."

Thick sugar-syrup and candy are undoubtedly the best bee-food, especially when the bees are to be confined a long time and no brood is to be raised.

614. An experiment of De Layens has proved that bees can use water to dissolve sugar (**272** *bis*). The same writer relates how a French bee-keeper, Mr. Beuzelin, feeds his bees in Winter:

"He saws into slices a large loaf of lump-sugar, and places these slices upon the frames under a cloth. Another bee-keeper

21

told me several years ago of having saved colonies in straw hives by simply suspending in them, with wires, lumps of sugar weighing several pounds."—(*Bulletin de la Suisse Romande.*)

While such methods succeed in a mild and damp climate, like that of France, they are not advisable in the Northern part of the United States, unless the bees are wintered in cellars (**646**).

615. The prudent Apiarist will regard the feeding of bees—the little given by way of encouragement excepted— as *an evil to be submitted to only when it cannot be avoided*, and will much prefer that they should obtain their supplies in the manner so beautifully described by him whose inimitable writings furnish us, on almost every subject, with the happiest illustrations :

> " So work the honey-bees,
> Creatures that, by a rule in Nature, teach
> The art of order to a peopled kingdom.
> They have a king and officers of sorts,
> Where some, like magistrates, correct at home,
> Others, like merchants, venture trade abroad ;
> Others, like soldiers, armed in their stings,
> Make boot upon the Summer's velvet buds ;
> Which pillage they, with merry march, bring home
> To the tent royal of their emperor,
> Who, busied in his majesty, surveys
> The singing masons building roofs of gold ;
> The civil citizens kneading up the honey ;
> The poor mechanic porters crowding in
> Their heavy burdens at his narrow gate ;
> The sad-eyed justice, with his surly hum,
> Delivering o'er, to executors pale,
> The lazy, yawning drone."
>
> SHAKESPEARE'S *Henry V, Act I, Scene 2.*

616. All attempts to derive profit from selling cheap honey or syrup, fed to bees, have invariably proved unsuccessful. The notion that they can change *all sweets*, however poor their quality, into *honey*, on the same principle that cows secrete milk from any acceptable food, is a complete delusion.

It is true that they can make white comb from almost every liquid sweet, because wax being a natural secretion of the bee, can be made from all saccharine substances, as fat can be put upon the ribs of an ox by any kind of nourishing food. But the quality of the comb has nothing to do with its contents; and the attempt to sell, as a prime article, inferior sweets, stored in beautiful comb, would be as truly a fraud as to offer for good money, coins which, although pure on the outside, contain a baser metal within.

Different kinds of honey or sugar-syrup fed to the bees can be as readily distinguished, after they have sealed them up, as before.

The Golden Age of bee-keeping, in which bees are to transmute inferior sweets into such balmy spoils as were gathered on Hybla or Hymettus, is as far from prosaic reality as the visions of the poet, who saw—

> "A golden hive, on a golden bank,
> Where golden bees, by alchemical prank,
> Gather gold instead of honey."

Even if cheap sugar could be " *made over* " by the bees so as to taste like honey, it would cost the producer, taking into account the amount consumed (**223**) in elaborating wax, almost if not quite, as much as the market price of white clover honey; and, if he feeds his bees after the natural supplies are over, they will suffer from filling up their brood cells.

617. The experienced Apiarist will fully appreciate the necessity of preventing his bees getting a taste of forbidden sweets, and the inexperienced, if incautious, will soon learn a salutary lesson. Bees were intended to gather their supplies from the nectaries of flowers, and, while following their natural instincts, have little disposition to meddle with property that does not belong to them; but, if their incautious owner tempts them with liquid food, at times when they can obtain nothing from the blossoms, they become so infatuated with such easy gatherings as to lose all discretion, and will perish by thousands if the vessels which contain the food are not furnished with floats, on which they can safely stand to help themselves.

As the fly was not intended to banquet on blossoms, but on substances in which it might easily be drowned, it cautiously alights on the edge of any vessel containing liquid food, and warily helps itself; while the poor bee, plunging in headlong, speedily perishes. The sad fate of their unfortunate companions does not in the least deter others who approach the tempting lure, from madly alighting on the bodies of the dying and the dead, to share the same miserable end! No one can understand the extent of their infatuation, until he has seen a confectioner's shop assailed by myriads of hungry bees. We have seen thousands strained out from the syrups in which they had perished; thousands more alighting even upon the boiling sweets; the floors covered and windows darkened with bees, some crawling, others flying, and others still, so completely besmeared as to be able neither to crawl nor fly—not one in ten able to carry home its ill-gotten spoils, and yet the air filled with new hosts of thoughtless comers.

We once furnished a candy-shop, in the vicinity of our Apiary, with wire-gauze windows and doors, after the bees had commenced their depredations. On finding themselves excluded, they alighted on the wire by thousands, fairly squealing with vexation as they vainly tried to force a passage through the meshes.* Baffled in every effort, they attempted to descend the chimney, reeking with sweet odors, even although most who entered it fell with scorched

---

* Manufacturers of candies and syrups will find it to their interest to fit such guards to their premises: for, if only one bee in a hundred escapes with its load, considerable loss will be incurred in the course of the season.

wings into the fire, and it became necessary to put wire-gauze over the top of the chimney also. (586).

618. As we have seen thousands of bees destroyed in such places, thousands more hopelessly struggling in the deluding sweets, and yet increasing thousands, all unmindful of their danger, blindly hovering over and alighting on them, how often have they reminded us of the infatuation of those who abandon themselves to the intoxicating cup! Even although such persons see the miserable victims of this degrading vice falling all around them into premature graves, they still press madly on, trampling, as it were, over their dead bodies, that they too may sink into the same abyss, and their sun also go down in hopeless gloom.

The avaricious bee that, despising the slow process of extracting nectar from "every opening flower," plunges recklessly into the tempting sweets, has ample time to bewail her folly. Even if she does not forfeit her life, she returns home with a woe-begone look, and sorrowful note, in marked contrast with the bright hues and merry sounds with which her industrious fellows come back from their happy rovings amid " budding honey-flowers and sweetly-breathing fields."

CHAPTER XIII

WINTERING AND SPRING DWINDLING.

Wintering.

619. Bees can be wintered safely in nearly all climates, where the Summer is long enough to enable them to store a Winter supply.   In the natural state, the vital heat of the live hollow trees in which they dwell, helps to maintain a higher temperature than that of the outside air, and bees Winter so well in such abodes, that travelers, who visit Northern Russia, wonder how so small an insect can live in such inhospitable countries.

620. As soon as frosty weather arrives, bees cluster compactly together in their hives, to keep warm.   They do not assemble on combs full of honey, but on the empty comb just below the honey.   They are never dormant, like wasps and hornets, and a thermometer pushed up among them will show a Summer temperature, even when, in the open air, it is many degrees below zero.

The bees in the cluster are *imbricated*, like the shingles of a roof, each bee having her head under the abdomen of the one above her, and so on, to the ones who are in reach of the honey.   These pass the honey to those below them, which pass it to the next, and so on, to the bottom of the mass.

621. When the cold becomes intense, they keep up an incessant tremulous motion, in order to develop more heat*

---

* Everybody knows that motion transforms itself into heat, and that heat is but a form of motion. . . .   whether the motion comes from a large body or from a small one, whether this motion be suddenly or gradually stopped, the result is the same, it is transformed into heat.—(Flammarion, ''Le Monde Avant la Création de l'Homme.'')

by active exercise; and. as those on the outside of the cluster become chilled, they are replaced by others. Besides, the fanning of wings, which causes this roar, sends the warm air from the top of the cluster to the bottom of the hive— thus warming the bees placed at the lowest part of the cluster; and these, if not too chilled, take advantage of a warmer day, to climb above the mass, and get honey in their turn.

When the weather is very cold, their humming can often be heard outside of the hive; and, if the hive be jarred, at any time, there comes a responsive murmur, which is longer or shorter in duration, and lower or higher in tone, according to the strength of the colony.

622. As all muscular exertion requires food to supply the waste of the system, the more quiet bees can be kept, the less they will eat. It is, therefore, highly important to preserve them as far as possible, in Winter, from every degree, either of heat or cold, which will arouse them to great activity.

When all the food which is in their reach is consumed, they will starve, if the temperature is too cold to allow them to move their cluster to the parts of the combs which contain honey; hence, if the central combs of the hive are not well stored with honey, they should be exchanged for such as are, so that, when the cold compels the bees to recede from the outer combs, they may cluster among their stores. In districts where bees gather but little honey in the Fall, such precautions, in cold climates, will be specially needed, as, often, after breeding is over, their central combs will be almost empty.

623. It is impossible to say how *much* honey will be needed to carry a colony safely through the Winter. Much will depend on the way in which they are wintered, whether in the open air or in special depositories, where they are protected against the undue excitement caused by sudden and severe atmospheric changes; much, also, on the length of the Winters, which vary so much in different latitudes, and the forwardness of the ensuing Spring. In some of our Northern States, bees will often gather nothing 'or more than six months, while, in the extreme South, they are seldom deprived of all natural supplies for as many weeks. In all our Northern and Middle States, if the colonies are to be wintered out of doors, they should have *at least* twenty-five pounds of honey.

In movable-frame hives, the amount of stores may be easily ascertained by actual inspection. The weight of hives is not always a safe criterion, as old combs are heavier than new ones, besides being often over-stored with bee-bread. (**263**.)

624. Practical bee-keepers usually judge of the amount of stores by sight. The majority of combs in an ordinary Langstroth hive should be about half full of honey, for outdoor wintering, in this latitude. Remember that food is needed, not only to carry them through the Winter, but also to help them to rais: brood largely, during the cold days of early Spring. Bees do not waste their stores, and the wealthy colonies w ll usually be found stronger, and better prepared for the following harvest.

Enthusiastic beginners. in Apiculture, are apt to overdo extracting (**753**), leaving too little honey in the brood-chamber for Winter. If the bees a:e not actually crowded with honey, we would advise them to leave, to strong colonies, all the honey that the brood-chamber contains. Some may think that nine or ten heavy Quinby frames, are too many for a colony, for they *may be wintered on six or seven.* We will here give a bit of our experience on that point:

625. Some 18 years ago, in an Apiary away from home, where we were raising comb-honey (**719**), we had a number of swarms. which, in the rush of the honey-crop, we did not examine until their combs were built. At that time, the

triangular bar (**319**) was the guide principally used, and
the combs of some of these swarms were joined together in
a way that rendered the frames immovable. In the Fall,
we extracted (**751**) from the brood-chamber of nearly
every colony, as was then our practice, leaving only seven
Quinby frames on an average—for Winter. The colonies,
that had crooked combs, were left with all their stores—ten
frames.—because we could not disturb them without breaking combs, and causing leakage and robbing, and it was not
the proper season to transfer (**574**) them. These colonies
did not have to be fed, the following Spring, became very
strong, and yielded the largest crop. This untried-for
result caused us to make further experiments, which proved
that *there is a profit in leaving, to strong colonies, a large
quantity of honey, so that they will not limit their Spring
breeding.*

626. The quality of the bee-food is an important matter
in wintering bees. Protracted cold weather compels them
to eat large quantities of honey, filling their intestines
with fecal matter which they cannot void, for bees never
discharge their fæces in the hive (**73**), unless they are
confined too long, or greatly disturbed.

Unhealthy food in prolonged confinement, sooner or later
causes diarrhea (**784**), not only in wintering out of doors,
but in cellar wintering (**646**), and in shipping bees long
distances (**587**).

Diarrhea, or as some call it, dysentery, in bees, is not
properly a disease, since it is only caused by the retaining
in the abdomen, of a large amount of excrements, which in
ordinary circumstances would be voided regularly.* These
excrements or fæces, from a reddish yellow to a muddy
black in color, according to the quality of the food eaten,

---

* Whenever bees have been confined for two weeks or more, they discharge
in flight excrements which soil everything about the Apiary. The house-
keeper avoids hanging clothes out to dry on such days

have an intolerably offensive smell. In excessive confinement, with a large consumption, from any cause, of more or less healthy food, when bees can no longer retain the excrements in their distended abdomen, they void them upon one another, upon the combs, upon the floor, and at the entrance of the hive, "which bees in a healthy state are particularly careful to keep clean."

If bees can void them, in flight (73), before it is too late, they experience no bad effects, hence it is indispensable, that bees wintered out of doors should be enabled to fly, at intervals, during the Winter.

627. From numerous experiments made, it is evident that *the purest saccharine matter will feed them with the least production of fœces.* Hence watery, unripe, or sour honey, and all honey containing extraneous matter, are more or less injurious to confined bees. Dark honey containing a large proportion of *mellose* is inferior to clover-honey or sugar-syrup. Honey harvested from flowers, which yield much pollen (263), is likely to contain many floating grains of it, and will be more injurious than clear, transparent honey, in cases where bees will be confined to their hives by cold for five or six weeks. Honey-dew (255) seems worse yet. The juices of fruits, apples, grapes, etc. (877), are worst of all. In the Winter of 1880–81, we purchased the remains of some 90 colonies, that had been winter-killed, and in which the only food left was apple-juice, that had been carried in, during the preceding Fall, and had turned to cider. This unwholesome food in Winter confinement, by causing diarrhea, had killed bees everywhere around us (784).

628. Happily these instances, of bees storing apple-juice, are scarce, but the practical bee-keeper will not allow such food to remain in the hive. It can be extracted (749), boiled, and fed back in Spring, for bees do not suffer from this food when not confined to their hives. The same may be said of inferior or unripe honey (**261**).

Much unsealed honey in the comb is injurious for Winter, even if the honey is ripe. This unsealed honey gathers moisture on account of its hygrometric properties, and becomes thin and watery. In addition to this peculiarity, honey, when cold, condenses the moisture or steam of the bees, in the same manner that a pitcher of cold water condenses the moisture of the air in a warm room. In some Winters, we have seen unsealed honey gather so much of the moisture of the bees that it overflowed, and ran out of the cells to the bottom-board. Luckily the bees usually consume this honey first, before Winter begins.

629. To avoid the accidents caused by poor honey, some Apiarists have suggested that all the honey might be extracted every Fall, and sugar-syrup fed in its place. This system is even carried farther by the inverting process, which (**726**) compels the bees to place all their honey in the surplus sections (**721**), leaving dry all the combs of the brood-chamber. At the first glance, this course seems profitable, when the difference between the price of comb-honey (**783**) and the cost of sugar-syrup is considered, but when we take into acccount the trouble of feeding, and the poor results obtained in wintering the bees, we see much labor for a small profit. Having ascertained that bees winter better on Spring or light-colored honey (**782**), we no longer extract from the brood-chamber, avoiding the annoyance and the extra labor of feeding. Our experience has convinced us that, unless the Spring crop has failed, or the food is decidedly bad, such as unripe honey (**249**), or honey-dew (**255**), or fruit-juice (**877**), it is cheaper to winter bees on natural stores. When sugar-syrup is needed, none but the best sugar should be used. (See Feeding. **605.**)

630. All empty combs, whether brood-combs or surplus-combs, should be removed from the hive previous to cold weather, as the bees, which may cluster in them, would starve at the first cold spell without being able to join the cluster. We have seen a whole colony perish, during a cold fortnight in December, because they had occupied an extracting story (which had but little honey in it, and had been left on by neglect), although there was plenty of honey in the hive, a few inches below them. The space, left empty by the removal of the combs, should be filled with a warm material placed between the side of the hive and the division-board.

631. As some bees which cluster on the outside combs are often unable to join the others in cold weather, it would be well to have holes, or Winter passages, through the combs, such as will allow them to pass readily, in cold weather, from one to another; but if these holes are made before they feel the need of them, they will frequently close them. It is suggested that small tubes made of *elder*, the pith of which has been removed, would make permanent Winter-passages, if inserted in the comb, at any time. On a cold November day, Mr. Langstroth found bees, in a hive without any Winter-passages, separated from the main cluster, and so chilled as not to be able to move; while, with the thermometer many degrees below zero, he repeatedly noticed, in other hives, at one of the holes made in the comb, a cluster, varying in size, ready to rush out at the slightest jar of their hive.

It has been found quite practical to give them a passage above the combs, or between the combs and the straw-mat, or quilt, above them. The Hill device is very good for this purpose, although we find that the bees often have *bridge-combs* in sufficient quantity above the frames to give them the necessary passage.

OUT-DOOR WINTERING.

632. The usual mode of allowing bees to remain all Winter on their Summer stands, is, in cold climates, very objectionable. In those parts of the country, however, where the cold is seldom so severe as to prevent them from flying, at frequent intervals, from their hives, no better way, all things considered, can be devised. In such favored regions, bees are but little removed from their native climate, and their w nts may be easily supplied, without those injurious effects which commonly result from disturbing them when the weather is so cold as to confine them to their hives.

If the colonies are to be wintered in the open air, they should all be made populous, and rich in stores, even if to do so requires their number to be reduced one-half or more. The bee-keeper who has ten strong colonies in the Spring, will, by judicious management with movable-frame hives, be able to close the season with a larger Apiary than one who begins it with thirty, or more, feeble ones.

632 (*bis*). Small colonies consume, proportionally, much more food than large ones, and then perish from inability to maintain sufficient heat.

Bees, in small or contracted hives, especially when deprived of all the honey gathered in Spring, as stated before (**629**), have too scanty a population for a successful wintering, especially out of doors; for, as it is by eating that bees generate warmth, the abdomens of a small number are soon filled with residues, and if the cold continues for weeks the bees get the diarrhea (**784**). We have often seen colonies in small hives perishing side by side with large ones whose bees were very healthy.

Such facts abound, and we have but to open the bee-journals to find the confirmation of our statement,

In the *American Bee-Journal* for February 8, 1888, page 83, Mr. J. P. Stone of Holly, Mich., asks why a colony, which was hived in 1859 in a large box, is prospering yet, while others have perished. The size given, $16 \times 16 \times 22$, which shows that the box has twice the capacity of an 8-frame Langstroth hive, answers his question.

In the following number of the same journal, page 107, Mr. Heddon mentions a colony which had wintered safely for seven years in a box ten times larger than the Langstroth, while many others died by its side. "The colony, when transferred, contained about double the number of bees usually raised from one queen."

Fig. 104.
COMMON HIVES PROTECTED BY STRAW.
(From Hamet.)

Yet small colonies can sometimes be safely wintered, if their combs and honey are not spread over a large space, and if they are sheltered so as to maintain the proper heat. It is therefore indispensable to reduce the combs of a hive to the amount of room which the bees can best keep warm, by the use of the division or contracting board (**349**), without forgetting to leave a sufficient supply of good honey.

UNITING.

633. A queenless colony, in the Fall, should always be united to some other hive.

If two or more colonies, which are to be united in the Fall, are not close together, their hives must be gradually drawn nearer, and the bees may then, with proper precautions, be put into the same hive. For this purpose, it is well to kill the poorest queen (if both have queens) and keep the best. This may be dispensed with, but the prudent bee-keeper will never neglect an opportunity to improve his stock. On a cool November day, the combs of the weakest colony that bear the cluster, should be lifted all together, and inserted in the other hive, after the bees of the latter have been thoroughly frightened with smoke. (**382.**)

634. If, when two colonies are put together, the bees in the one on the old stand are not gorged with honey, they will often attack the others, and speedily sting them to death, in spite of all their attempts to purchase immunity, by offering their honey. The late Wm. W. Cary, of Coleraine, Massachusetts, who has long been an accurate observer of the habits of bees, united colonies very successfully, by alarming those that were on the old stand; as soon as they showed by their notes, that they were subdued, he gave them the new-comers. The alarm which causes them to gorge themselves with honey, puts them, doubtless, upon their good behavior, long enough to give the others a fair chance.

They can also be made to unite peaceably, by sprinkling a little sweet-scented water on them (**485**). It is well to put a slanting board in front of the entrance (**603** *bis*) to show the moved bees that their location is changed. The empty **hive** should be removed from its place to prevent the bees from returning to it. The number of combs in the united colony can be reduced as soon as the bees have all clustered together.

In this manner a strong colony with little honey, and a weak one with plenty of stores, can be united to form a good hive of bees.

OUT–DOOR SHELTERING.

635. The moving of a colony to a warmer or better sheltered place, just before Winter, is not advisable, for a great many bees, not having noticed their new location, would perish of cold, while searching for their home, and the population would be greatly decreased.

In our Northern, Middle and Western States, the style of hive used has a considerable influence on the safety of out-door wintering.

With hives that are single-walled all around, great care should be taken to shelter the bees from the piercing winds, which in Winter so powerfully exhaust their animal heat; for, like human beings, if sheltered from the wind, they will endure a low temperature far better than a continuous current of very much warmer air.

In some parts of the West, where bees suffer much from cold winds, their hives are protected, in Winter, by sheaves of straw, fastened so as to defend them from both cold and wet. With a little ingenuity, farmers might easily turn their waste straw to a valuable account in sheltering their bees.

Not only can straw be used for this purpose with much service, but also forest leaves, corn fodder, and rushes. Snow is found to be a very good shelter, provided its successive melting and freezing does not interfere with the necessary ventilation. It must be removed from the entrance on the approach of a warm day.

Mr. Geo. H. Beard, of Winchester, Mo., safely wintered ninety-three colonies out of ninety-six, in the severe Winter of 1884-5, in two-story Simplicity hives, (**324**) by removing the oil-cloth and replacing it with coarse sack-cloth, filling the upper story with maple leaves, and covering the hives, on all sides, except the front, with what is commonly known as slough-grass. This success is worthy of notice, for in that memorable Winter, more than two-thirds of the bees in the Northern States died, some Apiarists losing all they had. Like that of 1855-6, it will long be remembered, not only for the uncommon degree and duration of its cold, but for the tremendous winds, which, often for days together, swept like a Polar blast over the land.

We have, for years, wintered part of our bees on the Summer stand, by sheltering them on all sides but the front, with forest leaves closely packed, and held with a framework of lath.

636. One of the most important requirements for successful out-door wintering, is the placing of warm absorbents, immediately over the cluster, to imbibe the excess of moisture that rises from the bees, without allowing the heat to escape.

In March, 1856, we lost some of our best colonies, under the following circumstances : The Winter had been intensely cold, and the hives, having no upward ventilation, were filled with frost, — in some instances, the ice on their glass sides being nearly a quarter of an inch thick. A few days of mild weather, in which the frost began to thaw, were followed by a severely cold spell with the thermometer below zero, accompanied by raging winds, and in many of the hives, the bees, which were still wet from the thaw, were *frozen together in an almost solid mass.*

As long as the vapor remains congealed, it can injure the bees only by keeping them from stores which they need ; but, as soon as a thaw sets in, hives which have no upward ventilation are in danger of being ruined.

22

Mr. E. T. Sturtevant, of East Cleveland, Ohio, widely known as an experienced Apiarist, thus gives his experience in wintering bees in the open air:

"No extremity of cold that we ever have in this climate, will injure bees, if their breath is allowed to pass off, so that they are dry. I never lost a good colony that was dry, and had plenty of honey."

The absorbents generally used are chaff in cushions, straw, forest leaves (maple leaves preferred), corn cobs, woolen rags, or wool waste, etc. Mr. Cheshire uses cork-dust, which he claims gives fourteen times as much protection as a dead-air space. The oil-cloth, which makes an air-tight covering, must be first removed, and if no straw-mat is used, the cushion of absorbents may be placed right over the frames. We use the straw-mat, and fill the upper half-story with dry leaves, these being the cheapest and best absorbent at our command.

In the coldest parts of our country, *if upward absorbents are neglected*, no amount of protection that can be given to hives, in the open air, will prevent them from becoming damp and mouldy, *even if frost is excluded*, unless a large amount of lower ventilation is given. Then they need as much air as in Summer. Often, the more they are protected, the greater the risk from dampness. A very thin hive *unpainted*, so that it may readily absorb the heat of the sun, will dry inside much sooner than one painted white, and in every way most thoroughly protected against the cold. The first, like a *garret*, will suffer from dampness for a short time only; while the other, like a *cellar*, may be so long in drying, as to injure, if not destroy, the bees.

637. *If the colonies are wintered in the open air, the entrance to their hives must be large enough to allow the bees to fly at will.* Many, it is true, will be lost, but a large part of these are diseased; and, even if they were not, it is better to lose some healthy bees than to incur the risk of losing, or greatly injuring, a whole colony by the excitement created by confining them when the weather is warm enough to entice them abroad.

If the sun is warm and the ground covered with new-fallen snow, the light may so blind the bees, that they will fall into this fleecy snow, and quickly perish. Even at such times, it is hardly advisable to confine them to their hives. A neighbor of ours killed four colonies, all he had, by

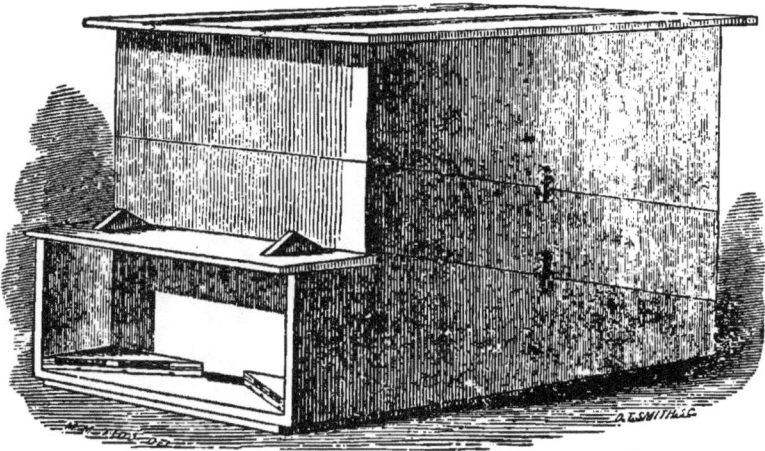

Fig. 105.

TWO-STORY DOUBLE-WALLED LANGSTROTH HIVE, OLD STYLE.

closing the entrances with wire-cloth for Winter. We had advised him to remove it, but he did not do so because some one had told him that his bees would get lost in the snow.

638. Great injury is often done by disturbing a colony of bees when the weather is so cold that they cannot fly. Many that are tempted to leave the cluster, perish before they can regain it, and every disturbance, by rousing them to needless activity, causes an increased consumption of food. On the other hand, it is of the utmost importance that they be allowed to fly and void their excrements (**73**) whenever the weather is warm enough. At such times it will be advisable to clean the bottom-boards of hives, of dead bees, and other refuse.

639. To show the advantages derived by the bees from a Winter flight, we will give our experience during one of

Fig. 106. INSIDE VIEW OF TWO-STORY DOUBLE-WALL LANGSTROTH HIVE.
Old style.

a, b, c, double bottom-board   *d*. stationary outer-case. *f*, portico   *g*, entrance thro double wall. *h, i*, front and back of lower hive   *j z*, rabetted pieces *l*, lower honey-bo *m*, lower part of cover. *o, q*, cover. *r*, upper honey-board. *u, u, t*, frames. *w*, front rear of upper story.

the coldest Winters, that of 1872–3. From the beginning of December to the middle of January, the weather was cold and the bees were unable to leave the hive. The 16th

of January was a rather pleasant day. We took occasion
of this to examine our weak colonies, being anxious in
regard to their condition. To our astonishment, they were
found alive, and our disturbing them caused them to fly
and discharge their excrements. Being convinced that all
our bees were safe, we did not disturb the strong colonies,
and a few of the latter remained quiet. The next day, the
cold weather returned, and lasted three weeks longer. Then
we discovered that the weak colonies, that had had a cleansing flight, were alive and well, while the strong ones which
had remained confined, were either dead or in bad condition.

640. In order to shelter bees more efficiently. in out-
door wintering, against climatic influences, Apiarists have
devised hives, with double walls, filled at the sides, as well
as on top, with some light material non-conductor of heat.
Some are made on the same principle as the old two-story
double-wall L. hive (fig. 106) without packing.

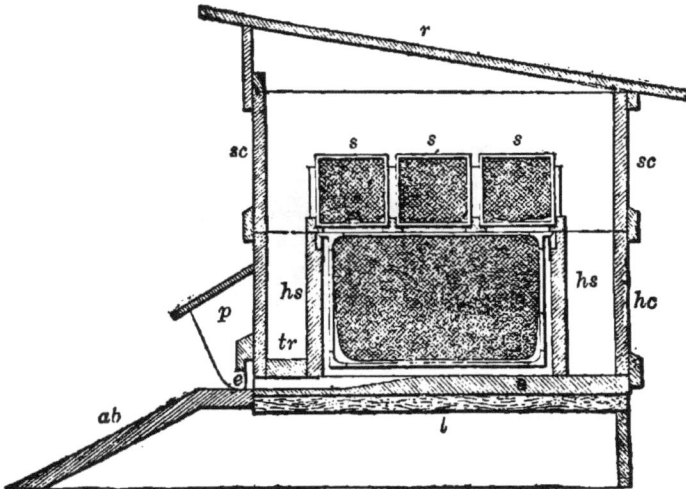

Fig 107, (From Cheshire.)
DOUBLE-WALL COWAN HIVE.

ab, apron-board. *e*, entrance. *p*. portico *hs*. hollow space *tr*, tun-
nel-roof or cover to entrance. *hc*, hive case. *sc*, surplus case. *r*, roof.

The most wide-spread style, is the chaff-hive, of A. I.
Root. This hive is far superior to single-wall hives for out-
door wintering. It is made in two stories, but all in one
piece. This renders it rather inconvenient to reach down
to the lower story, when handling bees. We, therefore,
made our chaff hives of a single story with half-story cap,
like that of fig. 69. This single-wall cap can be filled
with a cushion, dry leaves, or any other absorbents. Some
Apiarists also use *one-story chaff-hives* with loose bottom-
boards that can be taken off to remove the dead bees in
Spring.

641. After having used some eighty chaff-hives during
six or eight years, we find two disadvantages in them:
1st. They are heavy and inconvenient to handle, especially
when made to accommodate ten Quinby, or twenty Simplic-

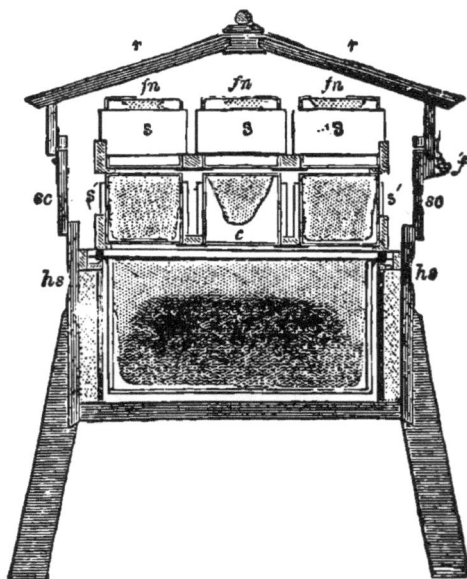

Fig. 108
INSIDE OF THE CHESHIRE HIVE.
hs, hives sides with cork-dust for packing. *sc*, section case. *'s*, section.
s, separators *fn*, foundation.

ity frames. *2d.* As they do not allow the heat or cold to pass in and out readily, the bees in these hives may remain in-doors, in occasional warm Winter days, while those of thin-front hives will have a cleansing flight. Thus, in hard Winters, these bees suffer as much from diarrhea (**626–784**) as others, unless the Apiarist takes pains to disturb them and *make them fly*, when necessary.

Fig. 109.

OUTER COVERING.

As used by J. G. Norton and others. One side is removed to show the hive within.

642. But we highly recommend the use of these hives, to the bee-keepers who do not wish to go to the trouble of sheltering their bees every Winter. With the chaff-hive, it is a matter of only a few minutes to put into Winter-quarters a colony, that has sufficient stores and bees. As to the advantage, claimed for these hives, of keeping weak colonies warm, in the Spring, we found it counterbalanced by the loss of the sun's heat during the first warm days,

and we found that bees bred as fast, in our ordinary
hives (double only on the windward sides), owing to the
quick absorption of the sun's rays by the boards.

643. To obtain the advantages of the chaff-hive without
any of its disadvantages, and at the same time retain in use
the single-wall Langstroth or Simplicity hives, some bee-
keepers have devised outer-boxes to be placed over the colonies during Winter, and removed in Spring. These can be
filled with absorbents, and make the best and safest out-
door shelters (Fig. 109). They are only hooked together
by nails partly driven, and are taken off in pieces, in the
Spring and put away, under shelter. The roofs may be
used over the hives all Summer, if desirable. The only
disadvantage of outer-boxes is that they may harbor mice
or insects. Some use them, without any packing, and we
know by experience, that even in this way, very small colonies may be wintered safely. If the hive has a portico, the
front of the box is made to fit around it. In any case, the
portico itself can be closed, during the coldest weather, by
a door fitting over it, but it must be opened on warm days.
In the extraordinary Winter of 1884–5, several bee-keepers
of McDonough County, Illinois, among whom, we will cite
Mr. J. G. Norton, of Macomb, safely **wintered** their Simplicity hives with this method, while their neighbors lost
all, or nearly all, their bees.

644. *If the colonies are strong in numbers and stores, have
upper moisture absorbents, easy communication from comb to
comb, good ripe honey, shelter from piercing winds, and can
have a cleansing flight once a month, they have all the conditions essential to wintering successfully in the open air.*

In-door Wintering.

645. In some parts of Europe, it is customary to winter all the bees of a village in a common vault or cellar. Dzierzon says:

"A *dry* cellar is very well adapted for wintering bees, even though it is not wholly secure from frost; the temperature will be much milder, and more uniform than in the open air; the bees will be more secure from disturbance, and will be protected from the piercing cold winds, which cause more injury than the greatest degree of cold when the air is calm.

" Universal experience teaches that the more effectually bees are protected from disturbance and from the variations of temperature, the better will they pass the Winter, the less will they consume of their stores, and the more vigorous and numerous will they be in the Spring. I have, therefore, constructed a special Winter repository for my bees, near my Apiary. It is weather-boarded both outside and within, and the intervening space is filled with hay or tan, etc.; the ground and plat enclosed is dug out to the depth of three or four feet, so as to secure a more moderate and equitable temperature. When my hives are placed in this depository, and the door locked, the darkness, uniform temperature, and entire repose the bees enjoy, enable them to pass the Winter securely. I usually place here my weaker colonies, and those whose hives are not made of the warmest materials, and they always do well. If such a structure is to be partly underground, a very dry site must be selected for it."

In Russia, bee-keepers dig a well from twenty to twenty-five feet deep, and six or eight feet wide. The hives, which, *there*, are hollow trees, are then piled horizontally upon one another, *like cord-wood*, with one end open. The well is filled to within six feet of the top, and a shed, made of straw, is built above. The bees are left there during the five or six months of Winter.

In some other countries, they are kept in caves, abandoned mines, or any under-ground place near at hand.

646. In the North of the United States, and in Canada, they are generally wintered in cellars, and remain there in quiet from November till April, sometimes till May.

In all localities, where the bees cannot fly at least once a month, in the Winter, it is best to follow this method of wintering.

As Dzierzon says, a *dry* cellar is the best, although bees can be wintered in a *damp cellar*, but with more danger of loss, especially if the food is not of the best. The honey of Northern countries is generally of finer quality than that of the South.

647. In the first place, the bees should be moved to the cellar, just after they have had a day's flight, at the opening of cold weather. We take only the brood-apartment leaving the cap, and sometimes the bottom-board, on the Summer stand, being careful to mark the number of each hive inside of its cap* so as to return it to the same location in Spring (**32-33**). In the cellar, the hives are piled one upon another. An empty hive or a box is put at the bottom of each pile, so that the bees will be as high up from the damp ground as possible. If the bottom-board is brought in with the hive, the entrance should be left open. It is well to raise the lower tier of hives from their bottoms with entrance-blocks. Some upper ventilation had better be given also, for the escape of moisture. If the cellar is damp, the combs will mould more or less; if it is dry, they will keep in perfect order.

648. After the bees are put in, they should be left in darkness, at the temperature that will keep them the quietest. We find that from 42 ° to 45 ° is the best. Every Apiarist should have a thermometer, and use it. The cost is insignificant, and it will pay for itself many times.

The fact that bees, in Russia (**645**), are confined in

---

*In a well-regulated Apiary, each hive bears a number painted on the body.

deep wells, for six months, shows that a total deprivation of light cannot be injurious. It prevents them from flying out of their hives, to which they would be unable to return, after flying to the windows, allured by the light, when the temperature of the cellar rises occasionally and unexpectedly to 50 or 60 degrees.

As bees, wintered on their Summer stands, begin to fly out when the temperature reaches about 50 degrees, and are in full flight at about 55, one can imagine how restless they become when the temperature of the cellar rises to 55 or 60 degrees. They wait impatiently for the dawn of the day which will afford them the opportunity for flying out. But as the days pass and darkness continues they are uneasy and tired.

Fig. 110.
CELLAR BLIND, TO GIVE AIR WITHOUT LIGHT.

The warmth incites them also to breed, and as they need water for their brood (271), some leave the hive in quest of it and are lost. This happens more or less every Winter.

To cool the air of the cellar, ice may be brought in and allowed to melt slowly over a tub.

The Apiarist must guard against cold, also, but in wintering a large number of colonies, the heat which they generate will usually keep the cellar quite warm in the coldest weather. In our experience, we have had to keep the cellar windows open, often, in cold weather.

649. To allow cold air to enter without giving light, we have devised cellar blinds (figs. 110 and 111). When the

Fig. 111.
CELLAR BLIND IN PLACE.

window, inside, is raised, a wire-cloth frame is put in its place to keep mice out, and there is a slide on the inside of the shutter which can be used to give more or less air as the case requires. Besides, the windows of our bee-cellar are made with double panes, to exclude cold or heat more efficiently, when they are shut. A slight quantity of pure air is needed at all times.

As we have said above, when the warmer days of Spring come, with alternates of cold, the bees will breed a little, and if this is not begun too early, it will be a help to them rather than an injury, for they will become strong, all the sooner, after being taken out.

650. A small number of colonies can be wintered in any ordinary cellar, quite safely, when their food is of good quali:y, and the temperature does not vary too much, but they *must* be *quiet* and *in the dark.*

651. If the temperature of the cellar is too low, or too high, or if the food is unhealthy, the bees will have a large amount of fecal accumulation in their intestines, and will show their anxiety by coming out of the hive in clusters, during the latter part of their confinement. If, in addition to this, the cellar is damp, the comb will mould; and when taken out, some colonies *may* desert (**407, 663**) their hives.

652. Great loss may be incurred in replacing, upon their Summer stands, the colonies which have been kept in special depositories. Unless the day when they are put out is very favorable, many will be lost when they fly to discharge their fæces. In movable-frame hives, this risk can be greatly diminished, by removing the cover from the frames, and allowing the sun to shine directly upon the bees; this will warm them up so quickly, that they will all discharge their fæces in a very short time.*

---

* The following is an extract from Mr. Langstroth's journal:

. '' Jan. 31st, 1857.—Removed the upper cover, exposing the bees to the full heat of the sun, the thermometer being 30° in the shade, and the atmosphere *calm.* The hive standing on the sunny side of the house, the bees quickly took wing and discharged their fæces. Very few were lost on the snow, and nearly all that alighted on it took wing without being chilled More bees were lost from other hives which were not opened, as few which left were able to return: while, in the one with the cover removed, the returning bees were able to alight at once among their warm companions.''

653. If more than one hundred colonies are wintered in the cellar, and it is desired to remove them all the same day, enough help should be secured to put them all on their stands before the warm part of the day is over. It is far better to keep them in the cellar even one week longer, than to take them out when the weather is so cold that they cannot cleanse themselves immediately; to our mind, 45 ° in the shade, or 55 ° in the sun, is the lowest temperature in which it is best to put bees out.

654. As bees remember their location, it is important to return each colony to its own place. If this is not done, the confusion may cause some colonies to abandon their hives. Dzierzon also advises placing them on their former stands, as many bees still remember the old spot. If it is desirable to remove some hives to a new location, a slanting board (**603** *bis*) should be placed in front of the hive. All the bottom-boards should be cleaned of dead bees or rubbish, without delay.

655. If the hives of an Apiary are all removed from the cellar on the same day, there will be but little danger of robbing, for they are somewhat bewildered when first brought out; but if some are taken out later than others, the last removed will be in danger, unless some precautions are taken.

656. If the bees that are wintering in the cellar, are found to be restless, it may be good policy to give them some water (**271**), or to take them out on a warm day when the temperature is at least 45 ° in the shade, to let them have a flight, and return them to the cellar afterward. We do not advise it as a practice however. On the contrary, if they are quiet, it is better to keep them indoors, till the early Spring days have fairly come, to avoid what is called Spring-dwindling (**659**).

657. Those, who have no cellar, can successfully winter their bees in clamps or silos as advised by the Rev. Mr,

Scholz, of Lower Silesia, widely known in Germany for his skill in bee-keeping. These clamps are made similar to

Fig. 112. (From L'Apicoltore, of Milan.)
BEE CLAMP FOR WINTERING.
l, air draft. *d*, roof.

those in which farmers place apples, potatoes, turnips, etc., to preserve them during cold weather. The only objection to

Fig. 113.
HOW TO PILE THE HIVES.

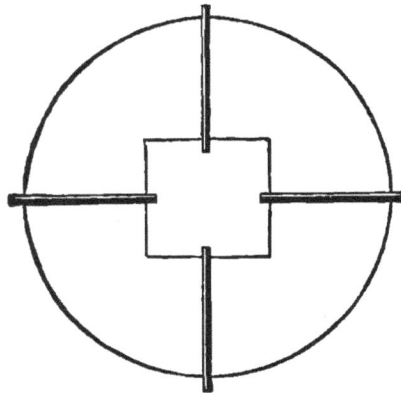

Fig. 114.
GROUND PLAN OF A BEE CLAMP.

this mode, is the dampness of the ground in wet and warm Winters. The hives are put, on a bed of straw, in a pyramidal form (fig. 113), and covered, first with old boards, then with a thick layer of straw, and another, of earth. Wooden pipes are placed at the bottom (fig. 114), and one in the shape of a chimney, at the top, for an air-draft. The requisites are the same as in cellar wintering, an equal temperature, sufficient ventilation, a fairly dry atmosphere, and quiet.

658. We must warn novices against the wintering of bees in any repository in which the temperature descends below the freezing point. In such places the bees consume a great deal of honey, and they soon become restless, for want of a flight. Their Summer stand even without shelter, is far safer than any such place, because they can at least take advantage of any warm Winter day to void their excrements. These facts are demonstrated beyond a doubt.

Spring Dwindling.

659. When the conditions necessary to the successful wintering of bees are not complied with, and they have suffered from diarrhea (**784**), many colonies may be lost by Spring dwindling, especially if the Spring is cold and backward. Even colonies, which appeared to have gone through the Winter strong in numbers, may slowly lose bee after bee till the queen alone remains in the hive. This is sometimes mistaken for desertion (**407**), as will be seen in the following paragraph, which we quote from *The London Quarterly Review*, and in which the author attributes to lack of loyalty in the bees, that which evidently must have been due only to Spring dwindling:

"Bees, like men, have their different dispositions, so that even their loyalty will sometimes fail them. An instance not long ago came to our knowledge, which probably few bee-keepers will credit. It is that of a hive which, having early exhausted its store, was found, on being examined one morning, to be utterly deserted. The comb was empty, and the only symptom of life was the poor queen herself, 'unfriended, melancholy, slow,' crawling over the honeyless cells, a sad spectacle of the fall of bee-greatness. Marius among the ruins of Carthage—Napoleon at Fontainebleau—was nothing to this."

Several such instances, caused by Spring dwindling, with subsequent robbing of the honey, were observed by us. Colonies are thus destroyed as late as April and May.

660. In some instances, the enlarged abdomen of the bees will show that they are suffering from constipation—or inability to discharge their fæces, even though they may have voided their abdomen since their long confinement. Probably their intestines are in an unhealthy condition. In the worst cases of Spring dwindling, sometimes, even the queens show signs of failing, and eventually disappear. This may occur also with colonies that were wintered in the cellar, if they have suffered from diarrhea, or have been removed too early.

There is another sort of Spring dwindling caused by the loss of working bees in cold Springs, while in search of water (**271**), or pollen (**263**), for the brood.

661. To avoid losses, or to check them as far as possible, after a hard Winter, it is indispensable that the following be observed :

1st. The hives should be located in a warm, sunny, well-sheltered place. All Apiaries that are placed in exposed windy situations, or facing North, suffer most from Spring dwindling.

2d. The number of combs in the hive should be reduced in early Spring, with the division-board or contractor, to suit the size of the cluster (**349**). This helps the bees to keep warm and raise brood. The space must again be enlarged gradually, when the colony begins to recruit.

We consider this contraction of the hive as altogether indispensable. Let us suppose that, in early Spring, we

23

have a colony whose population is so much reduced that it
cannot warm, to the degree needed for breeding, more than
500 cubic inches of space.   If we leave the brood-chamber
without contraction, as its surface, in a 10-frame Langstroth
hive, will be about 270 square inches, the cubic space
heated will have about two inches in thickness at the top,
since heat always rises.   If, on the contrary, we have
reduced the number of frames to three, the depth of the
space warmed at the top will amount to more than three
times as much, or to more than six inches.   Thus, the
bees will not only be more healthy, but the laying of the
queen, not being delayed by the cold, and the number of
the bees increasing faster, they will be able to repay the
bee-keeper for the care bestowed, instead of dwindling, or
remaining worthless for the Spring crop.

3d. The heat should be concentrated in the brood apartment, by all means, and not allowed to escape above.   The
entrance also must remain reduced.

4th. The bees should be provided with sufficient stores
of honey, pollen, and water.

662. Apiarists in general, do not attach enough importance to the necessity of furnishing water (**271**) to bees in
cold Springs, in order that they may stay at home in quiet.
Although Berlepsch laid too much stress on the question of
water, the lack of which he even said was the cause of dysentery, yet he was right in calling our attention to the need
of it for breeding:

" The Creator has given the bee an instinct to store up honey
and pollen, which are not always to be procured, but not water,
which is always accessible in her native regions.   In Northern
latitudes, when confined to the hive, often for months together,
they can obtain the water they need only from the watery particles contained in the honey, the perspiration which condenses
on the colder parts of the hive, or the humidity of the air which
enters their hives.

" In March and April, the rapidly-increasing amount of brood causes an increased demand for water; and when the thermometer is as low as 45°, bees may be seen carrying it in at noon, even on windy days, although many are sure to perish from cold. In these months, in 1856, during a protracted period of unfavorable weather we gave all our bees water, *and they remained at home in quiet*, whilst those of other Apiaries *were flying briskly in search of water*. At the beginning of May, our hives *were crowded with bees*; whilst the colonies of our neigbors *were mostly weak*.

"The consumption of water in March and April, in a populous colony, is very great, and in 1856, one hundred colonies required eleven Berlin quarts per week, *to keep on breeding uninterruptedly*. In Springs where the bees can fly safely almost every day, the want of water will not be felt.

"The loss of bees by *water-dearth*, is the result of climate, and no form of hive, or mode of wintering, can furnish an absolutely efficient security against it."—(Translated from the German, by S. Wagner.)

That bees cannot raise much brood without water, unless they have fresh-gathered honey, has been known from the times of Aristotle. Buera of Athens (Cotton, p. 104), aged 80 years, said in 1797:

"Bees daily supply the worms with water; should the state of the weather be such as to prevent the bees from fetching water for a few days, the worms would perish. These dead bees are removed out of the hive by the working-bees if they are healthy and strong; otherwise, the stock perishes from their putrid exhalations."

In any movable-frame hives, water can be given to the bees, by pouring it into the empty cells of a comb.

DESERTING.

663. We have shown (**407**) that bees sometimes desert their hives, when the colony is too weak, or short of stores, or suffering from dampness, mouldy combs, etc., etc. This desertion, which differs from natural swarming in

this, that it may take place in any season, and that the
deserting bees do not raise any queen-cells previously, is
more frequent in cold backward Springs than at any other
time.

At different times we have seen bees deserting their
hives and forsaking their brood for lack of pollen (**264**).
A comb containing pollen having been put in their hive
and the bees returned they remained happy. But the
worst of these desertions is when the bees have suffered
while wintered in-doors (**651.**) These colonies abandon
their hives very soon after being replaced on their Summer stands. When such desertion is feared, it is better
not to put out more than one dozen colonies at one time,
and to prepare a few dry combs, in clean hives, to hive
the swarm as soon as possible; for, too often some other
colonies following the example, mix with the first, the
queens are balled (**538**), causing great annoyance and
loss to the bee-keeper. Such swarms should be hived on
clean dry comb, and furnished with honey and pollen.
The capacity of the hive in which they are put should be
reduced to suit the size of the swarm, and increased very
cautiously, from time to time, when the bees seem to be
crowded; for warmth is indispensable to bees in Spring.
The condition of such colonies must be regularly ascertained and their wants supplied.

We would refer those who think that "*it is too much
trouble*" to examine their hives in the Spring, to the practice of the ancient bee-keepers, as set forth by Columella:
"The hives should be opened in the Spring, that all the
filth which was gathered in them during the Winter may be
removed. Spiders, which spoil their combs, and the worms
from which the moths proceed, must be killed. When the
hive has been thus cleaned, the bees will apply themselves
to work with the greater diligence and resolution." The
sooner those abandon bee-keeping, who consider the proper

care of their bees as "too much trouble," the better for themselves and their unfortunate bees.

In making this thorough cleansing, the Apiarist will learn which colonies require aid, and which can lend a helping hand to others; and any hive needing repairs, may be put in order before being used again. Such hives, if occasionally re-painted, will last for generations, and prove cheaper, in the long run, than any other kind.

CHAPTER XIV.

ROBBING, AND HOW PREVENTED.

An ounce of prevention is worth a ton of cure.

664. BEES are so prone to rob each other, in time of scarcity, that, unless great precautions are used, the Apiarist will often lose some of his most promising colonies. Idleness is, with them, as with men, a fruitful mother of mischief. They are, however, far more excusable than the lazy rogues of the human family; for they seldom attempt to live on stolen sweets, when they can procure a sufficiency by honest industry.

As soon as they can leave their hives in the Spring, they may begin to assail the weaker colonies. In this matter, the morals of our little friends seem to be sadly at fault; for, those colonies which have the largest surplus are—like some rich oppressors—the most anxious to prey upon the meagre possessions of others.

If the marauders, who are ever prowling about in search of plunder, attack a strong and healthy colony, they are usually glad to escape with their lives from its resolute defenders. The bee-keeper, therefore, who neglects to watch his needy colonies, and to assist such as are weak or queenless, must count upon suffering heavy losses from robber-bees.

665. It is sometimes difficult, for the novice, to discriminate between the honest inhabitants of a hive, and the robbers which often mingle with them. There is, however, an air of roguery about a thieving bee which, to the expert, is as characteristic as are the motions of a pickpocket to a skillful policeman. Its sneaking look, and nervous, guilty agitation, once seen, can never be mistaken. It does not, like the laborer carrying home the fruits of honest toil, alight boldly upon the entrance-board, or face the guards, knowing well that, if caught by these trusty guardians, its life would hardly be worth insuring. If it can glide by without touching any of the sentinels, those within—taking it for granted that all is right—usually permit it to help itself.

Bees which lose their way, and alight upon a strange hive, can readily be distinguished from these thieving scamps. The rogue, when caught, strives to pull away from his executioners, while the bewildered unfortunate shrinks into the smallest compass, submitting to any fate his captors may award.

These dishonest bees are the *"Jerry Sneaks"* of their profession, and after following it for a time, lose all taste for honest pursuits. Constantly creeping through small holes, and daubing themselves with honey, their plumes assume a smooth and almost black appearance, just as the hat and garments of a thievish loafer, acquire a "seedy" aspect.

Dzierzon thinks that these black bees, which Huber has described as so bitterly persecuted by the rest, are nothing more than thieves. Aristotle speaks of " a *black* bee which is called a *thief.*"

Some bee-keepers question whether a bee that once learns to steal ever returns to honest courses. The writer has known the value of an Apiary to be so seriously impaired by the bees beginning early in the season to rob each other, that the owner was often tempted to wish that he had never seen a bee.

666. Yet, we should hardly blame them for their robbing propensities. With them, as with men, much depends on the education which they are allowed to receive. Their nature teaches them to hunt for sweets industriously, wherever they can find them, and any sweet, which they can reach, by the most strenuous efforts, is considered by them, at once, as their private property. Were it not for this disposition of the bee, to hunt for sweets everywhere, and take them home, the honey of those colonies that dwell in the woods, and frequently perish during the Winter, would be wasted. The propensity to rob is acquired only during a dearth of honey in the flowers; for bees have a much greater relish for fresh honey, as produced in the blossoms, than for any other sweet on earth. This is so true, that in a day of abundant harvest, honey may be left exposed where bees can reach it, without being touched, or even approached, by a single bee, for hours; while, if placed in the very same spot during a dearth of honey, it will be covered with bees in very few minutes.

If the bee-keeper would not have his bees so demoralized that their value will be seriously diminished, he will be *exceedingly careful* in time of scarcity to prevent them from robbing each other. If the bees of a strong colony once get a taste of forbidden sweets, they will seldom stop until they have tested the strength of every hive. Even if all the colonies are able to defend themselves, many bees will be lost in these encounters, and much time wasted; for bees, whether engaged in robbing, or battling against robbers, lose both the disposition and the ability to engage in useful labors.

667. An experienced bee-keeper readily perceives when any robbing is going on in his Apiary. Bees are flying vagrantly about, hunting in nooks and corners, and at all the hive-crevices. Extensive robbing causes a general uproar, and the bees of all the hives are much more disposed to sting. The robbers sally out with the first peep of light, and often continue their depredations until it is so late that they cannot find the entrance to their hive. Some even pass the night in the plundered colony.

The cloud of robbers arriving and departing need never be mistaken for honest laborers (**174**) carrying, with unwieldy flight, their heavy burdens to the hive. These bold plunderers, as they *enter* a hive, are almost as hungry-looking as Pharaoh's lean kine, while, on *coming out*, they show by their burly looks that, like aldermen who have dined at the expense of the city, they are stuffed to their utmost capacity.

668. When robbing-bees have fairly overcome a colony, the attempt to stop them—by shutting up the hive, or by moving it to a new stand—if improperly conducted, is often far more disastrous than allowing them to finish their work. The air will be quickly filled with greedy bees, who, unable to bear their disappointment, will assail, with almost frantic desperation, some of the adjoining hives. In this way, the strongest colonies are sometimes overpowered, or thousands of bees slain in the desperate contest.

How to Stop Robbing.

When an Apiarist perceives that a colony is being robbed, he should contract the entrance (**339**), and, if the assailants persist in forcing their way in, he must close it entirely. In a few minutes the hive will be black with the greedy cormorants, who will not abandon it till they have attempted to squeeze themselves through the smallest openings. Before they assail a neighboring colony, they should be thoroughly sprinkled with cold water, which will somewhat cool their ardor.

Unless the bees, that were shut up, can have an abundance of air, they should be carried to a cool, dark place, after the Apiarist has allowed the robbers to escape out of it. Early the next morning they must be examined, and, if necessary, united to another hive.

" In Germany, when colonies in common hives are being robbed, they are often removed to a distant location, or put in a dark cellar. A hive, similar in appearance, is placed on their stand, and leaves of wormwood and the expressed juice of the plant are put on the bottom-board. Bees have such an antipathy to the odor of this plant, that the robbers speedily forsake the place, and the assailed colony may then be brought back.

" The Rev. Mr. Kleine says, that robbers may be repelled by imparting to the hive some intensely powerful and unaccustomed odor. He effects this the most readily by placing in it, in the evening, a small portion of *musk*, and on the following morning the bees, if they have a healthy queen, will boldly meet their assailants. These are nonplussed by the unwonted odor. and, if any of them enter the hive and carry off some of the coveted booty. on their return home, having a strange smell, they will be killed by their own household. The robbing is thus soon brought to a close."—S. WAGNER.

It will often be found that a hive which is overpowered by robbers has no queen, or one that is diseased.

669. One of the best methods which we have found to stop the robbing of one hive by another, when the robbed colony is worth saving, is to exchange them; *i. e.* to place the robbed colony on the stand of the robbing colony, and *vice versa*. The robbing colony can usually be found by sprinkling the returning bees with flour, as they come out of the robbed hive, and watching the direction which they take. It can also often be detected by the activity of its bees, if the neighboring hives are idle, especially after sunset.

This method, however, cannot be practiced when the robbing and the robbed colonies do not belong to the same person; or when the robbing is carried on by many hives at one time, although, in the latter case. the exchange of stands between the strongest of the robbing hives and the weak robbed colony, in the evening, and the reducing of the entrances of both, usually has a good result. The old robber bees, bewildered by this exchange, make their home in the robbed colony, since they find it on the stand where they are accustomed to bring their honey; and they defend it with as much energy as they used in attacking it before. See Quinby's "Mysteries of Bee-Keeping" N. Y., 1866.

670. We read in the *British Bee-Journal* that a carbolized sheet **(384)** can be used to stop robbing, if spread in front of the robbed hive. This same sheet, spread on the hive as soon as opened while extracting **(749)**, and on the surplus box where the combs are placed **(768)**, displeases the robbers and protects the comb,

671. There is a kind of pillage which is carried on so secretly as often to escape all notice. The bees engaged in it do not enter in large numbers, no fighting is visible, and the labors of the hive appear to be progressing with their usual quietness. All the while, however, strange bees are carrying off the honey as fast as it is gathered. After watching such a colony for some days, it occurred to us one evening, as it had an unhatched queen, to give it a fertile one. On the next morning, rising before the rogues were up, we had the pleasure of seeing them meet with such a warm reception, that they were glad to make a speedy retreat.

This is another proof that discouragement caused by queenlessness often leads to the loss of a colony.

PREVENTION.

672. *If the Apiarist would guard his bees against dishonest courses, he must be exceedingly careful, in his various operations, not to leave any combs or any honey where bees can find them, for, after once getting a taste of stolen honey, they will hover around him as soon as they see him operating on a hive, all ready to pounce upon it and snatch what they can of its exposed treasures.*

In times of scarcity, food should never be given to the bees in the day time, but only in the evening, always inside of the hive and above the combs. The feeding of bees (605) in the day time causes robbing in two ways. It excites the bees which are fed, and induces them to go out to hunt for more, and the smell of the food given attracts the bees of the other hives. Hence follows fighting and trouble. But, above all things, the Apiarist must try to keep his colonies strong.

When there is a scarcity of blossoms, the entrance of the hive should be lessened, to suit the needs of the colony, by moving the entrance blocks (339). If the hive contains more combs than the bees can well defend, the number of the combs should be reduced by the use of the division board (349).

673. It is especially with weak colonies that care should be taken, in Spring or Fall. The strong hives being better able to keep warm, their bees fly out earlier in the day and will readily discover the weaker ones, which, unless their honey is protected, they will soon overpower.

When the above instructions are carried out, if thieves try to slip into a feeble colony they are almost sure to be overhauled and put to death; and if robbers are bold enough to attempt to force an entrance, as the bottom-board slants forward (327) it gives the occupants of the hive a decided advantage. Should any succeed in entering, they will find hundreds standing in battle-array, and fare as badly as a forlorn hope that has stormed the walls of a beleaguered fortress, only to perish among thousands of enraged enemies.

Cracks and openings in disjointed hives, should be securely closed with yellow clay, until the bees can be transferred into better abodes.

When the hives are opened, the work must be performed speedily and carefully; and, if any great number of robbers show themselves during the operation, it is well, after closing the hive, and reducing the entrance, to place a bunch of gra-s (fine grass or fine weeds preferred) over it, for an hour, or till the temporary excitement has subsided. The guardian bees station themselves in this grass and chase out robbers much more easily than they could otherwise. The robbers themselves recognize that their chances of "dodging in" are slim, and give up the undertaking. We have never had any trouble with robbers after closing a hive in this way.

When the robbed colony is weak, the robbing may be abated by preventing any bees from entering it till evening, when other colonies have stopped flying; allowing, at the same time, any bee that wishes to depart from it, and closing the entrance till late in the morning. By this course most of the robbers will be tired of their useless attempts, while the remaining workers of the robbed hive will be ready to repel the attacks.

When none of these methods succeed, a small comb of hatching Italian bees (**551**) may be given, with the necessary precautions (**480**), to the weak colony, and the hive placed in the cellar for a few days. The hatched Italians will receive the intruders warmly when the hive is brought back.

The Italian bees (**551**) defend their hives much better than the black (**549**) against the intrusion of robbers, and the Cyprians and Syrians (**559**) surpass even the Italians.

When a comb of honey breaks down in a hive from any cause, it should be removed promptly, and the bottom-board should be exchanged for a clean one at once. If any drops of honey fall about the Apiary, it is best to cover them up with earth promptly. In short, no honey should be left exposed, where bees can plunder it.

CHAPTER XV.

COMB FOUNDATION.

674. The invention and introduction of comb foundation, with the use of movable frames (**286**), marked an important step in the progress of practical bee-culture. The main drawback to the perfect success of movable-frame hives was the difficulty of always obtaining straight combs in the frames (**318**). Although the bevelled top bar (**319**) often secured this object, yet, in many instances, the bees deviated from this guide and fastened their combs from one frame to another; and if the matter was not promptly attended to, the combs of the hive became as immovable as those of box hives. One frame slightly out of place was a sufficient incentive for the bees to fasten two frames together. In the management of four large Apiaries, previous to the introduction of comb foundation, we found that, in spite of our efforts, a certain number of colonies would so build their combs, that only a part of the frames were movable without the use of a knife. Even the combs that were built in the right place were made somewhat waving, or bulged in spots, and were thus rendered unfit for such interchanges as are daily required in ordinary manipulations.

675. Another drawback to success was the building of drone comb (**225**). We have had colonies in which nearly one-fourth of the combs were drone-comb. In such hives the number of drones that *might* be raised would be sufficient to consume the surplus honey. To be sure, with movable-frame hives, such combs can be removed, but the difficulty consists in procuring straight and neat worker-combs to replace them; for if we simply remove the drone-combs, the bees often replace them with the same kind (**233**).

676. *Good straight worker-comb, not too old, is the most valuable capital of the Apiarist* (**442**). For years, before the introduction of comb-foundation, we had been in the habit of buying all the worker-comb from dead colonies that we could find, but we never had enough.

The consideration of the above important points, and of the great cost of comb to the bees (**233**), had long ago drawn the attention of German Apiarists to the possibility of manufacturing the base, or *foundation*, of the comb.

677. In 1857, Johannes Mehring invented a press to make wax *wafers*, on which the rudiments of the cells were printed. Those only, who experienced the obstacles which this industry presents, can form an idea of the energy and perseverance that were required to succeed as he did.

The foundation made by him then, was far from being equal to what is now made. The projections of the cell-walls were too rudimentary, sometimes not printed, and the bees often built drone-cells instead of worker-cells; but these imperfect efforts were the beginning of an industry which has proved of immense advantage to bee-keepers, and has spread like wild-fire wherever bees are kept.

678. Another Apiarist, Peter Jacob, of Switzerland, improved on the Mehring press. and in 1865, some of his foundation was imported to America, by Mr. H. Steele, of Jersey City (*Am. Bee-Journal*, Vol. 2, page 221), and tried by Mr. J. L. Hubbard, who reported favorably upon it. In 1861, Mr. Wagner had secured a patent in the United States, for the manufacture of *artificial honey comb-foundation by whatever process made*. His patent was never put to use, and rather retarded the progress of this industry in America.

679. The first comb-foundation made in America, was

manufactured in 1875, by a German, Mr. F. Weiss, very probably on an imported machine. Mr. A. I. Root, to whom

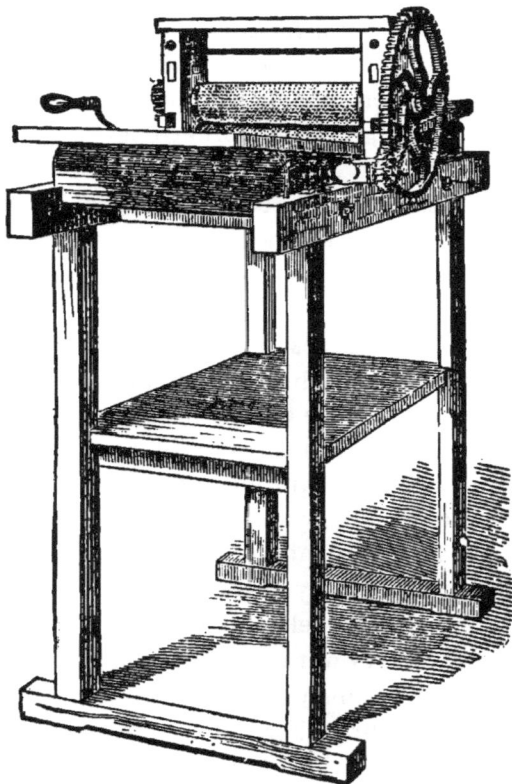

Fig. 115.
THE ORIGINAL "ROOT" MILL.
(From Root's "A. B. C.")

the credit is due* of popularizing the invention the world over manufactured a large roller-mill, in February, 1876,

---

* Some people think that when a man has made money by putting in practice the ideas of another, he is not entitled to any credit for it. But he, whose inquisitiveness has discovered the value of an invention, and whose energy has put it into practice, is almost as necessary and useful to the world as the original inventor himself.

PLATE 12.

JOHANNES MEHRING,

Inventor of Comb–Foundation.

This Apiarist is mentioned pages 51 and 367.

with the help of a skilled mechanic, A. Washburne. He sold hundreds of these mills afterwards.

680. In the practical use of comb-foundation, the most sanguine expectations were realized:

1. Every comb that is built on foundation is as straight as a board, and can be moved from one place to another, in any hive, without trouble.

2. The combs built on worker-foundation are exclusively worker-combs, with .the exception of occasional patches, when the foundation sags slightly.

3. All the wax produced by the bees, and gathered by the Apiarist from scraps, old combs, or cappings, is returned to the bees in this shape, instead of being sold at the commercial value of beeswax, which is several times less than its actual cost (**223**). The cost of foundation for brood-combs is not very great, especially if we consider that this capital is not consumed, but only employed; as the wax contained in the combs represents at least one-half of the primary value of the foundation, and can be rendered again, after years of use, none the worse for wear.

681. Different machines are in use in the United States. The flat-bottom foundation has the reputation of being the most regular, and thinnest; its main defect being the unnatural flat base of the cells, which renders it easier to manufacture, but objectionable to the bees, who have to remodel its base in using it (**213**). It is manufactured with or without wires imbedded in it, to help fasten it in the frames.

The Pelham-mill also makes an unnaturally-shaped foundation, the base of the cells being two instead of three-sided. This mill has the advantage of being very cheap, and is more easily manipulated than some of the others.

682. The Given-press makes foundation similar to that of the old European presses. It has been highly praised by

24

a number of Apiarists. As it is the easiest working of all foundation-machines, a great many, who could not succeed in making foundation on the mills, succeeded on this press. Another advantage claimed for it, is that it can make foundation in wired-frames by pressing it right over the wires. But a press has the disadvantage of leaving in the sheets all the irregularities, which they may have, when dipped; while in the roller-mills, these irregularities are "laminated out." Hence, pressed-foundation can never be as regular as rolled-foundation.

683. Plaster moulds and other utensils have been tried for foundation-making, but these cheap implements are almost entirely discarded.

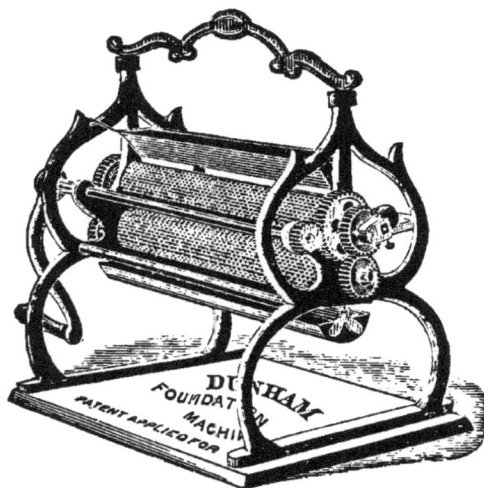

Fig. 116.
THE DUNHAM MILL.

684. The Root-mills,—the most practical—have been improved upon in different ways, by C. Olm, by Mrs. Dunham of Wisconsin, and by J. Vandervort of Pennsylvania. The latter gentleman, one of America's eminent machinists, makes most superior mills for any grade of foundation.

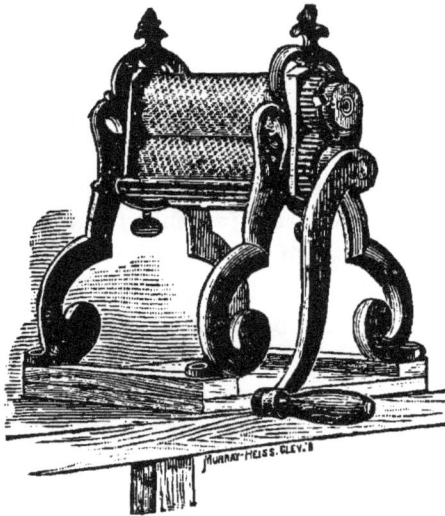

Fig. 117.
VANDERVORT MILL.

685. The wax used for thin surplus-foundation is a selected grade. Wax from cappings (**772**) and Southern wax are the best for this purpose. In every case, whether the foundation is to be used for surplus (**728**), or for brood-combs (**223**), the wax should be thoroughly cleaned by heating it to a high temperature and allowing it to cool slowly in flaring vessels, from which the cold wax can be easily removed. Wax, that is allowed to retain impurities, has less consistency, and will sag more readily. The method used by wax-bleachers of purifying with acids should not be resorted to, as the bees have a dislike for any disagreeable smell or taste.

686. *Nothing but pure wax should be used in any grade of foundation.* Paraffine, ceresine, etc., have been tried with disastrous results. Aside from the fact that these compounds melt at a lower degree than beeswax * and *break down* in

*"Paraffine melts at 110° Fahr., Beeswax at 162."—(Bloxam's Chemistry.)

the hive, the bees readily discover the imposition and show a decided preference for pure foundation.

The most common adulteration of crude bees-wax is made with tallow. Luckily, this is easily recognized by the soft, dull appearance of the cakes. The smell of tallow is also noticeable in freshly broken fragments.

687. The machines used for thin foundation should not be the same as those used for brood foundation. The latter, made on a light wall machine, would be too weak to stand the weight of the bees, in a full-sized brood frame, and would not contain wax enough for the bees to build their comb; for it is a remarkable fact that the bees *"thin out"* their foundation to a certain extent and make it considerably deeper out of the same material. When it has been made, *with a thin base and a heavy wall, the bees draw it out more readily into comb.*

On the other hand, foundation for surplus (**719**) must be made as light as the finest machine can make it, to avoid what is called the "fish-bone," a central rib found in the honey-comb that has been built on too heavy foundation. There is no "fish-bone," if the proper grade has been used, and even an expert in comb-honey hesitates in deciding whether the base is natural or artificial.

At the present day, nearly every section (**721**) of comb-honey that is sold, has been built on such foundation. The daintiest and most fastidious ladies can have no objection to it, and on visiting a well-managed foundation shop, they declare that the tender sheets are "nice enough to eat."

689. To prepare the wax sheets, we use soft wood boards $\frac{3}{8}$ of an inch thick, bathed in tepid water. They are wiped with a sponge, and dipped in melted wax, two or three times. The lower part of the board is then dipped in cold water, when it is turned bottom side up, and the other end is treated in the same manner. After the board has been put in water to cool for a little while, it is taken out; its

edges are trimmed with a sharp knife, and two smooth sheets
of wax are peeled off. If the sheets are intended for heavy
foundation, twice as many dips are necessary. The wax
should be liquid but not hot. If it is too hot, the sheets
will crack. To secure rapid work, you must have a room
arranged purposely for the dippers, with a zinc or tin floor
to catch the drips of water and wax.

690. The illustration, here given, (see plate) shows one
of the moulding tables in our foundation factory. The sheet
wax, after a few days' cooling in a deep and dry cellar, is
tempered, in the moulding tank with warm water, and run
through the rollers. The latter are lubricated with starch,
or soapsuds. When soapsuds are used, it is very important that the sheets be pressed so tightly in the rollers, as
to come out dry. This also makes a better print. The
foundation, as fast as it comes from the rollers, is laid upon
a hard wood block — a dozen sheets or more, at a time. A
wooden pattern is laid over them, and they are trimmed to
the proper size, by a knife made for the purpose, whose
blade has been wet with soapsuds. The projecting edges
are trimmed off, and the damaged sheets are melted over
for future use.

For the thin grades of foundation, the narrower the
sheets are, the thinner the foundation can be made. A
wide sheet spreads the rollers by springing the shafts to a
certain extent, and is heavier.

691. The manufacture of foundation, which at first
seemed likely to be undertaken by every Apiarist, has
become an industry of itself, owing to the greater skill and
speed acquired by those who make it daily. It might be
compared to cigar making. Any Apiarist can make wax
into sheets and run it through rollers, and any farmer can
raise tobacco and roll its leaves into cigars, but, to the
uninitiated, a neat sheet of foundation is as difficult to
make as an elegant cigar.

Acting upon ·ocialistic principles, in our manufacture of comb-foundation, we have interested our workmen, as we did the farmers on whose lands we have out-Apiaries (**584**). Pursuing the same principles, our workmen have associated together, dividing their earnings and electing their foreman. Such arrangements produce not only harmony, but many other results. Our laborers get better wages; and there is less need of close watching; for the work is always done with the view of increasing the business by satisfying the customers.

Well-made foundation will keep for years, in a dry place. It should never be handled when cold; and when too much softened by heat, should be cooled in a cellar, a few hours before it is handled.

692. The best grade of foundation for brood or extracting (**749**) combs is that which measures about five square feet to the pound; that for sections, ten to twelve feet. On this latter grade, the comb is not so readily built, for the bees have to add their own wax to it.

693. The foundation is fastened in sections by different machines, the most simple of which is the Parker-Fastener, sold by all dealers in bee-implements.

In his " Management of Bees" Mr. Doolittle describes his method. as follows:

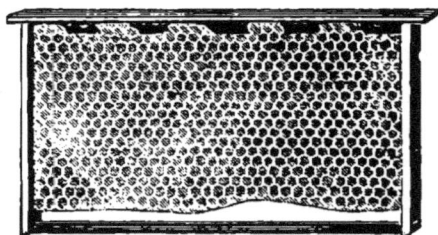

Fig. 118.

FOUNDATION FASTENED ON A
TRIANGULAR BAR.

From "Bees and Honey."

" Turn your sections top side down, hold a hot iron close to the box, and after holding the starter immediately above and touching the iron. draw the iron out quickly and press the starter gently on to the wood. when it is a fixture."

To fasten the foundation on a triangular top-bar, it can be pressed to each side of it as per engraving (fig. 118) taken from the *American Bee-Journal*. But, on a flat top-bar, it is much more readily fastened by the use of the

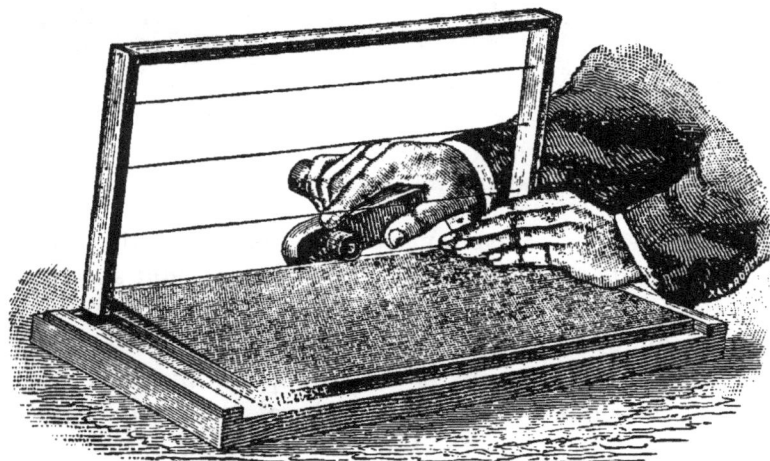

Fig. 119.
HAMBAUGH ROLLER.

roller (fig. 119) invented by our friend, Mr. Hambaugh, a successful Apiarist of Illinois.

694. In brood-frames, it may be fastened with or without wires. The wire used is malleable tinned wire, No. 30. A shallow frame needs no wires at all, but in brood-combs, —to insure safety and prevent warping—it is as well to use two or three horizontal wires as in fig. 119. This method of horizontal-wiring was first given us by Mr. Vandervort, to whom the world is also indebted for the spur for imbedding the wire in the foundation (fig. 120). The excessive wiring resorted to by some is worse than useless.

695. As comb-foundation is generally bought in long strips, it may be well to give directions to cut it into pieces of the right size for sections. This may be done with almost

Fig. 120.

VANDERVORT IMBEDDING SPUR

any sharp knife. Have a pattern of the size of the pieces wanted, made of hard wood. Take six or eight sheets at one time, arranged in an even pile. Lay your pattern on them, holding it down firmly; dip your knife in strong soap-suds, and if the wax is at the proper temperature, you will cut the eight pieces at one stroke of the knife. If the sheets have a tendency to *slip* from under the pattern, you may nail cleats on three sides of it, to *encase* the pile as in a box.

696. Are there a right and a wrong way, to suspend foundation in the frames? Or, in other words, should two of the six sides of the cell be perpendicular or horizontal? Huber, and Cheshire after him, call our attention to the fact, that the bees always build their combs, with two sides of the cells perpendicular. Mr. Cheshire explains, at length, the adaptation and advantages of this natural fact, and its bearing on the strength of the comb. From his explanations, it results that foundation suspended thus: *i. e.* with two perpendicular sides, would be properly fastened, while if suspended thus: *i. e.* .with two horizontal sides, it would be   improperly fastened.

Most of the machines that are made turn out foundation-sheets, which are to be hung horizontally, when the cells

PLATE 13.

MOULDING TABLE FOR COMB FOUNDATION.

a, foundation mill; *b*, brush; *c*, comb; *d*, clamp; *e*, sponge; *f*, tempering pan; *g*, apron boards.

are in the proper position. The Dunham-machine, however, makes sheets which should hang vertically, if the proper position is wanted. As the sheets principally used, are for frames of the Langstroth pattern (**299**), from eight to ten inches in depth, and sixteen to eighteen inches in length, and as the machines are all under fourteen inches in width, the Dunham foundation-sheets must be cut in two, or else must be fastened wrong in the frames, owing to the position of the cells in the rollers. In ninety-nine cases out of every hundred, the latter method has been followed, and as the Dunham heavy-brood foundation has given universal satisfaction, it proves that the position of the cells cannot have a great importance, practically, whenever a heavy grade is used. It is well, however, to place foundation in the correct position, whenever practicable, especially with the light grades for sections, which are more in danger of stretching under ordinary circumstances.

697. It is astonishing, as well as pleasing, to see how quickly a swarm will build its combs, when foundation is used. The enthusiasm, with which it is used by bee-keepers, is only exceeded by that of the bees, "in being hived on it." This invention certainly deserves to rank next to those of the *movable-frames* (**282**) and of the *honey-extractor.* (**749.**)

CHAPTER XVI

Pasturage and Overstocking.

Pasturage.

698. The quantity of honey yielded by different flowers varies considerably; some give so little, that a bee has to visit hundreds to fill her sack, while the corolla of others overflows with nectar.

In the vicinity of the Cape of Good Hope, there is a blossom, the *Protea mellifera,* which probably surpasses all others in the abundance of its nectar. Indeed, so abundant is it, that it is said, the natives gather it by dipping it from the flowers, with spoons. Mr. De Planta, in a lengthy and scientific article published in the *Revue Internationale d'Apiculture,* gives an account of his analysis of some samples of this honey, which he had received through the "Moravian United Brothers." He reports it to have the scent and the taste of ripe bananas, and considers it very sweet and good.

699. The same plants yield nectar in different quantities in different countries. The Caucasian Comfrey, from which the bees reap a rich harvest in Europe, is of little account here.

700. Every bee-keeper should carefully acquaint himself with the honey-resources of his own neighborhood. We will mention particularly some of the most important plants from which bees draw their supplies. Since Dzierzon's discovery of the use which may be made of flour, early blossoms producing pollen *only,* are not so important.

All the varieties of willow abound in both pollen and honey, and their early blossoming gives them a special value.

> "First the gray willow's glossy pearls they steal,
> Or rob the hazel of its golden meal,
> While the gay crocus and the violet blue,
> Yield to their flexible trunks ambrosial dew."—EVANS.

The sugar-maple (*Acer saccharinus*) yields a large supply of delicious honey, and its blossoms, hanging in graceful fringes, will be alive with bees.

In some sections, the wild gooseberry is a valuable help to the bees, as it blossoms very early, and they work eagerly on it.

Of the fruit trees, the apricot, peach, plum, cherry and pear, are great favorites ; but none furnishes so much honey as the apple.

The dandelion, whose blossoms furnish pollen and honey, when the yield from the fruit trees is nearly over, is worthy of rank among honey-producing plants.

Fig. 121.
BLOSSOM OF TULIP TREE.

The tulip tree (*Liriodendron*) (Fig. 121), is one of the

greatest honey-producing trees in the world. As its blossoms expand in succession, new swarms will sometimes fill their hives from this source alone. The honey, though dark, is of a good flavor. This tree often attains a height of over one hundred feet, and its rich foliage, with its large blossoms of mingled green and yellow, make it a most beautiful sight.

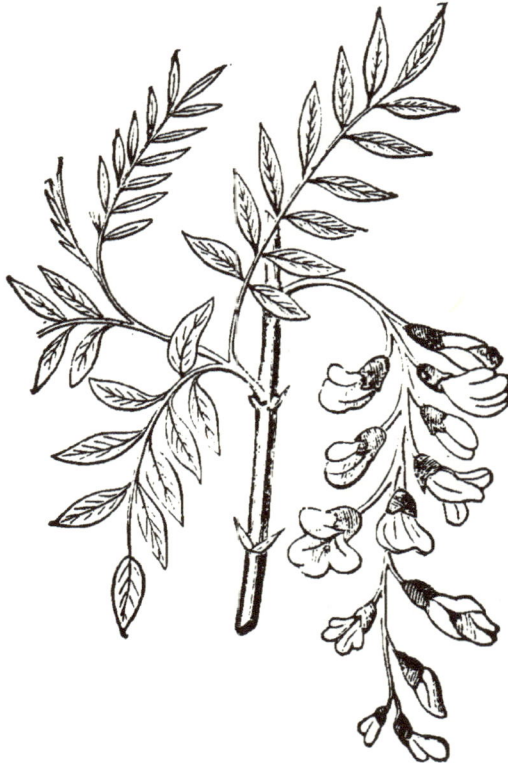

Fig. 122.
LOCUST BLOSSOMS

The common locust (Fig. 122), is a very desirable tree for the vicinity of an Apiary, yielding much honey when it is peculiarly needed by the bees.

The wild cherry blooms about the same time.

701. Of all the sources from which bees derive their supplies, white clover (Fig. 123), is usually the most

Fig. 123.
WHITE CLOVER.

important. It yields large quantities of very pure white honey, and wherever it abounds, the bee will find a rich harvest. In most parts of this country it seems to be the chief reliance of the Apiary. Blossoming at a season of the year when the weather is usually both dry and hot, and the bees gathering its honey after the sun has dried off the dew, it is ready to be sealed over almost at once.

It is at the blossoming of this important plant that the main crop of honey usually begins, and that the bees prop-agate in the greatest number.

The flowers of red clover (fig. 124) also produce a large quantity of nectar; unfortunately its corollas are usually too deep for the tongue of our bees. Yet sometimes, in Summer, they can reach the nectar, either because its corollas are shorter on account of dryness, or because they are more copiously filled.

Fig. 124.
RED CLOVER.

702. The linden, or bass-wood (*Tilia Americana*, fig. 125), yields white honey of a strong flavor, and, as it blossoms when both the swarms and parent-colonies are usually populous, the weather settled, and other bee-forage scarce, its value to the bee-keeper is great.

> " Here their delicious task, the fervent bees
> In swarming millions tend: around, athwart,
> Through the soft air the busy nations fly,
> Cling to the bud, and with inserted tube,
> Suck its pure essence, its etherial soul."—THOMSON.

This majestic tree, adorned with beautiful clusters of

Fig. 125.
LINDEN OR BASSWOOD.

fragrant blossoms, is well worth attention as an ornamental shade-tree. By adorning our villages and country residences with a fair allowance of tulip, linden, and such other trees as are not only beautiful to the eye, but attractive to bees, the honey-resources of the country might, in process of time, be greatly increased. In many districts, locust and bass-wood plantations would be valuable for their timber alone.

703. We have also a variety of clover imported from Sweden, which grows as tall as the red clover, bears many blossoms on a stalk, in size resembling the white, and, while it answers admirably for bees, is preferred by cattle to almost any other kind of grass. It is known

Fig. 126.
From "Bees & Honey.")
ALSIKE CLOVER.

by the name of Alsike, or Swedish clover (Fig. 126).

The objection made to this clover is that its stem is so light that it falls to the ground. This is remedied by sowing it with timothy. The latter helps it to stand. It is as good for honey as white clover.

704. The raspberry furnishes a most delicious honey. In flavor it is superior to that from the white clover. The sides of the roads, the borders of the fields, and the pastures of much of the "hill-country" of New England, abound with the wild red raspberry, and, in such favored locations, numerous colonies of bees may be kept. When it is in blossom, bees hold even the white clover in light esteem. Its drooping blossoms protect the honey from moisture, and they can work upon it when the weather is so wet that they can obtain nothing from the upright blossoms of the clover. As it furnishes a succession of flowers for some weeks, it yields a supply almost as lasting as the white clover. The precipitous and rocky lands, where it most abounds, might be made almost as valuable as some of the vine-clad terraces of the mountain districts of Europe.

The borage (*Borago officinalis*), (Fig. 143), blossoms continually from June until severe frost, and, like the raspberry, is frequented by bees even in moist weather. The honey from it is of a superior quality.

The *Canada thistle*, the *viper bugloss* yield good honey after white clover has begun to fail. But these plants are troublesome, for they cannot easily be gotten rid of.

705. *Melilot*, or sweet clover (figs. 127 and 136), which grows on any barren or rocky soil without cultivation, is one of the most valuable honey-plants. It will not thrive, however, where cattle can graze on it, as they soon destroy it. If cut early to be used as forage, it blooms later than white clover and till frost. It is a biennial.

The different varieties of smart-weeds (*Persicaria*), golden rod, buckwheat, asters, iron-weed, Spanish-needles in low

Fig. 127. (From L'Apicoltore.)
YELLOW OR OFFICINAL MELILOT.

.ands and marshy places, give a very abundant honey-crop
in the latter part of the Summer.   They form the bulk of
what is called the " Fall crop " in this latitude.

In California the sage, in Texas the horse-mint, in Florida the mangrove, form the main honey-harvests of those
countries.

706. We here present a list of the flowers known as
being visited by the bees for their nectar or for their pollen.
We have grouped them in Families, and we give engravings
of their most prominent types, in order to help the Apiarist

25

in his investigations. But our list is far from being com plete, and every day brings some new discoveries.

Compositæ:—Dandelion, Thistle, Chamomile, (Fig. 128), Sunflower, Ox-eye Daisy, Goldenrod (Fig. 129), Coreopsis,

Fig. 128.
CHAMOMILE.

Fig. 129.
GOLDEN ROD OR SOLIDAGO.

Lettuce, Chicory, Boneset, Iron-weed, Indian Plantain, Fire-weed, Aster (Fig. 130-131), Burr Marigold, Spanish Needles, Coneflower, Isatis tinctoria, Star Thistle, Thoroughwort, Butterweed, Sneeze-wort, Blue Bottle, Ragweed, several varieties of Echinops, one of which, the Spherocephalus, was introduced here by Mr. Chapman. The Echinops ritro (smaller in size) (Fig. 132), is cultivated in Europe on ac-

Fig. 130.
ASTER ROSEUS.

Fig. 131.
ASTER TRADESCANT.

count of its beautiful blue heads. This family includes also the Helenium tenuifolium (Fig. 133), whose honey is poisonous.—(Dr. J. P. H. Brown.)

Fig 132.
ECHINOPS RITRO.

Fig. 133.
HELENIUM TENUIFOLIUM.

Leguminous:—Judas tree (Fig. 134), which blooms very early, Locust tree, Honey-locust (Fig. 135), Wistaria,

Fig. 134.   (From L'Apicoltore.)
JUDAS TREE.

white, red and alsike Clover, Melilot (Fig. 136), Lucerne
or Alfalfa, Peas, Beans, Vetches, Lentils, False-Indigo,
Partridge pea, Wild senna, Milk vetch, Yellow-Wood,
Mesquit-tree of Texas, Cleome-integrifolia, and pungens
(Fig. 137).

Labiate:— (from *Labium*, a lip.) Sage (Fig. 139),
Mint (Fig. 140), Ground Ivy (Fig. 138), Horehound,
Catnip, Motherwort, Horse-Mint, Basil, Hissop, Bergamot,
Marjoram, Thyme, Melissa, Dead Nettle, Brunella, Penny-
Royal.

Rosaceous:— Wild Rose, Cherry (Fig. 142), Plum,
Peach, Apricot. Apple, Pear, Quince, Hawthorne, Blackberry, Raspberry, Strawberry, Juneberry, Cinquefoil,

Fig. 135.
HONEY-LOCUST LEAVES & THORNS.

Fig. 136.
MELILOTUS ALBA.

Fig. 137.
CLEOME PUNGENS.

Fig. 138.
GROUND IVY.

Bowmansroot, Queen of the Prairie, Meadow Sweet, Pyracantha.

Fig. 139.
SAGE.

Fig. 140.
MINT.

Fi:. 141.
KNOT-WEED.

Fig. 142.
CHERRY BLOSSOMS.

Polygonous:— (Knot-Weed) Buckwheat, Lady Thumb, Rhubarb, Sorrel, and a variety of Knot-Weeds or Persicarias (Fig. 141).

Fig. 143. (From L'Apicoltore.)
BORAGE.

Fig. 144.
VERONICA OFFICINALIS.

Borage Family:—Borage (Fig. 143), Viper-bugloss, Comfrey, Phacelia, Virginia Lungwort, Hound tongue, Gromwell, False Gromwell.

Scrophularia:—Scrophularia nodosa (Simpson's honey-

Fig 145.

ASCLEPIAS TUBEROSA.   PLEURISY ROOT.

Fig. 146.

ASCLEPIAS SYRIACA.

plant), Veronicas (Fig. 144), Yellow Jessamine of the South, whose honey is poisonous. (DR. J. P. H. BROWN.)

Asclepiadaceæ:—The common Milk-weed (Fig. 146), or Silkweed, Asclepias Syriaca, is much frequented by bees, but these visits are often fatal to them. All the grains of pollen of the Silkweed. in each anther, are collected in a compact mass, inclosed in a sack; these sacks are united in pairs (*a*. Fig. 147) by a kind of thread, terminated by a small, viscous gland. These threads stick to the feet (*b*. Fig. 147) and often to the labial palpi (**46**) of the bees, who cannot easily get rid of them, and perish. In some parts of Ohio and Western Illinois, a variety

Fig. 147.

POLLEN OF MILKWEED.

a, sacs of pollen in pairs; *b*, the same attached to a bee's foot.

(From "A B C of Bee Culture.")

of the common kind, the *Alsclepias Sullivantii*, does not present to bees these difficulties to the same degree. We have seen bees gathering honey freely on four or five different varieties which grow in our neighborhood, and especially on the Tuberosa or Pleurisy root (Fig. 145), fitly recommended by James Heddon. This kind is noticeable by its orange flowers.

Fig 148.
RAPE.

Fig. 149.
BLACK MUSTARD.

Cruciferous:—Rape (Fig. 148), Mustard (Fig. 149), Cabbage, Radish, Candy Tuft, Stock, Wall-Flower, Moonwort, Sweet Alyssum, Cress.

Ericaceæ:—This family, on the Old Continent, includes the numerous varieties of Heath, on which bees reap a large harvest of inferior honey, so thick that it is impossible to extract it. Blueberry, Sour Wood, Laurel, Clethra alnifolia, Cowberry (Fig. 150), Huckleberry, Whortleberry, Gaultheria Procumbens, or Creeping wintergreen,— which is indicated, by some English bee-keepers, as pre-

Fig. 150.
COWBERRY.

Fig. 151.
VALERIAN.

Fig. 152.
ŒNOTHERA GRANDIFLORA.

Fig. 153.
EPILOBIUM SPICATUM.

F 154
LILY.

Fig. 155.
HYACINTH.

Fig. 156.
LILY OF THE VALLEY.

Fig. 157.
SOLOMON'S SEAL.

venting bees from stinging the hands when they are rubbed with its leaves,—belong to this family.

Valerianaceæ:—Valerian (Fig. 151), Corn salad or Lamb lettuce, belong to this family.

Onagraceæ:—(Evening Primrose family) Gaura, Fuschia, Œnothera (Fig. 152) Epilobium (Willow Herb, Fig. 153).

Liliaceæ:—Lilies (Fig. 154), Asparagus, Wild Hyacinth (Fig. 155), Star of Bethlehem, Lily of the Valley (Fig. 156), Solomon's Seal (Fig. 157), Dog's-tooth Violet, three-headed Night-shade, Garlic, Onion, Crocus.

Malvaceæ:—Common Mallows, and others, Hollyhock, Cotton (Fig. 158), Abutilon.

Fig. 15⁰.
COTTON.

Caprifoliaceæ:—Honeysuckle, Snow and Coral berries, Arrow-wood.

Fig. 159.—MELON.

Cucurbitaceæ:—Cucumber, Melon (Fig. 159), Squash, Gourd.

Fig, 160.
FENNEL.

Fig. 161.
PINK.

Umbelliferæ:—Parsley, Angelica, Lovage, Fennel (Fig. 160), Parsnip, Coriander, Cow-parsnip.

Fig. 162.
PLANTAIN.

Fig. 163.
BARBERRY.

Caryophyllaceœ ;— Pink (Fig. 161), Licknis, Chickweed, Saponaria.

We can name also: Rib-Grass, or Plantain (Fig. 162), Goosefoot, Blue-eyed grass, Corn-flag, Buckthorn, Barberry (Fig. 163), Sumac, Grape-vine, Polanysia, Button weed, . Mignonette, or Reseda (Fig. 164), Teasel, Skunk cabbage,·

Fig. 164.
MIGNONETTE.

Fig. 165.
WILLOW.
Male Blossom.

Fig. 166
WILLOW.
Female Blossom.

Waterleaf, Hemp, Touch-me-not, Amaranth, Crowfoot, St. John's wort, and among the trees: Willow (Figs. 165-166), Poplar, which have their sexual organs on different trees; Oak (Fig. 167), Walnut, Hickory, Beech, Birch, Alder, Elm, Hazel-nut (Fig. 168), Maple whose organs of reproduction are separated, although on the same tree.

Horse chestnut, Persimmon, Gum-tree, Dogwood, Button-bush, Cypress, Liquidambar, Linden.

We should mention also, Aylanthus glandulosus (Varnish tree of Japan), a large, ornamental tree, which gives an abundance of honey so bad in taste, as to compel the bee-keepers who have some in their neighborhood to extract it as soon as it is gathered, that it may not injure the quality of their crop.

Fig. 167.
OAK. FEMALE BLOSSOM.

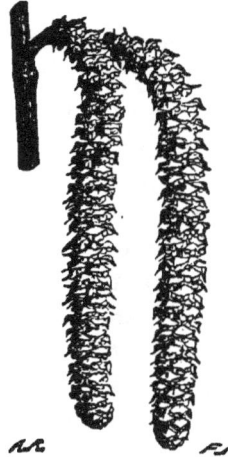

Fig. 168.
HAZEL NUT. MALE BLOSSOM.

Bees also visit some of the plants of the grass family, such as corn and sorghum. A plant of this family, the Setaria, or bristly fox-tail grass, is known in France under the name of *accroche-abeilles*, (bee-catcher). Its curved hairs grasp the bees' legs, and the poor insects, unable to free themselves, are soon exhausted, and die.

Overstocking.

OUR COUNTRY NOT IN DANGER OF BEING OVERSTOCKED WITH BEES.

707. If the opinions, entertained by some, as to the danger of overstocking were correct, bee-keeping in this country, would always have been an insignificant pursuit.

It is difficult to repress a smile when the owner of a few hives, in a district where hundreds might be made to prosper, gravely imputes his ill-success to the fact, that too many bees are kept in his vicinity. If, in the Spring, a colony of bees is prosperous and healthy, it will gather abundant stores, in a favorable season, even if many equally strong are in its immediate vicinity ; while, if it is feeble, it will be of little or no value, even if it is in " a land flowing with milk and honey," and there is not another colony within a dozen miles of it.

As the great Napoleon gained many of his victories by having an overwhelming force at the right place, in the right time, so the bee-keeper must have strong colonies, when numbers can be turned to the best account. If they become strong only when they can do nothing but consume what little honey has been previously gathered, he is like a farmer who suffers his crops to rot on the ground, and then hires a set of idlers to eat him out of house and home.

708. Although bees can fly, in search of food, over three miles, still, *if it is not within a circle of about two miles in every direction from the Apiary, they will be able to store but little surplus honey.** If pasturage abounds within a quar-

---

* " Judging from the sweep that bees take from the side of a railroad train in motion, we should estimate their pace at about thirty miles an hour. This would give them four minutes to reach the extremity of their common range." *London Quarterly Review.*

ter of a mile from their hives, so much the better; there is no great advantage, however, in having it close to them, unless there is a great supply, as bees, when they leave the hive, seldom alight upon the neighboring flowers. The instinct to fly some distance seems to have been given them to prevent them from wasting their time in prying into flowers already despoiled of their sweets by previous gatherers.

"Mr. Kaden, of Mayence, thinks that the range of the bee's flight does not usually extend more than three miles in all directions. Several years ago, a vessel, laden with sugar, anchored off Mayence, and was soon visited by the bees of the neighborhood, which continued to pass to and from the vessel from dawn to dark. One morning, when the bees were in full flight, the vessel sailed up the river. For a short time, the bees continued to fly as numerously as before; but gradually the number diminished, and, in the course of half an hour, all had ceased to follow the vessel, which had, meanwhile, sailed more than four miles." —*Bienenzeitung*, 1854, p. 83.

Our own experience corroborates the statements of Kaden. We have known strong colonies of bees to starve upon the hills in a year of drouth, while the Mississippi bottoms, less than four miles distant, which had been overflowed during the Spring, were yielding a large crop. It is evident that districts, where the honey blossoms are scarce, can be much more readily overstocked than those rich bottom lands which are covered wit h blossoms, the greater part of the Summer. A great amount of *land in cultivation*, is not always a hindrance to honey production, for cultivated lands often grow weeds, which yield an abundance of honey. Heartsease and Spanish needle grow plentifully in cornfields and wheat stubble in wet seasons. Pasture lands abound with white clover.

709. It is impossible to give the exact number of colonies that a country can support profitably. In poor locations, a few hives will probably harvest all the honey to be

26

found, while some districts can support perhaps a hundred
or more to the square mile. The bee-keeper must be his
own judge, as to the honey capacity of his district.

"When a large flock of sheep, says Oettl, is grazing on a
limited area, there may soon be a deficiency of pasturage. But
this cannot be asserted of bees, as a good honey-district cannot
readily be overstocked with them. To-day, when the air is
moist and warm, the plants may yield a superabundance of
nectar; while to-morrow, being cold and wet, there may be a
total want of it. When there is sufficient heat and moisture, the
saccharine juices of plants will readily fill the nectaries, and will
be quickly replenished when carried off by the bees. Every cold
night checks the flow of honey, and every clear, warm day re-
opens the fountains. *The flowers expanded to-day must be visited
while open; for, if left to wither, their stores are lost.* The same
remarks will apply substantially in the case of honey-dews.
Hence, bees cannot. as many suppose. collect to-morrow what is
left ungathered to-day, as sheep may graze hereafter on the pasturage they do not need now. Strong colonies and large Apiaries
are in a position to collect ample stores when forage suddenly
abounds, while, by patient, persevering industry. they may still
gather a sufficiency, and even a surplus, when the supply is
small, but more regular and protracted."

Although we believe that a district can be overstocked,
so as to make bee-culture unprofitable, yet the above extract
gives a correct view of the honey harvest, which depends
much on the weather, and must be gathered when produced.

The same able Apiarist, whose *golden rule* in bee-keeping
is, to keep none but *strong colonies*, says that in the lapse
of twenty years since he established his Apiary, there
has not occurred a season in which the bees did not
procure adequate supplies for themselves, and a surplus
besides. Sometimes, indeed, he came near despairing, when
April, May, and June were continually cold, wet, and un-
productive; but in July, his strong colonies speedily filled
their garners, and stored up some treasure for him; while,
in such seasons, small colonies could not even gather enough
.to keep them from starvation.

710. According to Oettl (p. 389), Bohemia contained 160,000 colonies in 1853, from a careful estimate, and he thought the country could readily support four times that number. This province contains 19,822 square English miles.

We say square English miles, and we insist on the word English, for we have read of reports from Germany, showing incredible figures as to the number of bees, and the amount of beeswax and honey gathered on areas of a few square miles; and yet, some of these reports may have been true, for there are different sized miles, in Germany. The German geographical mile is equal to $4.\frac{611}{1000}$ English miles; the German short mile, to $3.\frac{897}{1000}$; and the German long mile to $5.\frac{753}{1000}$, &c. ; the shortest German square mile being as about 15 of the English, and the long being about equal to 33 of our square miles. This we glean from "Chambers Encyclopedia."

According to an official report, there were in Denmark, in 1838, eighty-six thousand and thirty-six colonies of bees. The annual product of honey appears to have been about 1,841,800 lbs. In 1855, the export of wax from that country was 118,379 lbs.

In 1856, according to official returns there were 58,964 colonies of bees in the kingdom of Wurtemberg.

In 1857, the yield of honey and wax in the empire of Austria was estimated to be worth over seven millions of dollars.

Doubtless, in these districts, where honey is so largely produced, great attention is paid to the cultivation of crops which, while in themselves profitable, afford abundant pasturage for bees.

711. California, which seems to be the Eldorado of bee-culture, can probably support the greatest number of bees to the square mile, and yet in some seasons the bees starve there in great numbers owing to the drouth.

We have no official statistics of the honey crops of the United States, but the following extract from the *American Bee-Journal* (1886), will give an idea of the immensity of our honey resources, considering the comparatively small areas of this country now occupied by Apiarists.

"The California *Grocer* says that the crop of 1885 was about 1,250,000 pounds. The foreign export from San Francisco during the year was approximately 8,800 cases. The shipments East by rail were 360,000 pounds from San Francisco, and 910,000 pounds from Los Angeles, including both comb and extracted. We notice that another California paper estimates the crop of 1885 at 2,000,000 pounds, and the crop of the United States for 1885 was put down at 26,000,000 pounds. We do not think these figures are quite large enough, though it was an exceedingly poor crop."

But former years have given still better results. Through the courtesy of Mr. N. W. McLain, of the U. S. Apicultural Station, we have received the following statistics from "The Resources of California, 1881":

The honey shipped from Ventura County, California, during 1880 amounted to 1,050,000 lbs. The Pacific Coast Steamship Company of San Diego shipped 1,191,800 pounds of honey from that county in the same year.

The crop of the five lower counties in California that year, was estimated by several parties at over three million pounds.

According to a report of S. D. Stone, Clerk of the Merchants' Exchange of San Francisco, the actual amount of honey shipped to that city from different parts of California in the sixteen months ending May 1, 1881, was 4,340,400 pounds, equal to two hundred and seventeen car-loads.

One hundred tons of honey, in one lot, were shipped during the same year, from Los Angeles to Europe on the French bark Papillon. This had *all* been purchased from Los Angeles Apiarists.

712. In the excellent season of 1883, the honey crop of

Hancock County, Illinois, was estimated at about 200,000 pounds, which made an average of less than half a pound per acre. 36,000 pounds of this was our own crop, and the county did not contain one-tenth of the bees that could have been kept profitably on it. Yet, at this low rate, the crop of Illinois alone, with the same percentage of bees, would have been 15,000,000 pounds. We cannot form an adequate idea of the enormous amount of honey, which is wasted from the lack of bees to harvest it.

713. In our own experience in the Mississippi Valley, we have found eighty to one hundred colonies to be the number from which the most honey could be expected in one Apiary. Dr. C. C. Miller in his interesting work "A Year Among the Bees," says also that one hundred colonies is the best number in one location. Mr. Heddon strongly urges bee-keepers not to locate within any area already occupied by an Apiary of one hundred colonies or more. The extensive experience of both these Apiarists confirms ours, but we must remember that locations differ greatly.

714. In all arrangements, aim to save *every step* for the bees that you possibly can. With the alighting-board properly arranged, the grass kept down, or better still, coal-ashes or sand **(568)** spread in front of the apron-board **(343)**, bees will be able to store more honey, even if they have to go a considerable distance for it, than they otherwise could from pasturage nearer at hand. Many bee-keepers utterly neglect all suitable precautions to facilitate the labors of their bees, as though they imagined them to be miniature locomotives, always fired up, and capable of an indefinite amount of exertion. A bee *cannot* put forth more than a certain amount of physical effort, and a large portion of this ought not to be spent in contending against difficulties from which it might easily be guarded. They may often be seen panting after their return from labor, and so exhausted as to need rest before they enter the hive.

715. With proper management, at least fifty pounds of surplus honey may be obtained from each colony that is wintered in good condition. This is not a "guess" estimate, it is the average of our crops during a period of over twenty years in different localities.

Such an average may appear small to experienced bee-keepers, but we think it large enough when we consider that we have very few linden trees in our neighborhood.

A careful man, who, with Langstroth hives, will begin bee-keeping on a prudent scale, enlarging his operations as his skill and experience increase, will succeed in any region. But, in favorable localities, a much larger profit may be realized.

Bee-keepers cannot be too cautious in entering largely upon new systems of management, until they have ascertained, not only that they are good, but that *they* can make a good use of them. There is, however, a golden mean between the stupid conservatism that tries nothing new, and that rash experimenting, on an extravagant scale, which is so characteristic of many people.

CHAPTER XVII.

Honey Production.

716. History does not mention the first discovery of honey, by human beings. Whether it became known to primitive man by accident, from the splitting of a bee-tree by lightning, or by his observation of the fondness of some animals for it,—certain it is that when he first tasted the thick and transparent liquid, the fear of stings was overcome, and the bee-hunter was born. Since that time, the manner of securing honey has undergone a great many changes, improving and retrograding, as we can judge from writings now extant.

Killing bees (**276**) for their honey was, unquestionably, an invention of the dark ages, when the human family had lost—in Apiarian pursuits, as well as in other things—the skill of former ages. In the times of Aristotle, Varro, Columella, and Pliny, such a barbarous practice did not exist. The old cultivators took only what their bees could spare, killing no colonies, except such as were feeble or diseased.

The Modern methods have again done away with these customs among enlightened men, and the time has come when the following epitaph, taken from a German work, might properly be placed over every pit of brimstoned bees:

HERE RESTS,
CUT OFF FROM USEFUL LABOR,
A COLONY OF
INDUSTRIOUS BEES
BASELY MURDERED
BY ITS
UNGRATEFUL AND IGNORANT
OWNER.

To the epitaph should be appended Thomson's verses:

> " Ah, see, where robbed and murdered in that pit,
> Lies the still heaving hive! at evening snatched,
> Beneath the cloud of guilt-concealing night,
> And fixed o'er sulphur! while, not dreaming ill,
> The happy people, in their waxen cells,
> Sat tending public cares.
> Sudden, the dark, oppressive steam ascends.
> And, used to milder scents, the tender race,
> By thousands, tumble from their honied dome
> Into a gulf of blue sulphureous flame!''

717. The present methods are as far ahead of the old ways, as the steel rail is ahead of the miry road; as the palace car is ahead of the stage coach.

It is to the production of surplus honey that all the efforts of the bee-keeper tend, and the problem of Apiculture is, how to raise the most honey from what colonies we have, with the greatest profit.

718. In raising honey, whether comb or extracted, the Apiarist should remember the following:

1st. His colonies should be strongest in bees at the time of the expected honey harvest (**565**).

2d. Each honey harvest usually lasts but a few weeks.

If a colony is weak in Spring, the harvest may come and pass away, and the bees be able to obtain very little from it. During this time of meagre accumulations, the orchards and pastures may present

> "One boundless blush,one white empurpled shower
> Of mingled blossoms;''

and tens of thousands of bees from stronger colonies may be engaged all day in sipping the fragrant sweets, so that every gale which ''fans its odoriferous wings '' about their dwellings, dispenses

> "Native perfumes, and whispers whence they stole
> Those balmy spoils."*

---

* The scent of the hives, during the height of the gathering season, usually indicates from what sources the bees have gathered their supplies.

By the time the feeble colony becomes strong — if at all — the honey harvest is over, and, instead of gathering enough for its own use, it may starve, unless fed. Bee-keeping, with colonies which are feeble, except in extraordinary seasons and locations, is emphatically nothing but " vexation of spirit."

3rd. Colonies that swarm **(406)** cannot be expected to furnish much surplus, in average localities and seasons. (See Artificial Increase **469.**)

4th. A hive containing or raising many drones **(189)** cannot save as much surplus as one that has but few, owing to the cost of production of these drones.

We have insisted on this point already, but it is of such importance, that we cannot refrain from recalling it. The hives should be overhauled every Spring, and the drone comb, cut out and replaced by neat pieces of worker comb, or of comb foundation **(674)**. Every square foot of drone comb, replaced with worker comb, represents an annual saving, in our estimation, of at least one dollar to the colony.

COMB HONEY.

719. Although the production of comb honey is less advantageous than that of extracted honey **(746)**, yet a newly made and well sealed honey comb is unquestionably most attractive, and, when nicely put up, will find a place of honor, even on the tables of the wealthy. White comb honey will always be a fancy article, and will sell at paying prices.

Dark honey in the comb never finds ready sale. Hence, the bee-keepers, in districts where white honey is harvested, are mostly producers of comb honey; while those in the districts producing dark honey, in the South mainly, rely more on extracted honey.

720. We have not the space to describe the different evolutions, through which the production of comb honey has passed since box-hive times ; production in large frames, in glass boxes, in tumblers, etc.

Honey in large frames does not sell well, and cannot be safely transported. Were it not for this, its production in this way would be advisable. The experienced bee-keeper well knows that bees will make more honey in a large box, than in several small ones whose united capacity is the same. In small boxes, they cannot so well maintain their animal heat in cool weather and cannot ventilate so readily in hot weather.*

The bees have another important and natural objection to the small receptacles, mentioned by a noted Apiarist, as will be seen farther (**741**). Practically, there is more labor for the bees in small receptacles, as the joints and corners of the combs require more time and more wax.

721. But to produce salable comb honey, we have no choice. We *must* produce it in as small a receptacle as possible. The *Adair* section boxes, which we used as early as 1868, marked the first progressive step, so far as we know.

These sections forming a case by the overlapping of their top and bottom bars, and furnished with glass at each end, were much admired, and we sold several tons of honey, in this shape, in St. Louis, at the now fabulous prices of from 25 to 28 cents per pound.

722. But the one and two pound sections, as now made, have been universally adopted of late years.

The one pound sections sell best, but, at the difference of only one cent per pound, we would prefer to use the two pound sections.

---

* In the exceedingly hot season of 1878, the colonies that were provided with glass boxes yielded on an average, less than one-fourth of the average yielded by others.

These sections are made of two kinds, dovetailed in four pieces, or in one piece and bent. The first can be made

Fig. 169 (From '' Bees and Honey.'')
ONE PIECE SECTION.

of any kind of white wood, while the latter are made of bass wood only. When the one piece sections are made by the splitting process, they are less apt to break in bending, but sawed sections can be safely put up by wetting the V notches, before bending them.

723. Sections are usually made $\frac{1}{8}$ inch thick and $1\frac{1}{2}$ to 2 inches wide. The standard section for Langstroth hives is $4\frac{1}{4}$ x $4\frac{1}{4}$ inches, with openings at the bottom and top.

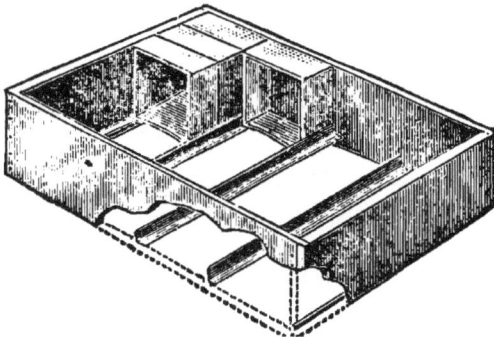

Fig. 170. (From ''Gleaning in Bee Culture.'')
MILLER'S CRATE OR SUPER.

724. They are given to the bees in the upper story, like the extracting cases (fig. 178). Storage room, *on the sides of the brood chamber*, has been periodically advised by inventors of new hives, but bees never fill and seal sections placed at the side as fast as if put above the brood chamber.*

* There are few noted bee-keepers who are successful with a combination of top and side storage. We will cite one of the leaders of American Apiculture, G. M. Doolittle.

Sections are either crated. in cases (fig. 170), or hung in broad frames (fig. 171), of full depth, or half depth. Both

Fig 171 (Fro.n ''Bees an l Honey.'')
FULL DEPTH SECTION FRAME.

ways have their friends, and both are good, as long the main principles are adhered to.

725. These principles are based on the difficulties, that have to be overcome in comb-honey production, as follows:

1st. Inducing the bees to work in small receptacles;

2d. Forcing them to build the combs straight and even, without bulge, so that the sections can be interchanged without being bruised against one another, when taken off and crated for market;

3d. Keeping the queen in the brood apartment, and preventing her from breeding in the sections;

4th. Preventing swarming as much as possible;

5th. Arranging the sections so as to have as little propolis put on them as possible (**237**);

6th. Getting the greatest number of sections thoroughly sealed, as unsealed honey is unsalable.

726. *1st.* INDUCING BEES TO WORK IN SMALL RECEPTACLES.

Rather than work in small, empty receptacles, the bees *sometimes* crowd their honey in the brood chamber, till the queen can find no room to lay in, and swarming, or a smaller crop of honey, is the consequence. To remedy this evil, some of our leading bee-keepers have resorted to an old, discarded, French practice, ''reversing.'' Reversing consists in turning the brood chamber upside down and placing hives containing empty combs, whose bees died the preceding Winter, or empty supers, over it. The honey contained in the brood chamber, which is always placed above and behind the brood, safe from pilfering intruders, is now at the bottom, near the entrance. The cells are wrong side up (fig. 172), and the most watery honey is in danger of leaking out. Hence an uproar in the hive, and the immediate result is, that the bees promptly occupy the upper story, and store in it all this ill-situated honey. The result is so radical, that "reversing bee-keepers" admit that their bees have to be fed in the Fall, as too little honey is left in the brood chamber for the hives to winter on.* In the box-hive times, the following was already the almost unanimous report of bee-keepers on the results of "reversing." The recruiting and feeding for Winter of reversed colonies being considered too costly and risky, the Apiaries were supplied every year with new colonies bought from bee-keepers whose business was to raise swarms to sell.

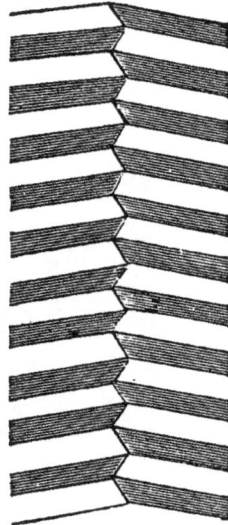

Fig. 172.
SLOPE OF THE CELLS
WHEN INVERTED.

"If you want the greatest quantity of honey, reverse your colonies; but if reversing was practiced everywhere, we would diminish the number of our colonies, and would finally even

* In reference to this, Mr. Shuck says: "This is not necessarily true. Stop inverting, and the frames fill just the same as they do in any non-invertible hive, of course. I attach importance to the system in preparation for the harvest, and getting the workers started right. After that, the hive may be used as a non-inverter. If you practice inversion weekly, the whole gather is likely to be in the supers, and you will be obliged to feed for Winter. If you cease inverting about the middle of basswood, you will have surplus, and the bees will have Winter stores, provided the flowers yield honey."

destroy the race of bees, for as far as *bee reproduction* is concerned the '*reversing Apiarist*' reaches the same result as the '*brimstoning Apiarist.*'"—French Apiarian Congress, Paris, 1861. *L'Apiculteur,* Volume 6, page 175.

Fig. 173. SCHUCK'S REVERSIBLE HIVE. (PATENTED.)

In the present state of progress in bee culture, "reversing" is less damaging, but its disadvantages to the bees cannot overbalance its advantages, unless it is practiced **very** cautiously and sparingly.

727. Yet this practice is sufficiently enticing — as it forces the bees to occupy the supers so quickly — to have caused the invention of a number of reversible hives or

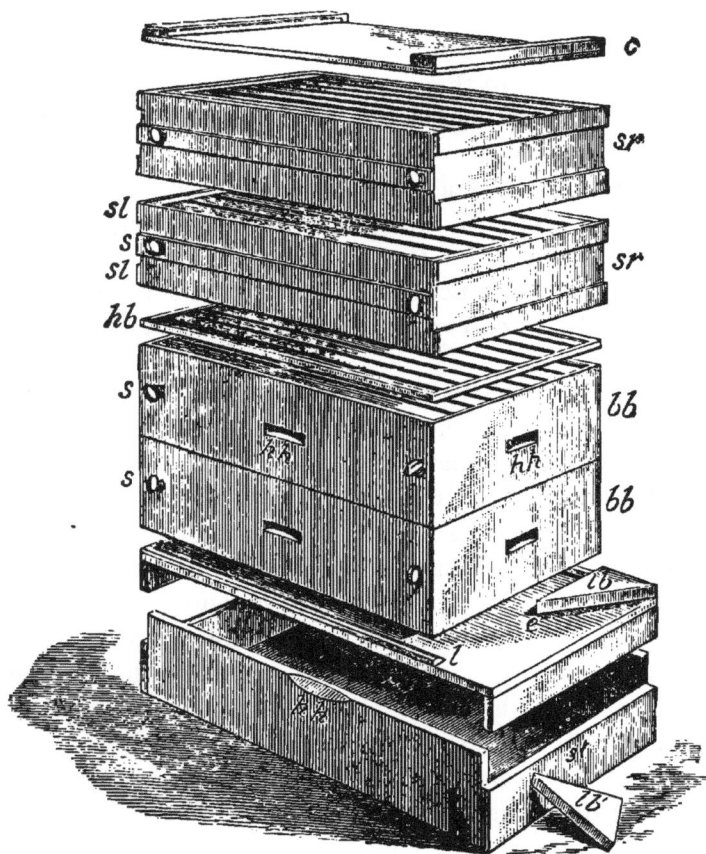

Fig 174. (From Cheshire.)
HEDDON'S REVERSIBLE HIVE. (PATENTED.)

st, stand; *bb*, body; *hb*, honey board; *sr*, section racks; *c*, cover; *hh*, hand holds; *lb*, entrance blocks; *e*, entrance; *l*, cleat to give bee space; *s*, screws to hold frames.

frames. If our readers desire to try "reversible hives," they will have but to choose among the many. The most popular reversible hives of the present day are the Shuck

and the Heddon, both patented. The former has frames of
the same size as the regular Langstroth pattern, and is
quite popular in Iowa.

728. Reversing during the harvest does not cause the
bees to gather any more honey ; nay, they harvest even a
little less, owing to the time occupied in transporting the
honey, *but it is all placed in the surplus apartment* at the
mercy of their owner.

A much safer method to induce the bees to work in the
supers, is to place in them. nearest the brood, a few unfinished sections from the previous season.* The supers
should be located as near the brood apartment as possible,
with as much direct communication as can be conveniently
given.

729. But, with the greatest skill, it is impossible to
attract the bees into the supers, as long as there are empty
combs in the brood-chamber.

If the queen is unable to occupy all the combs with
brood, the empty ones should be removed at the beginning
of the honey harvest, and either given to swarms or divided
colonies, or placed outside of the division board (**349**).
This is called "contraction." We would warn our readers
against excessive contraction, for, after the honey season
is over, a hive which has been contracted to, say, two-
thirds, of its capacity, has become dwarfed in honey,
brood, and bees, and will run some risks through the Winter.   Besides, that part of the super, which is above
the empty space, is but reluctantly occupied by bees.

"If the reader has ever constructed a hive. whose surplus
department was wider than the brood chamber, jutting out over
the same, he has noticed the partial neglect paid by the bees, to
the surplus boxes which rested over wood instead of combs.

* This is what Dr C. C. Miller calls a " bait." These unfinished sections
have been emptied of their honey by the extractor, and cleaned by the bees
the previous Fall.

" Now this same difference made by the bees, between wood and comb, they will also make between combs of honey and combs of brood, and with our 8-frame Langstroth hive, we notice far less neglect of the side surplus combs than we noticed when using the 10-frame hives. This is one objection to the method of contracting by replacing the side combs of brood chambers with fillers or dummies."—J. Heddon " Success in Bee-Culture."

730. A method which avoids contraction, and makes the best honey-producing colonies still better, consists in taking brood combs from colonies that are not likely to yield any surplus, and exchanging them, for empty combs from the best colonies, just before the honey harvest. This method requires too many manipulations to be very advantageous, and prevents the poorest colonies from becoming stronger.

731. *2d.* SECURING STRAIGHT, EVEN COMBS, IN SECTIONS. With thin comb foundation, in strips filling ½ to ¾ of the section, the combs are always straight, but their surface, when sealed, is not always *even*. Some cells are built longer than others, and, in packing the honey, these bulged combs might come in contact with one another and get bruised. To prevent this occurrence, many Apiarists use "separators," made of tin, wood, or coarse wire cloth, placed between the rows of sections, as in fig. 171. This invention, claimed by Mr. Betsinger, of New York, was first tried in the brood chamber, by Mr. Langstroth in 1858. It was suggested by Mr. Colvin. (See former edition, page 374.)

Let the reader bear in mind that these separators although useful, are not indispensable. They are to a certain extent an annoyance to the bees. Some Apiarists of ability, among whom we will cite Mr. Geo. II. Beard, of Missouri, manage to secure very nice honey in sections without them ; but if we were to produce large quantities of comb honey, we should use them, and would give the preference to those made of tin.

27

732. *3d.* KEEPING THE QUEEN IN THE BROOD APART-
MENT. If the supers have been put on just previous to the
opening of the honey crop, with sufficient bait to attract the
bees in them, there will be but little danger of the queen's
moving up into them, unless her breeding room is too much
cramped by honey, or by the exiguity of the brood nest.

The condition of the honey crop has something to do
with her propensity to move out of the brood apartment.
When the honey crop is heavy, and of short duration, there
is no danger on this score, as the honey combs are filled as
fast as they are built, and the
queen, should she move to the
super, would soon leave it, owing
to her inability to lay there. In
localities where the crop is lasting
and intermittent, much advantage
has been derived from the use of
the Collin perforated zinc (**191**).
The only obstacle to its use, is
that it hinders ventilation and free
access for the bees.

Fig. 175.
PERFORATED ZINC.
(From Root's "Gleanings.")

733. *4th.* SWARMING WITH COMB-HONEY PRODUCTION.
As the directions given by us elsewhere (**465**) do not
altogether prevent swarming, when comb-honey is raised,
and as the swarming of a colony usually ends its surplus
production for the season, it has been found advisable to
give the surplus cases to the swarm, instead of leaving them
on the old hive. To further strengthen the swarm, which
is thus depended upon for surplus, it is placed on the stand
of the old hive, and the latter is removed to a new location.
This is a very practical method. It is due to Messrs.
Heddon and Hutchinson, — at least *they* have popularized
it. But the prudent Apiarist, who follows this course, will
keep a vigilant eye on the old colony, thus deprived of all
its working force, and will help it, if needed.

734. *5th.* PREVENTING THE BEES FROM "PROPOLIZING."*

" Propolis on sections is a nuisance, be the same little or much, and a plan which will allow of the filling of the section with nice comb honey without changing the clean appearance which they present when placed upon the hive, will be heralded with delight by all, and give great honor to him who works out the plan."—G. M. Doolittle, " Gleanings," page 171. 1886.

We have shown (**238**) that bees " propolize" every crack, and daub with this yellowish or brownish glue every thing inside of their hive. This is very hard to clean, and it can never be removed sufficiently to restore to the sections their original whiteness.

" All four sides of the sections are scraped clean of propolis, and the edges as well. It is not a difficult job for a careful hand, but a very disagreeable one. The fine dust of the bee-glue is very unpleasant to breathe. A scraper should be a careful person, or in ten minutes' time he will do more damage than his day's work is worth. Even a careful person seems to need to spoil at least one section, before taking the care necessary to avoid injuring others. But when the knife makes an ugly gash in the face of a beautiful white section of honey, that settles it that care will be taken afterward."—Dr. C. C. Miller : " A Year Among the Bees."

To prevent propolizing, the sections should be fitted tightly together, and as little of their outside as possible exposed to the bees. The honey should be removed promptly, when sealed, before the bees have time to do much gluing (**237**).

735. *6th.* SECURING SEALED COMB HONEY. For this purpose no more cases should be given than the bees are likely to fill. The second case should not be added until the first is nearly filled. The outside sections, being the last filled, may not be sealed at all, unless the bees are somewhat crowded for room. To remedy this, many bee-keepers are

---

*This word " propolizing " is unauthorized by Webster, but it is needed as a technic word.

in the habit of "*tiering out*," instead of "*tiering up*;"
that is, they put the empty or unfinished sections in the
middle of the super, removing all that are filled, or placing
them on the outside. This is an increase of labor, but some
hold that it pays. Mr. Doolittle, in his practical pamphlet,
"My Management," explains that, at the close of the honey
season, he reduces the number of sections on the hive, by
narrowing up the surplus room, with a division board, which
he calls a "follower." Mr. Doolittle uses both side and
top-storing in his hives.

"As the cases are raised from the sides at this time, the follower is moved up, so as to shut the bees out of half the side
cases, unless in case of some extremely populous colony. By
this means the working force is thrown into a more compact
space, the result of which is a tendency toward completing the
sections they have commenced work in, rather than building
comb in more. After a week I go over the whole yard again, this
time shutting the bees out of the side boxes entirely, which
throws the full force of the bees into the top boxes, and, although
the honey-season may now be over, by getting this force of bees
all together they will cap the partly-filled boxes, where they
otherwise would not. This gives sections lighter in weight, but
makes much more of our crop in a salable form."

736. It very often happens that the bees fasten the comb
only at the top of the section. For safe transportation it
is very important that it should be fastened to the section
wall, all around. To secure this, not only do Apiarists use
foundation (**674**), but some have devised "reversible"
section cases. When the sections are turned over, the
empty space now at the top, seems unnatural to the bees,
and they hasten to fill it, making a solid comb in the section. But this is not the only method.

"Years ago my sections were always filled so full by the bees,
that they carried very securely in transportation. Afterwards I
began to have trouble from combs breaking down. It was due,
perhaps, mainly to the bees having too much surplus room.
Some sections would be filled with a nice comb of honey, not

very strongly attached at the top, very little at the side, and not at all at the bottom. Aside from depending upon crowding the bees to make them fill the sections, I wanted a plan whereby I could be sure of having the sections fastened at the bottom as well as at the top. I tried to take partly filled sections out of the supers and reversing them, and went so far as to invent a reversible super. I abandoned this however, and adopted the plan of putting a starter in the bottom as well as at the top of the section." ("A Year Among the Bees.")

Dr. Miller, who is an authority on comb honey production, further states that he uses a foundation "starter" one inch wide at the bottom, and wide enough at the top to leave only ¼ inch of room between the two. This allows for the slight stretching usual in comb foundat on.

737. To prevent the building of bridges between the upper and lower stories, some Apiarists use the Heddon skeleton or slatted honey board (fig. 76), which is separated from both the super and the brood chamber by a *bee space*, and in which the slats *break the joints* or passages of the bees thus ⎯⎯⎯⎯⎯⎯

This honey board answers its purpose, but we object to it, because it places the supers in less direct communication with the brood chamber.

We will now consider a few of the various cases and crates used in the production of comb honey.

738. THE DEEP BROAD FRAMES (fig. 171), have the decided advantage of allowing the Apiarist to use sections in a full size upper story. In limited comb honey production, they can probably be used with satisfaction. They also allow of a side storage as practiced by Mr. Doolittle.

THE HALF STORY BROAD FRAMES, are superior to the former, — though they require special cases, — because the bees can be confined to a shallow space, and when the crop is limited, or the weather cool, the sections are better and more promptly finished. We prefer half story comb honey supers, for the same reason we do half story extracting supers. Apiarists, who will follow our methods for extracting and raise but little comb honey, will see the benefit of using the same cases for both grades.

Mr. Heddon's invertible broad frames, in invertible section cases, are undoubtedly a good thing, especially as they are crowded together by the pressure of screws or offsets.

739. THE SECTION CRATE, invertible or not, is now used by the majority of specialists. Messrs. Miller, Shuck, Armstrong, Manum, Foster, all comb honey producers, have each a particular style of crate. Mr. C. C. Miller places his sections in crates without top or bottom, three-eighths of an inch deeper than the sections. To support the sections in these boxes he nails, under both ends, a strip of tin, which projects one fourth inch inside. Strips of tin, bent in the form of an L and soldered back to back, to form three inverted T's (fig. 170), are supported, across the box, by six small pieces of sheet iron, nailed at regular intervals under the sides of the box. Mr. A. I. Root improved these T as seen in the figure. These crates holding 28 or 32 sections, can be piled upon one another, leaving a bee space between them, while a similar bee space is provided between the sections and the slats of the skeleton honey board (fig. 76), by the shape of the latter.

740. Another way was contrived by Mr. Manum, of Bristol, Vermont, whose success in raising comb honey is well known.

He also uses a box without top or bottom, and holding only one row of 2-lb. sections, or two rows of 1 lb., eight to the row. These boxes, too, have strips of tin nailed under both sides and a band of sheet iron, for a cross-piece, running from end to end. A thumbscrew placed at one end, and acting on an offset, presses the sections against each other, and keeps the separators in place. Mr. Manum has used these clamps for several years and is well satisfied with them.

By the use of the Manum clamps, the sections are placed so closely that the bees cannot put any propolis between their edges. But their other parts are not protected.

741. To our mind, the implements invented by Mr. Oliver Foster, of Mount Vernon, Iowa, are worthy of notice and his conceptions of the general management of sections are so well explained, that we could not do better than copy a few pages of his small pamphlet.

"There should be *free communication* between the sections in *every direction.* They should have deep slots on all 8 edges as shown in Fig. 176 so that bees can pass freely over the combs from end to end of the case, as well as from side to side, and from top to bottom.

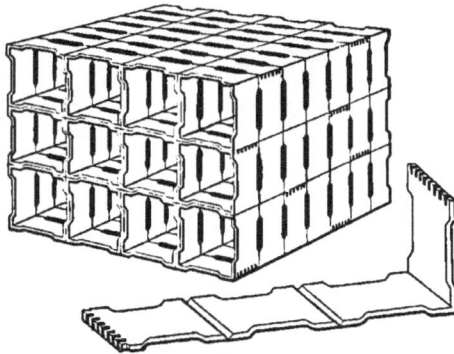

Fig. 176.
OPEN SECTIONS.
(From "How to Raise Comb Honey.")

"You may not appreciate the importance of this until you have tried them.

"*When we take into consideration that the object on the part of the bees in storing up honey in Summer, is to have it accessible for Winter consumption, and that in Winter, the bees collect in a round ball, as nearly as possible, in a semi-torpid sta'e with but little if any motion, except that gradual moving of bees from the center to the surface and from the surface to the center of this ball, we may imagine how unwelcome it is to them to be obliged to divide their stores between four separate apartments, each of which is four inches square and twelve inches long, with no communication between these apartments.*"

The italics are ours. This passage is most important.

742. "The case is made of four plane boards. B, B, C, C, (fig. 177). They are cut 1-16 in. narrower than the sections are high. A side and an end are nailed together in the form of a letter L. When two of these L shaped sections are placed together, they form the rectangular case, open at two opposite corners diagon-

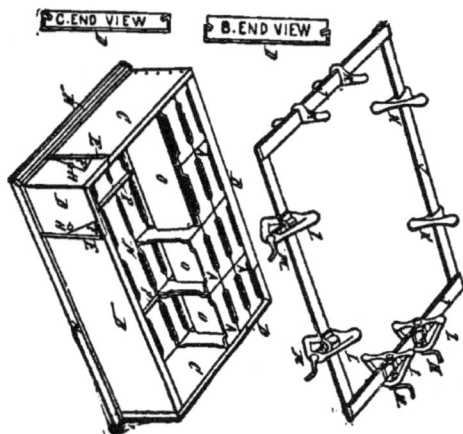

Fig. 177.

A, A, A, Section Boxes.

B, B, C, C, Plane side and end boards.

D, D, folded tin corner plates.

E, E, Flanges folded outward on ends of D.

F, F, Tin wedges which hold the case tight on the sections after clamping.

J, J, J, Iron clamp by which the case is drawn tight on the sections both ways.

II, II, Heads of nails through slots I.

O, O, O, Tin Separator in place.

P, P, Narrow tin strips supporting separators.

N, N, N, Slotted honey board,

ally. The boards are mitred together at these open corners and are clasped together by the tin angle plate D. These corner plates are also bent L shape.

"They are as high when folded as the sections, and $3\frac{1}{2}$ inches from the corner to each end. They have a small flange, bent outward on each end, E, and a double fold bent inward on each side, which forms sockets $\frac{3}{8}$ inch wide in which the end of the boards slide in and out, thus expanding or contracting the case in length and width.

"The folded side edges of the tin slide in saw grooves cut in the edges of the boards, are shown in the small figures, and the case is held rigid, whether open or closed. A small nail is driven through each of the slots I, into the wood, to prevent the case from opening farther than about $\frac{1}{2}$ inch larger each way than when closed.

"The case when closed is a little smaller than the tier of sections to be used.

"To fill the case it is placed on a level board and opened out. The sections are then carelessly arranged inside, and then drawn into position by pressing the case together. A wrought iron clamp, J, is then slipped over the case, and by operating the screws. M. the case is drawn so tight on the sections that all cracks between them are closed up, thus protecting the surface of the boxes from being soiled.

"To prevent the spreading of the case when the clamp is removed, four simple tin wedges, F, F, are slipped under the flange, and the nail head.

"This bottomless case of sections is then placed on the hive on a slotted honey board, which is level on top and has slots to correspond with those between the sections, save that the slots in the board are a little narrower, to secure perfect protection to the sections. If separators are used, they are simply dropped in between the rows of sections as each row is put in. (See O, fig. 177). They rest on the edges of two narrow strips of tin, P, P, that pass across each end of the case between the rows of sections at the bottom. These strips are movable, and securely held in place while handling, like the sections, by the lateral pressure of the case. The iron clamp is not a necessity, but it is very convenient where several colonies are kept. The case is equally adapted to use with or without separators. It can be used with or without an outer case. It can be ' tiered up ', ' reversed ', (inverted) or placed on end or on one side for ' side storing '."

743. In removing the cases from the hive, apply the clamp and lift all together, or open the case and take out one box at a time, using a little smoke, and shaking and brushing off the bees. Nearly all of the bees can be shaken from a single case-full before opening it; but the neatest way to get them out is to place the cases in an empty hive a little to one side of the front of the hive from which they were taken. Fasten a wire cloth tube over the only opening at the entrance of this empty hive. Make the tube 6 inches long, $\frac{3}{8}$ inch in diameter at the small end, and $1\frac{1}{2}$ inch at the end attached to the hive. Place the hive in position so that the point of the tube will touch the front end of the hive containing the colony. In a few moments, the bees will be marching ' double quick ' out through the tube, and in an hour or so every bee will be out." (Oliver Foster, " How to Raise Comb Honey." 1886.)

We advise every bee-keeper to procure this small pamphlet.

744. In support of what Mr. Foster wrote in behalf of the open-side sections, we may add that bees seem to consider a row of these sections as formed of a single comb, and that, in consequence, they attach each small comb to the sides, giving them more solidity. For the same reason bees are also less inclined to make bulged combs, and separators may be set aside with less risk of lack of uniformity.

Another and very important point, in favor of these sections, is the increased facility to ripen honey by evaporation, for the air can easily circulate from side to side, instead of from top to bottom only, as when closed-side sections are used.

745. Before closing our chapter on the production of comb-honey, in which we have tried to give our readers some of the best known methods, we must warn them against using too many contrivances, whenever they can possibly help it. All improvements that are made must be based on a full consideration of the instincts of the bees. Like Mr. Hutchinson ("Production of Comb-Honey" p. 18), we "have seen bees sulk for days during a good honey flow, simply because the present condition of things was not to their liking." Use as large sections as your market will allow. If you use separators and honey-boards, at all, let them be light and perforated. In a word, make your bees feel as natural and as much "at home" as possible.

Extracted Honey.

746. To separate the honey from the wax, the beekeepers of old used to melt or break the comb and drain the honey out.

Beeswax, as a sweet-scented luminiferous substance. far superior to oils or the crude grease of animals, was greatly appreciated by the priests. and placed among the best offerings required to please the gods. The custom of offering wax. or wax candles, continued to this day by some churches, especially by the Greek and Roman Catholic churches, caused for centuries the levy of heavy taxes, payable in beeswax, in countries where the inhabitants kept bees. Some countries, in Europe, had to pay to the church, every year, several hundred thousand pounds of beeswax. Such taxes compelled the bee-keepers to separate the honey from the wax with as little waste as possible.

Different grades of honey were harvested by the careful Apiarists. The light-colored combs produced a light-colored and pure honey; the combs which had contained brood produced turbid honey of inferior quality.

747. These primitive methods were afterwards greatly ameliorated, as for instance, in the French province of Gatinais, where the bee-keepers used the heat of the sun to melt the combs, and separate the honey from the melted wax. The choice honey obtained in Gatinais, from the *sainfoin*, cannot be excelled by our best extracted clover honey, as to color and taste, and it is sold in Paris altogether.

Owing to these causes, *strained* honey, of different grades, was a staple in Europe. But the demand being ahead of the supply, especially when the season was unfavorable for bees, Europe imported strained honey from Chili, and Cuba, and lately, extracted honey from California.

748. These causes did not exist in this country. Bees were scarce here at first. The American settlers had too much work on hand to care much for bees. The few who owned a limited number of colonies, brimstoned one of them occasionally, and consumed the honey at home. The more extensive bee owners could sell some broken combs to their neighbors, or a few pounds of strained honey to the druggist, who was not very hard to please, being accustomed to buy Cuba honey, harvested with the most slovenly carelessness. By and by, however, owing to very favorable conditions, the wild woods swarmed with bees in the "hollow trees," and the *bee-hunter* made his appearance. Thousands of trees fell under his ax, to yield the sweets that they contained. This rough-and-ready bee-keeping, or rather bee-killing, produced comparatively large quantities of honey ; but, as this honey was nearly always badly broken up and mixed with pollen, dead bees, and rotten wood, it became customary to boil the honey, so as to force the impurities and the wax to rise on top with the scum. Hence the cheap, liquid, dirty and opaque *strained* honey, dark in color and strong in taste. By the side of this unwholesome article, a little fancy comb honey was sold, that led to a national preference for comb honey.

Bu. in view of the cost of comb to the bees (**223**), in honey, time and labor, it was earnestly desired by progressive bee-keepers, especially after the invention of the movable frames, that some process be devised to empty the honey out of the combs without damaging the latter, so that they could be returned to the bees to be filled again and again.

749. In 1865 the late Major de Hruschka, of Dolo, near Venice, Italy, invented " *Il Smelatore*," THE HONEY EXTRACTOR.

It happened in this wise : He had given to his son, a small piece of comb honey, on a plate. The boy put the plate in his basket, and swung the basket around him, like a sling. Hruschka noticed that some honey had been drained out by the motion, and concluded that combs could be emptied by centrifugal force.

This invention was hailed, in the whole bee-keeping world, as equal to, and the complement of, the invention of movable frames ; and it fully deserved this honor.

750. As soon as we heard of the discovery, we had a machine made. It was not so elegant as those which are now offered by our manufacturers. It was a bulky and cumbersome affair; four feet in diameter and three feet high ; yet it worked to our satisfaction, and we became convinced, by actual trial, of the great gain which could be obtained, by returning the empty combs to the bees.

751. Let us say here, that the profit was greater than we had anticipated ; but we, together with a great many others, first committed the fault of extracting, before the honey was altogether ripened by evaporation. Like "Novice," who thought of emptying his cistern to put the overflow of his extracted honey, we had to go to town again and again, for jars and barrels, to lodge our crop. But experience taught us that we cannot get a good merchantable article, unless the honey is ripe.

752. If we give to bees empty combs, to store their honey, we will find, by comparing the products of colonies who have to build their combs, with those of colonies who always have empty combs to fill, that these last produce at least twice as much as the others.

A little consideration will readily show, to the intelligent bee-keeper, the great advantages given to the bees by furnishing them with a full supply of empty combs. To illustrate all these advantages, let us compare two colonies of bees, of equal strength, at the beginning of the honey season ; one with empty boxes, the other with empty comb in the boxes.

The two colonies have been breeding plentifully, and harvesting a large quantity of pollen, and a little honey, for several weeks past. The brood chamber is full from top to bottom. After perhaps one rainy day, the honey crop begins. The bees that have been given empty combs can go right up in them, and begin storing, as fast as they bring their honey from the fields. Not a minute is lost ; and as they have plenty of storing room, there is no need of crowding the queen out of her breeding cells.

In the other hive, there is indeed plenty of empty space in the upper story ; but before it can be put to any use, it has to be first partly filled with combs. Before a half day is over, the greater part of the bees have harvested, and brought, to their newly-hatched companions, all the honey that the latter can possibly hold in their sacks. What shall they do with the surplus? They have to go into that upper story, and hang there (205) for hours, waiting for the honey to be transformed into beeswax, by the wonderful action of these admirable little stomachs, whose work man cannot imitate, despite his science. But, while this slow transformation is going on, while the small scales of wax are emerging from under the rings of the abdomen (201) of each industrious little worker; while their sisters are slowly but busily carrying, moulding and arranging the warm little pieces of wax in their respective places, in order to build the frail comb (206); during all this time, the honey is flowing in the blossoms, and the other colony is fast increasing its supply of sweets. Meanwhile, the few bees, which have found a place for their load, go back after more, and, finding no room, they watch for the appearance of each hatching bee, from its cell, and at once fill that cell with honey; thus depriving the queen of her breeding-room, and forcing her to remain idle, at a time when she should be laying most busily.

The loss is therefore treble. First, this colony loses the present work of all the bees which have to remain inside to help make wax. Secondly, it loses the honey of which this wax is made. Thirdly, it loses the production of thousands of workers, by depriving the queen of her breeding-room, in the brood-chamber. All this, for what purpose? To enable the owner to eat his honey with the wax (719); when, as every one well knows, wax is tasteless and indigestible.

One word more in regard to the loss of production, by the crowding of the queen. This loss is two-fold in itself. When the bees find that the queen is crowded out of her breeding-room, they become more readily induced to make preparations for swarming (**406**).

It is then that a large number of young bees would be necessary to make up for the loss which the colony will sustain, in the departure of the swarm ; and yet the diminished number of eggs laid produces exactly the reverse of the desired result.

There is perhaps a fourth item of loss, in failing to furnish empty combs to this colony, and that when the season is not very favorable. Many practical bee-keepers have noticed that, in rather unfavorable seasons, it is difficult to induce bees to work in an empty surplus box, which they would work in readily if it were furnished with combs. It is a question which may remain doubtful, whether the bees do not sometimes, in such cases, remain idle for a day or two, rather than begin building comb in a box which they do not expect to be able to fill.

753. In view of the above *facts*, and after an experience of twenty years with the honey extractor, we strongly urge all beginners to produce extracted honey in preference to comb-honey, wherever they can sell it readily for half as much as comb honey. We have shown the advantages of its production *to the bees;* let us now show the advantages *to the Apiarist.*

754. *1st.* He can control, and take care of, a much greater number of colonies. The manipulations of an Apiary, run for extracted honey, occupy less than one half of the time required for the production of comb-honey. Our largest comb-honey producers acknowledge that one man cannot handle more than two hundred colonies successfully, when run for comb-honey (**719**), while as many as five hundred colonies, located in different Apiaries (**582**), are managed successfully by one Apiarist, when run for extracted honey During extracting time, of course, additional help is required, but this needs not be skilled labor, which is always hard to find.

755. *2d.* By the production of extracted honey, the surplus combs are saved, and given to the bees at the opening of the following harvest. This virtually does away with natural swarming, and enables the bee-keeper to control the increase of his colonies to suit his desires. One of the most successful comb-honey producers, Mr. Manum, of Vermont, who sold some 15 tons of comb-honey in 1885, acknowledged to us, that with his management in the production of comb-honey, it was nearly impossible to control swarming, and that the time was not far distant when he would have too many bees. He owned seven hundred colonies at the time.

756. The farmer, or merchant, who keeps only a few hives, to produce honey for his own use, will find it much preferable to produce extracted honey. With three colonies of bees and an extractor, in a very ordinary location, from 150 to 300 lbs. of honey can be produced on an average, every season.

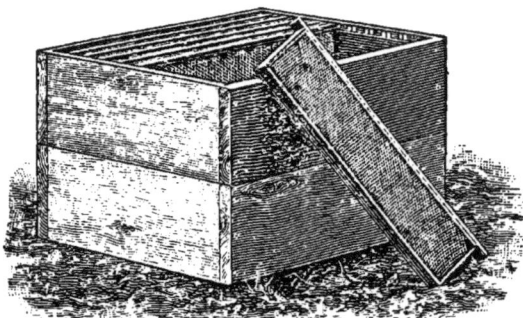

Fig. 178.
TWO HALF-STORY SUPERS FOR EXTRACTING.

757. For the production of extracted honey, we use half stories or cases (fig. 178) with frames 6 inches deep, and of the same length as the frames of the lower story. We also use full-story supers, but only on standard Langstroth

PLATE 14.

FRANCESCO DI HRUSCHKA,
Inventor of the Honey Extractor.

This Apiarist is mentioned page 428.

.

hives, and we decidedly prefer the half-story supers, for several reasons, after having used both kinds on a large scale for years.

The frames of the half-story supers are more easily handled when full, and the combs are less apt to break down from heat or handling. The half-story super is better suited for the use of an average colony, and in cool weather is more easily kept warm by the bees, than a full-story. Very strong colonies, in extraordinary seasons, can be readily accommodated with two and even three of these cases successively.

758. With the full-story supers, the queen and the bees are more apt to desert the lower story altogether, in poor honey seasons, and establish their brood-nest in the upper story, especially when the combs of the lower or brood chamber are old, and those above are new. The sole advantage of the full-story super is that the frames in it are exactly of the same size as those below, and can be interchanged with them if necessary ; but with large hives it will never be required to use upper story combs for feeding, and even if the queen should breed in these shallow cases, at times, she is soon crowded out of them by the surplus honey.

759. The upper story frames are filled with comb foundation (**674**), or even with old worker comb, and can be used indefinitely, since the honey is extracted from them, and they are returned unbroken to the bees. We have now several thousands of these combs, some of which have already passed fifteen or twenty times through the extractor and are now as good as at first, nay, even better ; for some, which were very dark, are lighter in color now, on account of the dark cells having been shaved by the honey knife and mended, by the bees, with new wax. These supers are given to the bees, a few days previous to the opening of the honey crop.

28

The mat (**352**), and cloth (**353**), are removed and the upper story is placed immediately over the frames (fig. 68).

760. One great advantage of this style of supers, lies in the facility, with which the bees can reach the upper story from any comb, or from any part of a comb, either to deposit their honey or for ventilation, during hot weather. Bees show their preference for these large receptacles very decidedly. For comparison, let two or three broad frames (**299**)—filled with sections which are of more difficult ventilation and access — be placed in the center of one of these supers with some extracting frames on each side, all equally filled with strips of foundation (**674**), and the small sections (**721**) will be filled last almost in every instance, even although placed nearest to the center of the brood-nest.

Mr. Langstroth was the first to call the attention of Apiarists to the loss incurred by compelling bees to store the surplus honey in small receptacles. The bee-keeper cannot afford to sell honey stored in small sections, except at a considerable advance over its value in large frames. Colonies, which do not have the breeding apartment nearly full of brood, honey and pollen, need not be supplied with supers (**757**), till they show a marked progress.

761. After the opening of the honey crop, which is very easily noticed by the greater activity of the bees and the *whitening* of the upper cells of *their combs*, a regular inspection of their progress is necessary. The season is short, but the daily yield is sometimes enormous.

762 Mr. A. Braun stated, in the *Bienenzeitung*, September, 1854, that he had a mammoth hive furnished with combs containing at least 184,230 cells,* and placed on a platform scale, that its weight might readily be ascertained

---

* Such a hive would hold about three bushels Mr. Wildman says that ''a clergyman set a well stocked hive of bees on a tub turned bottom up, after having made a hole through the bottom, and took from the tub four hundred and twenty pounds of honey.''

at stated periods. On the eighteenth of May it gained eighteen pounds and a half. On the eighteenth of June, a swarm weighing seven pounds issued from it, and the following day it gained over six pounds in weight. Ten days of abundant pasturage would enable such a colony to gather a large surplus, while five times the number of equally favorable opportunities would be of small avail to. a feeble one.

The largest yield of extracted honey, ever harvested by the colonies of one Apiary under our control, was 13,000 pounds in about fifty days, the most protracted honey crop we ever knew. This was harvested by eighty-seven colonies, making a daily average of three pounds a day per colony of evaporated honey. Such seasons are scarce.

As some colonies harvest much more than others, they need more attention.

763. *To secure the greatest possible amount of extracted honey, the colony should never be left without some empty comb.*

As soon as the combs of one of these supers are about three-fourths full, we put another rack under the first, and sometimes a third under the second. All this without waiting for the honey to be sealed; but we never remove the honey, to extract it, until the crop is at an end, for we want to get our honey *entirely ripened.*

Honey is evaporated, or ripened, by the forced circulation of air, caused by the fanning of the bees through the hive, in connection with the great heat generated by them. As honey evaporates, it diminishes in volume, and as long as the bees continue their harvest, they constantly bring in unripened, or watery honey, which they store in the partly filled cells that contain honey already evaporated. It is for this reason that unsealed honey, after the crop is over, is as ripe as honey sealed during the crop, and sometimes riper. If the crop is abundant, they often seal their combs too soon, and the honey thus sealed may afterwards ferment in the cell and burst the capping.

764. Some Apiarists extract the honey as fast as it is harvested by the bees, and afterwards ripen it artificially by exposing it to heat in open vessels. We do not like this method, and prefer to extract the whole crop at once. It is much more economical, for, with our system, one skilled man attends to as many as five or six Apiaries during the honey crop, and extracts at leisure afterwards, with almost any kind of cheap help. Since honey now has to compete in price with the cheapest sweets, the question of *economical production* is not to be disregarded.

" He who produces at maximum cost will fail. He who produces at minimum cost will succeed."—(Jas. Heddon.)

765. As some colonies do not begin work in the supers until very late, and do not fill all the space given them, the surplus of other colonies can be given them in such a manner that all will be equally filled. This can be done without brushing the bees off **(485)**.

The equalizing of empty combs in the surplus stories of different colonies, towards the end of the crop, will save time in extracting, as the supers will be found more evenly full. The giving of a few combs of honey to a colony that has not yet begun work in the supers also acts as an inducement, and gives the bees new energy.

HARVESTING.

766. The extracting, to be done swiftly, requires the work of four persons: three men and a boy. This work is done at a time when the bees have ceased to make honey, and the greatest care has to be exercised not to leave any honey within the reach of robber bees. The work of opening the hives, removing the combs and brushing off the bees, must be done quietly, but swiftly and carefully. The receptacles for combs should each have a cover, and the hive should be closed and its entrance reduced, as promptly as possible. In this way, there is not the least danger of robbing; but if robbing is once begun, by some carelessness or forgetfulness of the operator, the work has to be stopped until it has subsided. A basin of water and a towel, placed near at hand, are found to be very convenient, when the hives are very full; as the operator and his help sometimes get their fingers sticky with honey.

767. The utensils needed for neat extracting on a large scale are: In the Apiary, — a good smoker (**382**), one or two brushes made of asparagus tops, or some other light fibrous material, a wood chisel to loosen the cases, two tin pans, described farther on (**770**), one comb bucket, and two strong linen or cotton "*robber cloths*," which can be carbolized beforehand by the Raynor process (**384**).

768. The "robber cloths", so named by Dr. C. C. Miller, are used to cover the cases to keep away robbers. They are made of very coarse cloth or gunny, about a yard square.

"Take two pieces of lath, each about as long as the hive, and lay one upon the other, with one edge of the cloth between them. The cloth is longer than the lath, allowing 6 inches or more of the cloth to project at each end of the lath. Now nail the laths together with 1½ inch wire nails, clinching them. Serve the opposite end the same way, and the robber cloth is complete. You can take hold of the lath with one hand, lift the cloth from a hive or super, and with a quick throw, instantly cover up again your hive or super perfectly bee tight." ("A Year Among the Bees," 1886.)

769. The operator opens a hive, removes the super, places it in a tin pan (**770**), and covers it with a robber cloth. He then examines the brood chamber, from which one or two combs may be removed if advisable. We usually leave all the honey in the lower story, unless the bees are crowded out of breeding room, which will not happen, if they have plenty of room above.

When any comb is removed from the brood chamber, the bees are shaken off in front of the hive, the remaining ones are brushed off, and the comb is placed in the comb bucket (fig. 85). The hive is then closed, an empty surplus case (fig. 178) is placed in a second tin pan, and the combs of the filled case are, one after another, shaken and brushed in front of the hive and transferred to the empty case. The assistant, usually an apprentice, can help a great

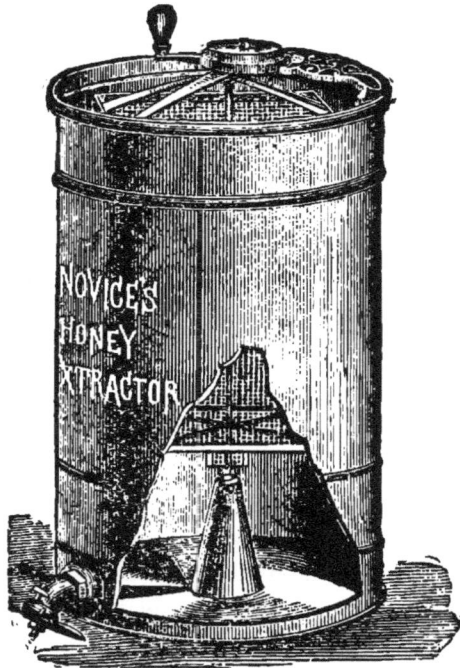

Fig. 179.
NOVICE'S EXTRACTOR.
(From the "A B. C. of Bee Culture.")

deal in this, and if the bees are handled according to rules, no one need be stung. When the combs are all transferred, the assistant carries the case to the honey room, while the Apiarist prepares to open another colony; the case which has just been emptied serving to transfer the combs of the next super.

When the harvest is large, a wheel-barrow may be used to bring the honey to the honey room.

770. In the honey house, there should be an extractor, a capping can, (fig. 183) a honey knife, a funnel with sieve, a pail, a barrel, and two tin pans like those used in the

Fig. 180.
MUTH'S HONEY KNIFE.

Apiary. The floor may be covered with painted, or oil-cloth, or strong enamel-cloth, in case any honey is spilled; each

Fig. 181.
THE MUTH'S EXTRACTOR SHOWING
HIS SLANTING BASKET.

Fig. 182.
EXCELSIOR EXTRACTOR.
(From "Bees & Honey.")

person may be provided with a good enamel-cloth apron, and all the windows furnished with wire cloth netting, to allow the bees to escape (**586**). The tin pans above mentioned are shallow, in the shape of bread pans, large enough to receive one of the supers freely, to keep the leaking honey from daubing anything, or from attracting robbers (**666**). They are supplied with strong handles.

771. We have said that we do not usually take honey from the brood chamber, but in an emergency we sometimes extract even from combs containing brood. We never noticed any loss of worker brood unless it was actually thrown out. If a few worker larvæ are displaced by the rotation, the bees push them back to the bottom of the cells. In all cases, when there is brood, the crank must be turned slowly.

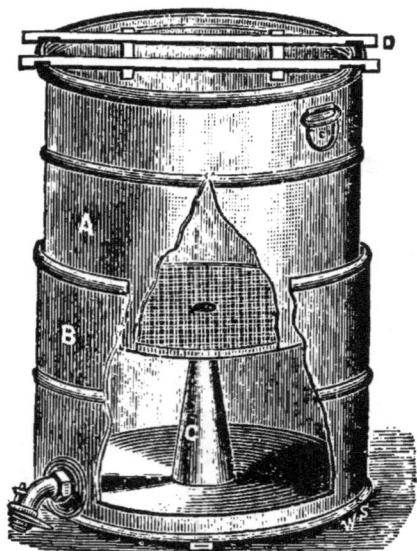

Fig. 183.
THE DADANT CAPPING-CAN.

772. In the extracting room, a man, *the shaver*, as we call him, uncaps the combs, as fast as they are brought. He stands before the *capping-can* (fig. 183). The capping can is formed of a lower can B, 24 inches wide and 14 inches high with a slanting bottom, a faucet and a central pivot C.

On this lower can is placed another can A, 23 inches wide and 22 inches high, with a coarse wire cloth bottom resting at the center on the pivot C. The upper can acts as a large sieve. On the top of it is placed a wooden frame D, notched, so as to fit on the edges of the can. It is on this frame that the combs are uncapped, and the cappings fall in the sieve, where the honey drains out of them, into the lower can. Our capping can is meant to hold the cappings of two days' extracting.

773. The all-metal extractors, of different makes, are the only ones now in use. Two-frame extractors are the most common, but we use four-frame extractors altogether, one in each Apiary. These extractors accommodate eight half-story frames.

774. In regard to the honey or uncapping knife, justice compels us to say that, so far, to our knowledge, there is but one which is really practical, the *Bingham honey knife.*

Fig. 184.
THE BINGHAM KNIFE.

This knife does away with the annoyance of having the cappings stick to the comb again, after having been shaved off, because it is made with a bevel, which causes *the shaver* to hold it in a slanting position, so that the cappings cannot stick to the comb again, unless purposely allowed to do so.

As fast as the combs are uncapped on both sides, they are put into the extractor, which may be turned by a boy. Care should be taken that the combs, that are placed opposite one another, be of nearly equal weight, as the unequal weight causes the extractor to swing right and left, fatiguing the boy and injuring the machine.

775. A quiet; regular motion is all that is necessary to throw the honey out, and, in warm weather, it fairly *rains* against the sides of the can with a noise similar to that of a shower on a tin roof.

776. Now is the time to invite the neighbors and their children to come to see the fun, and taste the golden nectar. Aside from the pleasure of making everybody happy, the present of a few pounds of honey proves an inducement to its use, and an advertisement for the producer.

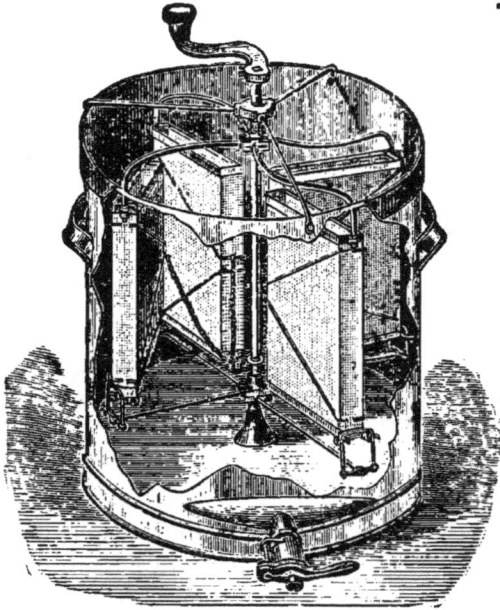

Fig. 185.
THE STANLEY AUTOMATIC REVERSING EXTRACTOR.

Extracting-day should always be understood to mean "free honey to all visitors." Let them visit the honey-room, and if the ladies get their dresses a little daubed while peeping in the extractor, they will soon find out that honey does not stain like grease, but will *wash* off in warm water.

777. After the combs are extracted on one side, they are turned over and extracted on the other. Mr. Stanley, of New York, invented an extractor (figure 185), in which, the combs are turned over by simply reversing the motion of the gear. This invention has not been sufficiently tried to be proclaimed decidedly superior; but it appears to have some advantages, the main drawbacks being the greater cost of the machine and its bulk. Similar extractors were introduced into England, by Mr. Cowan, several years ago.

778. The extractor is fastened on a high platform, so that the honey pail can be put under the faucet. A barrel is in readiness, with the large funnel and sieve over it. This sieve should be large enough to take a pailful of honey, so as to cause no delay.

A mark is made on the barrel, with a crayon, or chalk, as each pailful is poured in. In this way we know when the barrel is full, without having to gauge it, and we avoid having the honey run over and waste.

779. We would advise beginners, who extract for the first time, to go slowly and carefully. A little care, besides saving time, will save the waste of several pounds of honey, and make things more comfortable; for a pound of honey wasted goes a great way towards making everything sticky and dirty. If a splendid crop and neat work are pleasurable, a daubed honey-room and cross bees in the Apiary irritate both the Apiarist and his assistants, who soon become sick of the work. When things are rightly managed, the work is so delightful that more help can be found than is needed.

780. Of all manipulations, extracting is that which requires the greatest precautions against robbing (**664**). Carefully avoid all unnecessary exposure of comb or honey. Robbers not only annoy the Apiarist, but they cause the bees to get angry, and to sting.

781. All the cases, when extracted, are piled up on the oil-cloth carpet, till the day's work is done. The combs are never put back into the hive before evening, at sun down; to prevent too much excitement in the Apiary. In half an hour, every hand helping, the whole number is distributed on the hives; though we may have extracted as much as two thousand pounds in a day.

There are seasons, in which a very slight continuation of the honey crop, permits returning the combs, as fast as they are extracted. In such seasons it causes no excitement, and is much more convenient.

782. Within two or three days after extracting, the bees have cleaned the combs, and repaired them. But, to prevent the moths from injuring them, we keep them on the hives during the whole summer; the bees take care of them, and in the Winter, we pile up the cases, carefully closed, in cold rooms, where the cold of Winter destroys the eggs of the moth (**802**).

In localities, where there are two distinct crops of honey, each crop should be harvested separately. Thus, we always extract the the June crop in July, and the Fall crop in September.

783. Honey production, with the above methods, is so successful that the problem for practical Apiarists is no longer, how to produce large crops of honey, but *how to sell it* (**839**). Extracted honey can certainly be produced, at less cost, than the cheapest of cane sugar, and it can be truly said, that in the last thirty years, there has been more progress in bee-culture, than in any other branch of rural economy.

CHAPTER XVIII.

DISEASES OF BEES.

784. Bees are subject to but few diseases that deserve special notice. We have said (**626**) that we consider diarrhea as the result of an accumulation of fœces only, but Mr. Cheshire has examined some of the fœces of diarrhea, and found in some of them living organisms, which indicate that, sometimes, the distension of the abdomen is not caused by the overloading of the intestines alone. These organisms, when better known, will probably explain some of the losses of bees, after Winter, and the Spring dwindling (**659**), which reduces so many colonies.

785. We have said also (**665**), that those bees, who are in the habit of robbing, assume a smooth, black appearance. Mr. Cheshire thinks that this explanation of glossy black bees is inaccurate, and claims that an examination of such bees has shown, in them. the presence of living organisms, which he named *bacilli gaytoni*, after Miss Gayton, who found some of her colonies suffering from this disease, for three years in succession. These organisms have since received, in England, the name of *bacilli depilis*. This last term means hairless, the bees affected with the disease losing all, or nearly all, their hair. We do not question the accuracy of the examination of these shiny, hairless bees, but we know that bees who are habitual robbers lose their hair, and assume this sli.k, shiny appearance, without suffering any disease; for they belong to healthy colonies, and are only a small exception among other bees.

Foul-Brood.

786. There are other unimportant diseases, which have not yet been studied, but all are nothing, when com· ared to the dreadful contagious malady, already known thousands of years ago* and commonly called *foul-brood*, because it shows its effects mainly by the dying of the brood, but the denomin.tion is improper, for the brood is not alone diseased.

" When we remember that bees live in the closest contact in very numerous colonies; that their usual system of intercommunication is by actual touch ; that they habitually pass food from one stomach to another, while all the food they have has been carried either within or upon the bodies of their fellows ; that their very home is formed of one of their secretions, and that their beds, cradles and larder are all interchangeable, we shall admit that the circumstances are such as would appear to favor the development of contagious diseases."— (Cheshire.)

787. The scientific and indeed the true name of foul-brood is *bacillus*† *alvei*, "*small stick of bee-hives*" because it is composed of living organisms resembling small sticks. It develops very rapidly, and has been found, by Schonfeld and by Cheshire, not only in the brood, bu in the bees and queens. The rapid depopulating of the colonies infested, coupled with the fact that Mr. Bertrand has known several queens to die in diseased colonies, leaves no doubt as to the accuracy of the microscopical experiments made by Cheshire, on queens who were found with ba illi. not only in their organs, but also in the half developed eggs of their ovaries. According to the English microscopists, there are two kinds of bacilli alvei, the major and the minor, the

---

* As Aristotle (*History of Animals, Book IX , Chap.* 40) speaks of a disease which is accompanied by a disgusting smell of the hive, there is reason to believe that foul-brood was common more than two thousand years ago.

† *Bacillus*, plural *bacilli*, from the Latin, *a stick*.

larger and the smaller (*British Bee-Journal*), but are they equally to be feared?

These imperceptible "sticks" break successively into several parts, every one of which forms a colony of spores, that pass through divers shapes before developing into new bacilli. We can judge of the promptness of their reproduction, and of their minuteness, when we read in Cheshire, that a dead larva frequently contains as many as *one billion* of these spores (**28**).

788. In the *Bulletin Agricole du Département de l'Aube*, Mr. Brunet narrates the experiments made by Mr. Marcel Dupont, to breed the bacilli of foul-brood. Knowing that Pasteur used beef-broth in this kind of experiments, Mr. Dupont filled three glass-tubes with unsalted beef-broth, prepared according to the directions given by Pasteur, and after sealing and boiling them, to kill any living organisms that might have existed inside, he introduced into two of them, with a fine needle, a small quantity of a liquid, in which particles from the body of a diseased larva had been dissolved. One week after, the broth in both of these tubes, was cloudy and full of bacilli, while the liquid, in the third tube, had remained clear and unchanged.

789. Description. As we have never seen a case of bacillus alvei, we will borrow from those who have been more "lucky" (?) than ourselves, a description of the disease, for its detection in hives, and the remedies recommended by the best authorities.

" In most cases the larva is attacked when nearly ready to seal up. It turns slightly yellow, or grayish spots appear on it. It then seems to soften, settles down in the bottom of the cell, in a shapeless mass, at first white, yellow, or grayish in color, soon changing to brown. At this stage it becomes glutinous and ropy; then, after a varying length of time, owing to the weather, it dries up into a dark coffee-colored mass. Usually the bees make no attempt to clean out infected cells, and they will sometimes fill them with honey, covering up this dried foul-brood matter at the bottom.

Sometimes the larvæ do not die until sealed over. We have been told that such may be easily detected by a sunken capping perforated by a "pin-hole". This is by no means invariably the case. Such larvæ will often dry up entirely, without the cap being perforated or perceptibly sunken, although it usually becomes darker in color than those covering healthy larvæ.

The most fatal misapprehension has been in regard to the smell of the disease. In its first stages there is no perceptible smell, and it is not until the disease has made a considerable progress

Fig 186.   (From Cowan.)
APPEARANCE OF FOUL-BROODY COMBS.

that any unusual smell would be noticed by most persons. In the last stages, when sometimes half or more of the cells in the hive are filled with rotten brood, the odor becomes sufficiently pronounced, but the nose is not to be relied on to decide whether a colony has foul brood or not. Long before it can be detected by the sense of smell, the colony is in a condition to communicate the disease to others.

The eye alone can be depended on, and it must be a sharp and trained eye too, if any headway is to be made in curing the disease. (J. A. Green, in "Gleanings," January 1887.

790. "Foul-brood can be detected in the Spring, either through an unusual spreading of the brood, resulting from an unnoticed previous infection, of an indefinite number of cells, which contain sick or dead larvæ, or, if the disease is just beginning, by the presence, among the brood, of sick or rotten larvæ. The larvæ die and rot either before or after sealing. It is only when the disease has lasted for some time, that the cappings are punctured, and that the brood has an offensive odor.

" The spreading of brood in the Spring is not always caused by foul-brood. A defective queen, some old pollen in the cells, &c., may also cause it. The brood may die (we do not say rot) by other causes also, and we should regret to see our bee-keepers become unduly frightened, and make a useless inspection of all the brood in their hives, for such work is not an agreeable pastime. But if foul-brood has already appeared in the neighborhood, or in the Apiary, it is well to drive the bees from the brood-combs and to inspect the latter with a scrutinizing eye. We have sometimes diagnosticated foul-brood in hives which had but two or three sick larvæ, barely turning yellow. When the disease has already spread, it strikes the eye. The brood is shapeless, yellow, brown, black, and the cappings change color and sink."—(Bertrand, *Revue Internationale d'Apiculture.*)

791. CURE. Several methods of cure for foul-brood have been given, with more or less successful results. Mr. D. A. Jones, has written a small pamphlet, in which he gives his method. He removes all the broodless combs, from the infected colony, drives (**473**) or shakes the bees into a box covered with wire-cloth, leaving enough bees in the hive to take care of the brood, if it is worth saving ; and puts the driven bees in a dark cellar for three to six days, turning the box on its side so as to see the bees through the wire-cloth. He keeps them thus till he sees some of them dying from starvation. Then, he puts them into a clean hive, on comb-foundation, and feeds them with the honey that has been removed from their combs, after having boiled it with one-fifth of water. The bees that hatch from the brood receive the same treatment before being returned to their colonies; all the combs are melted, and the hives, frames, &c., are boiled for ten minutes before being used again. Although Mr. Jones has been successful with this

29

method, it has not proved effective in every case, for, since the bees and the queen may be contaminated in their organs (**798**), the disease, after a time, may reappear. Every means should be used to kill all the spores of the bacilli. Mr. Cheshire has kept some of them in a glass tube (''Bees and Beekeeping,'' page 560), and exposed them on several occasions to a temperature below frost, and they were alive after sixteen and a-half months. Mr. Jones reports having kept foul-brood combs exposed a whole winter to a temperature of 35° below zero, — in Canada, — without succeeding in killing the spores. (''Gleanings in Bee-Culture,'' 1884, page 767.)

792. We will now give the method of Hilbert, as practiced by Chas. F. Muth and described in his ''Practical Hints:''

"In April, I discovered two colonies in my Apiary, affected with the disease. I brimstoned the bees the same evening, burned up the combs and frames, and disinfected the hives. Another colony showed it in May. Feeling sorry to kill a beautiful queen, besides a very strong colony of pure Italians, I brushed them on ten frames of comb-foundation, into a clean hive, and placed over them a jar with food, as I shall describe hereafter. The old combs and frames were burned up, and the hives disinfected. This feeding was kept up until all the sheets of comb-foundation were built out nicely and filled with brood and honey. It was a beautiful colony of bees about four weeks afterwards, full of healthy brood, and with combs as regular as can only be made by the aid of comb-foundation. Four more colonies were discovered infected, one after another. All went through the same process, and every one is a healthy colony at present. I was so convinced of the completeness of this cure, that I introduced into one of these colonies my first Cyprian queen sent me by friend Dadant.

" All are doing finely now, and no more foul-brood. Should, however, another one of my colonies show signs of the disease, it would not be because it had caught it from its neighbor which I had attempted to cure, but because the germ of foul-brood was hidden somewhere in the hive, and of late had come in contact with a larva.

" The formula of the mixture is as follows:

> 16 gr. salicylic acid
> 16 gr. soda borax,
> 1 oz. water.

" I keep on hand a bottle of this mixture, so as to be always ready for an emergency ; also a druggist's ounce glass, so that I may know what I am doing. My food was honey, with about 25 per cent. water added. But we may feed honey or sugar syrup, adding to every quart of food an ounce of the above mixture. Bees being without comb and brood, partake of it readily, and by the time their comb-foundation is built out, you will find your colony in a healthy and prosperous condition.

" Thus you see foul-brood can be rooted out completely, and without an extra amount of trouble, provided you are sufficiently impressed with its dangerous, insidious character, and are prepared to meet it promptly on its first appearance.

" When an atomizer is used on combs and larvæ the medicine should be only half as strong as given in the formula."

793. Since our friend Muth wrote the above, Hilbert improved the method, by dispensing with soda borax, and adding to his treatment fumigations with evaporating salicylic acid. We give this new method, for it has been used successfully by Mr. Bertrand and several of his neighbors in a number of different apiaries.

Prepare:

Solution No. 1,

> Crystallized salicylic acid 1 oz,
> Pure alcohol, 8 oz.

With this mixture prepare:

Solution No. 2, for washing or sprinkling the combs with an atomizer, 20 drops of solution No. 1, mixed with 7 ounces of tepid rain water, or 200 drops in a pint of water.

Solution No. 3, to be used in the food of the bees, about 220 drops of solution No. 1 in a quart of syrup or honey boiled with about a fifth of tepid water. To avoid the trouble of counting the drops every time, it is advisable to put them, the first time in a graded vial, or in a small bottle in which a mark can be made for the repeated measurement of the solution. The water can be measured in the same way.

Describing the Hilbert process, Mr. Cowan, who has also succeeded in curing a number of cases, writes:

794. "One of the simplest and most rapid ways of curing the disease is by Hilbert's fumigating process, as the fumes of salicylic acid have the power of penetrating everything in the hive and destroying all the germs of foul-brood. The apparatus used for this purpose is the fumigator improved by Mr. Ed. Bertrand, (fig. 187).

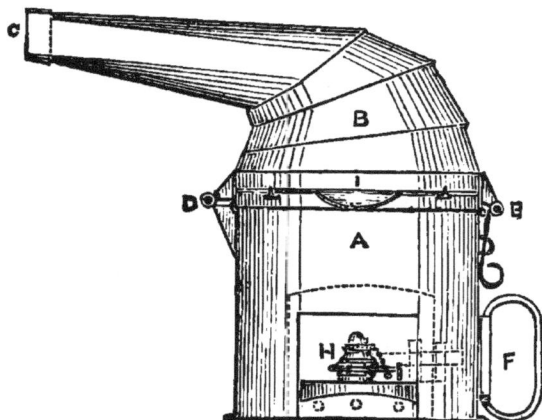

Fig. 187.
BERTRAND FUMIGATOR.

It consists of a cylinder A, to which is hinged, at D, a cover B, having a nozzle at C. This is 5 inches by $1\frac{1}{4}$, so as to be easily inserted between hive and floor board, and it is kept in position by the fastening E. A spirit lamp H, has the flame so regulated that the acid placed in the metal dish I, above it, is gently evaporated. The hive to be operated upon is not removed from its stand, but is raised up at the back off its floor-board by means of blocks of wood, and wedges are inserted at the sides, so as to leave only space for the insertion of nozzle, C, of fumigator. With hives on legs, the floor-board can be lowered. Fifteen and a half grains of salicylic acid are then placed in the dish I, and the flame of the lamp so regulated that the acid is gently evaporated. Too much flame will cause it to boil over, and waste ; too little would not melt it, so that just the right amount is found out by experiment. The nozzle of the fumigator in operation is now inserted in the opening at the bottom, and the corners of the quilt turned up, so as to allow the vapor of the acid to circulate freely. The fumiga tions should be performed early in the morning, or in the evening, when all the bees are at home. The entrance of the hive need not be closed. Any portion of the hive not reached by the fumes of the acid, the alighting-board and ground, near the hive, should be washed or syringed with salicylic acid 1 oz , soda borax 1 oz , water 2 quarts, or solution No. 3. It would be much better if the frames could be transferred to a clean hive after fumigation, and the infected hive scalded and painted over with the same solution, and with this view I have adapted my hives for easy separation and purification. Many hives, however, cannot be taken to pieces so readily, therefore they must be disinfected on the spot as well as possible, by the expenditure of a little more of the solution. Each hive should be fumigated from four to six times, at intervals of six days. The bees must receive every other evening a quarter of a pint of syrup containing 30 to 50 drops of solution No. 1. A foul-broody hive should be fumigated b fore being opened, as few frames left as the bees can well occupy, and if possible, the bees should be forced to build fresh combs, and rapid brood-rearing encouraged."

"All the hives in the Apiary should be fed with syrup contain-ing salicylic acid while the disease lasts.

"The honey from the infected combs can be removed and boiled for a short time, and by adding salicylic acid to it, can be used as food for the bees. All combs should be fumigated before being stored away, and sprayed with spray diffuser, on both sides and round the edges before being used again, with solution No. 1.

"All hives, floor-boards, frames, and utensils, used about an Apiary should be scalded and thoroughly cleansed when done with, and all woodwork painted over with the salicylic solution, to prevent the disease spreading any further.

"If the treatment above given be adopted in time, it will effect a cure, but if the disease is neglected and allowed to assume the worst type, much more trouble will be experienced in its eradica-tion. Some advise destroying the hives, but I never found any necessity to do this, as salicylic acid is sufficient to destroy any germs of the disease which may have adhered to the hive." (*British Bee-Keepers' Guide Book.*)

795. Mr. Cheshire, in turn, finding this process of evap-orating salicylic acid long and tedious, contrived a new

method in which he uses carbolic acid, otherwise called
phenol, after the suggestion of an Irish Apiarist, Mr. R.
Sproule.

As bees strongly dislike carbolic acid, since it is used to
frighten them (670), the quantity has to be very small, or
they will not touch the food containing it. The dose used
by Mr. Cheshire, in the food, is about one ounce for forty
pounds of syrup, amounting to 1-640th, but this proportion
may be changed according to circumstances. When there
is no honey in the fields, he says that the proportion may
be reduced to 1-750th.

"The carbolic acid should be added to the syrup when the
latter is cool and equally mixed by careful stirring."—(Cheshire.
Page 565.)

When the bees refused to touch the food thus prepared,
Mr. Cheshire succeeded in compelling them to use it, by
pouring it into the combs, in the cells immediately around
and over the brood. He advises the use of one part of
phenol in fifty parts of water, for spraying the infected
combs that are removed from the bees, but in no case does
he spray the inside of the brood-nest of the diseased colony with this solution.—(*British Bee Journal*, 1887, page
397.)

796. For our part, we should prefer the Bertrand-Cowan
method of applying Hilbert's recipe, to all others. It is
most likely, however, that either of these methods will be
successful if the Apiarist is careful and perseverant, but if
he neglects the minutest precautions, for instance, washing
his hands in a solution of phenol or of salicylic acid, before
going to some other hive, after handling a sick colony,
or if he does not apply a preventive treatment to all
his colonies during and after the treatment of the sick ones,
he may retain the disease in his Apiary indefinitely, for if
but a few of the spores escape, they will soon spread the
contagion again.

797. This reminds our Senior of an incident that happened in his younger days, while he lived with his father, who was a physician. A laborer had come to the old doctor for an ointment to cure the ·· itch ''. He had caught this — now uncommon and ever disgraceful — contagious skin disease, while working as a harvest hand, in the country. Directions were given him for using the ointment, and he was told that his wife should anoint with it also, as a preventive. But the woman, who did not have the disease, refused to use it, and two weeks afterwards the man came back for more ointment. He was cured, but his wife had the itch in her turn. The doctor gave him some, and told him that he should use it too, or he might catch the disease again ; but he did not mind the warning, and two weeks later, he had to call for more. '' Well,'' said the old doctor, '' I hope that these two experiments will convince you of the necessity of a thorough treatment for both, with a disease that is transmitted so readily, by contact.''

The case is exactly the same with the bacillus. While we are treating one colony, a few spores may be transmitted to a neighboring hive, by the contact of a single bee, and the disease is spread, unknown to us, while we are congratulating ourselves, in the firm belief that we have eradicated it.

798. The cure may be delayed, and may even fail altogether, if the queen is infected. Then the only resource is to kill her and give the colony another from a healthy hive.

799. When an Apiarist finds out that foul-brood exists in his vicinity, his best plan is probably to feed his bees regularly on salicycated food. A lump of camphor, placed inside of the hive on the bottom-board, is advised by some. Salt **(273)**, which improves the blood of all animals, by decreasing the number of white globules, shows its effects on the general health of all beings, and renders them more capable of battling against any disease, whether contagious or not.

800. Foul-brood is transmitted from one hive to another —like Asiatic cholera among men,—by different means. Robbing (**664**) is probably one of the main helps to contamination, as the robber bees may take the bacillus home, among their hair, unawares. Working bees may even gather the scourge from some sweet-scented blossom contaminated by previous visitors. The transportation, or shipping, of bees, from one part of the country to another, is often a mean of spreading the disease, and some of our State legislatures have made very stringent laws on the subject.

Contagious diseases were once the scourge of the land. Who has not heard of the plague, the dread disease of the dark ages ? According to Chambers' Encyclopedia, the plague of 1665 destroyed seventy thousand people, in London alone. Earlier still, in 1348, according to Sismondi, the plague destroyed three-fifths of the entire population of Europe, extending even up into Iceland. It was during that terrible scourge that the city of Florence lost over one hundred thousand people. If those dreaded diseases are now but little feared, we owe it to scientific discoveries. The microscope has shown that nearly all contagious diseases, which men or animals are subject to, are caused by living organisms, and medical science now teaches how they may be avoided by inoculation, or other means. More discoveries are daily made, and we can hope that the day is not far, when the advancement of science will have put an end to all these ills, and the bacillus alvei will be a thing of the past.

801. Aside from foul-brood, accidents may cause the brood to die, and even to rot in the cells, without special damage to the bees. Sudden and cold weather, in a promising Spring, when the bees have been spreading their brood, and are compelled to leave a part of it uncovered ; the neglect of the Apiarist, or his mismanagement, in placing back the brood, — after an inspection, — out of the reach of the cluster; or even the suffocation of a colony by heat (**367**), or by close confinement (**368**), may cause the death of the brood.

These accidents have none of the malignance of foul-brood, and nothing need be done in such occurrences besides removing the dead brood, and burying it carefully.

CHAPTER XIX.

ENEMIES OF BEES.

802. THE Bee-Moth (*Tinea mellonella*) is mentioned by Aristotle, Virgil, Columella and other ancient authors, as one of the most formidable enemies of the honey-bee. Even in the first part of this century, the bee-writers, almost without exception, regarded it as the plague of their Apiaries.

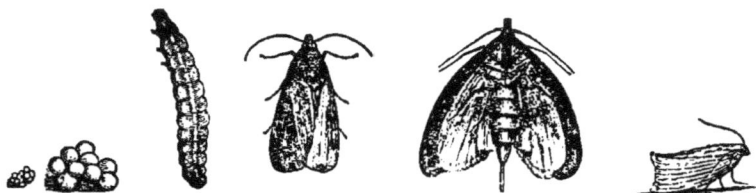

Fig. 188.
BEE-MOTH.
Eggs, natural size and magnified, larva and moths.

Swammerdam speaks of two species of the bee-moth (called in his time the "*bee-wolf*"), one much larger than the other. Linnæus and Réaumur also describe two kinds — *Tinea cereana* and *Tinea mellonella*.* Most writers supposed the former to be the male, and the latter the female of the same species. The following description is abridged from Dr. Harris' Report on the Insects of Massachusetts :

803. "Very few of the *Tineæ* exceed or even equal it in size. In its adult state it is a winged moth, or miller, measuring, from the head to the tip of the closed wings, from five-eighths to three-quarters of an inch in length, and its wings expand from one inch and one-tenth to one inch and four-tenths. The fore-wings

---

* Scientists do not agree exactly as to these species, nor their names, calling them, *galleria cereana, galleria alvearia, tinea cerella,* &c.

shut together flatly on the top of the back, slope steeply down-wards at the sides, and are turned up at the end somewhat like the tail of a fowl. The female is much larger than the male, and much darker-colored. There are two broods of these insects in the course of the year.* Some winged moths of the first brood begin to appear towards the end of April or early in May — ear-lier or later, according to climate and season. Those of the second brood are more abundant in August; but some may be found between these periods, and even much later."

No writer with whom we are acquainted has given such an exact description of the differences between the sexes, that they can always be readily distinguished. The wood-cuts of the moths, larvæ, and cocoons, which we present to our readers, were drawn from nature, by Mr. M. M. Tidd, of Boston, Mass., and engraved by Mr. D. T. Smith, of the

Fig. 189.—Female.

same city. Mr. Tidd seems first to have noticed that the *snout or palpus of the fe-male, projects* so as to resemble a beak, while that of the male is very short.

While some males are larger than some females, and

Fig. 190.—Male.

some females much lighter-colored than the average of males, and occasionally some males as dark as the darkest females, *the peculiarity of the snout of the female is so marked, that she may always be distin-guished at a glance.*

804. These insects are seldom seen on the wing, unless started from their lurking places about the hives, until to-wards dark. On cloudy days, however, the female may be noticed endeavoring, before sunset, to gain entrance into the hives.

"If disturbed in the daytime," says Dr. Harris, "they open their wings a little. and spring or glide swiftly away, so that it is very difficult to seize or hold them."

---

* Prof. Cook is of opinion (Gui le, page 315) that there may be three broods in a year and we believe he is correct. We have seen them most numerous in hot October weather.

They are surprisingly agile, both on foot and on the wing, the motion of a bee being very slow, in comparison. "They are," says Réaumur, "the most nimble-footed creatures that I know."

In the evening, they take wing, when the bees are at rest, and hover around the hive till, having found the door, they go in and lay their eggs.

"It is curious," says Huber, "to observe how artfully the moth knows how to profit by the disadvantage of the bees, which require much light for seeing objects, and the precautions taken by the latter in reconnoitering and expelling so dangerous an enemy."

"Those that are prevented from getting within the hive, lay their eggs in the cracks on the outside; and the little worm-like caterpillars hatched therefrom, easily creep into the hive through the cracks, or gnaw a passage for themselves under the edges of it." — Dr. Harris.

One afternoon, about twenty-five years ago, our Senior saw a female bee-moth on the front of an eke hive (**278**), and noticed that she was laying in the crack, between two ekes, through which the propolis could be seen; the ekes being rabetted to receive the comb-bars, their thickness there was reduced to about three-eighths of an inch.

The moth laid about ten eggs, then walked about, seeming satisfied with her work, and came back to lay about the same number, repeating the manœuver several times.

This shows that moths may lay eggs in the hive from the outside, if propolis is a food for their just-hatched larvæ. One of our objects, in preserving the strip around the hive to support the cap (fig. 68), and in incasing the bottom (**342**), was to hinder the moth.

805. "As soon as hatched, the worm encloses itself in a case of white silk, which it spins around its body; at first it is like a mere thread, but gradually increases in size, and, during its growth, feeds upon the cells around it, for which purpose it has only to put forth its head, and find its wants supplied. It devours its food with great avidity, and, consequently, increases so much in bulk, that its gallery soon becomes too short and narrow, and the creature is obliged to thrust itself forward and lengthen the gallery, as well to obtain more room as to procure an additional supply of food. Its augmented size exposing it to attacks from surrounding foes, the wary insect fortifies its new abode with additional strength and thickness, by blending with the filaments of its silken covering a mixture of wax and its own excrement, for the external barrier of a new gallery,* the *interior* and parti-

Fig. 191.
GALLERY OF MOTH WORM.

tions of which are lined with a smooth surface of white silk, which admits the occasional movements of the insects, without injury to its delicate texture.

"In performing these operations, the insect might be expected to meet with opposition from the bees, and to be gradually rendered more assailable as it advanced in age. It never, however, exposes any part but its head and neck, both of which are covered with stout helmets, or scales, impenetrable to the sting of a bee, as is the composition of the galleries that surround it."— BEVAN.

806. The worm is here given of full size, and with all its

Fig. 192.
THE WORMS.

peculiarities. The scaly head is shown in one of the worms; while the three pairs of claw-like fore legs, and

* This representation of the web, or gallery of the worm, was copied from Swammerdam.

the five pairs of hind ones, are delineated. The tail is also furnished with two of these legs. The breathing holes are seen on the back.

807. Wax is the chief food of these worms, but as Dr. Dönhoff says: "Larvæ fed exclusively on *pure* wax will die, wax being a non-nitrogenous (**221**) substance, and not furnishing the aliment required for their perfect development:" and his statement agrees with the fact that their larvæ prefer the brood-combs, which are lined with the skins cast away by the bee-larvæ (**167**), and which, in consequence, are more liable to be devoured than the new ones. In fact, they eat pollen and propolis, and while making their cocoons, they even seem to relish woody fibre, for they often eat into the wood of the frames or of the hives in which they are allowed to propagate, while comb-foundation remains almost untouched by them.

808. When obliged to steal their living among a strong colony of bees, they seldom fare well enough to reach the size which they attain when rioting at pleasure among the full combs of a discouraged population. In about three weeks, the larvæ stop eating, and seek a suitable place for encasing themselves in their silky shroud. In hives where they reign unmolested, almost any place will answer their purpose, and they often pile their cocoons upon one another, or join them together in long rows. They sometimes occupy the empty combs, so that their cocoons resemble the capping of the honey-cells. In Fig. 193, Mr. Tidd has given a drawing, accurate in size and form, of a curious instance of this kind. The black spots, resembling grains of gunpowder, are the excrements of the worms.

If the colony is strong, the worm runs a dangerous gauntlet, as it passes, in search of some crevice, through the ranks of its enraged foes. Its motions, however, are exceedingly quick, and it is full of cunning devices, being able to crawl backwards, to twist round on itself, to curl up almost into a knot, and to flatten itself out like a pancake. If obliged to leave the hive, it gets under some board or concealed crack, spins its cocoon, and patiently awaits its transformation.

Fig. 193.
COCOONS SPUN BY LARVÆ OF BEE-MOTHS.

809. The time required for the larvæ to break forth into winged insects, varies with the temperature to which they are exposed, and the season of the year when they spin their cocoons. We have known them to spin and hatch in ten or eleven days; and they often spin so late in the Fall, as not to emerge until the ensuing Spring.

810. In Northern latitudes where the thermometer, ranges for days and weeks below 10° the bee-moth-worm can winter only in the hive near the bee-cluster. It is a fact worthy of notice that Apiaries that are wintered *in the cellar* are more annoyed by the moth during the following Summer than those that are wintered out of doors, because none of the larvæ of the moth perish.

Dr. Dönhoff says that the larvæ become motionless at a temperature of 38° to 40°, and entirely torpid at a lower temperature. A number, which he left all Winter in his summer-house, revived in the Spring, and passed through their natural changes. This was in Germany where the Winters are milder than in our Northern and Middle States.

"If, when the thermometer stood at 10°, I dissected a chrysalis, it was not frozen, but congealed immediately afterwards. This shows that, at so low a temperature, the vital force is sufficient to resist frost. In the hive, the chrysalids and larvæ, in various stages of development, pass the Winter in a state of torpor, in corners and crevices, and among the waste on the bottom-boards. In March or April, they revive, and the bees of strong colonies commence operations for dislodging them." — DÖNHOFF.

Some larvæ which Mr. Langstroth exposed to a temperature of 6° below zero, froze solid, and never revived. Others, after remaining for eight hours in a temperature of about 12°, seemed, after reviving, to remain for weeks in a crippled condition.

"The eggs of the bee-moth are perfectly round, and very small, being only about one-eighth of a line in diameter. In the ducts of the ovarium, they are ranged together in the form of a rosary. They are not developed consecutively, like those of the queen bee, but are found in the ducts, fully and perfectly formed, a few days after the female moth emerges from the cocoon. She deposits them, usually, in little clusters on the combs. If we wish to witness the discharge of the eggs, it is only necessary to seize a female moth, two or three days old, with finger and thumb, by the head — she will instantly protrude her ovipositor, and the eggs may then be distinctly seen passing along through the semi-transparent duct.

Fig. 194.
WEBS AND REMNANTS OF COMBS DESTROYED BY MOTHS.

"Last Summer I reared a bee-moth larva in a small box It spun a cocoon, from which issued a female moth. Holding her by the head, I allowed her to deposit eggs on a piece of honey-comb. Three weeks afterwards, I examined the comb, and found on it some web and two larvæ. The eggs were all shrivelled and dried

30

up, except a few which were perforated, and from which, I suppose, the larvæ emerged. This appears to be a case of true parthenogenesis in the bee-moth." — *Translated from* Dr. Dönhoff *by* S. Wagner.

811. In Fig. 194, Mr. Tidd has faithfully delineated, and Mr. Smith skillfully engraved, the black mass of tangled webs, cocoons, excrements, and perforated combs, which may be found in a hive where the worms have completed their work of destruction.

The entrance of a moth into a hive and the ravages committed by her progeny. forcibly illustrate the havoc which vice often makes when admitted to prey unchecked on the precious treasures of the human heart. Only some tiny eggs are deposited by the insidious moth, which give birth to very innocent-looking worms ; but let them once get the control, and the fragrance* of the honied dome is soon corrupted, the hum of happy industry stilled, and everyt'ing useful and beautiful ruthlessly destroyed.

As a feeble colony is often unable to cover all its combs, the outside ones may become filled with the eggs of the moth. The discouraged aspect of the bees soon indicates that there is trouble of some kind within, and the bottom-board will be covered with pieces of bee-bread mixed with the *excrement* of the worms.

If a feeble colony cannot be strengthened so as to protect its empty combs, the careful bee-keeper will take them away until the bees are numerous enough to need them.

812. Combs having no brood, from dead colonies, or surplus combs, with or without honey, should be smoked with the fumes of burning sulphur, to kill the eggs or worms of the moth. when kept from the bees in the month; of June. July, August, and September. The box, hive, or room in which they are kept should be tightly closed to prevent the gas from escaping till it has done its work.

---

* The odor of the moth and larvæ is very offensive.

In smoking comb-honey in a room, the sulphur may be placed on hot coals in a dish, and care should be taken not to use too much of it, as the gas has the effect of turning the propolis to a greenish color, quite damaging to the looks of the beautiful sections. Enough smoke to kill the *flies*, in a room, will be found sufficient. Dry combs kept over Winter in a well closed room without a fire, are not in danger of the moth the following Summer, unless they are in some manner exposed. Combs, in which there have been moths, should be examined occasionally, to be smoked again if any worms are found.

A bee-keeper of Switzerland, Mr. Castellaz, keeps his combs in a closed box, in which he places some lumps of camphor. He says that bees accept these combs, even when impregnated with the odor of camphor.

In Italy where the moths are very troublesome, on account of the mildness of the Winters, some bee-keepers pile their combs flat in a box in which they have put about one inch of fine dry sand ; all the cells of every layer of comb are filled with sand, and the last one is entirely covered with it. The sand is shaken out, before the combs are melted or returned to the bees.

813. Italian bees, unless exceedingly weak and queenless, **(182)**, will defend a large number of combs against moths. One of our neighbors, who had, occasionally, helped us in the Apiary, after witnessing our success in bee culture, bought a colony of Italian bees and divided **(470)** it into three swarms, without regard to the scantiness of the crop. His swarms having dwindled to naught, he returned their combs to the impoverished colony, whose population was unable to cover more than two or three combs. But the returned combs had not been protected against moths, which hatched so numerous that our neighbor, surprised to see about as many moths as bees going out of the hive, came to us for advice. On opening the hive, we found three combs of brood crowded with bees, and seven others that were a perfect mass of webs, spotted with excrements. The bees were all on their combs and the moths on theirs; not one worm could be found on either of the three combs, protected by the Italians. Both populations, the one of bees, the other of moths, seemed to dwell harmoniously near each other.

814. The most fruitful cause of the ravages of the moth still remains to be described. *If a colony becomes hopelessly queenless* (**510**), *it must, unless otherwise destroyed, inevitably fall a prey to the bee-moth.* By watching, in glass hives, the proceedings of colonies purposely made queenless, we have ascertained that they make little or no resistance to her entrance, and allow her to lay her eggs where she pleases. The worms, after hatching, appear to have their own way, and are even more at home than the dispirited bees.

How worthless, then, to a hopelessly queenless colony, are all the traps and other devices which, formerly, have been so much relied upon. Any passage which admits a bee is large enough for the moth, and if a single female enters such a hive, she may lay eggs enough to destroy it, however strong. Under a low estimate, she would lay, at least, two hundred eggs in the hive, and the second generation will count by thousands, while those of the third will exceed a million.

The fact that hopelessly queenless stocks do not oppose any effectual resistance to the moths or worms, has for a long time been well known to the Germans. Mr. Wagner informed us "that their best treatises, for many years, speak of this as a settled fact, so that it has become an axiom that, if a colony is overpowered by robber-bees, its owner is not entitled to compensation, *as it was, in all likelihood, queenless, and would certainly have been destroyed by the moth.*"

In the *Ohio Cultivator* for 1849, page 185, Micajah T. Johnson says:— "One thing is certain—if bees, from any cause, should lose their queen, and not have the means in their power of raising another, the miller and the worms soon take possession. *I believe no hive is destroyed by worms while an efficient queen remains in it.*"

This seems to be the earliest published notice of this important fact by any American observer.

It is certain that a queenless hive seldom maintains a guard at the entrance after night, and does not fill the air with the pleasant voice of happy industry. Even to our dull ears, the difference between the hum of a prosperous hive and the unhappy note of a despairing one is often sufficiently obvious; may it not be even more so to the acute senses of the provident mother-moth?

Her unerring sagacity resembles the instinct by which birds that prey upon carrion, single out from the herd a diseased animal, hovering over its head with their dismal croakings, or sitting in ill-omened flocks on the surrounding trees, watching it as its life ebbs away, and snapping their blood-thirsty beaks, impatient to tear out its eyes, just glazing in death, and banquet on its flesh, still warm with the blood of life. Let any fatal accident befall an animal, and how soon will you see them,—

"First a speck and then a Vulture,"

speeding, from all quarters of the heavens, on their eager flight to their destined prey, when only a short time before not one could be perceived.

When a colony becomes hopelessly queenless, even should the bees retain their wonted zeal in gathering stores and defending themselves against the moth, they must as certainly perish as a carcass must decay, even if it is not assailed by filthy flies and ravenous worms. Occasionally, after the death of the bees, large stores of honey are found in their hives. Such instances, however, are rare ; for a motherless hive is almost always assaulted by stronger colonies, which, seeming to have an instinctive knowledge of its orphanage, hasten to take possession of its spoils; or, if it escape the Scylla of these pitiless plunderers, it is dashed upon a more merciless Charybdis, when the miscreant moths find out its destitution.

815. The introduction of movable-frame hives and Italian bees, with the new system of management, has done away with the fear of. the moth. It is no longer common to hear bee-keepers speak of having "good luck" or "bad luck" with their bees ; as bees are now managed, success or failure never depends on what is called "luck."

To one acquainted with the habits of the moth, the bee-keeper who is constantly lamenting its ravages, seems almost as much deluded as a farmer would be, who, after searching diligently for his cow, and finding her nearly devoured by carrion worms, should denounce these worthy scavengers as the primary cause of her untimely end.

The bee-moth has, for thousands of years, supported itself on the labors of the bee, and there is no reason to suppose that it will ever become exterminated. In a state of nature, a queenless hive, or one whose inmates have died, being of no further account, the mission of the moth is to gather up its fragments that nothing may be lost.

From these remarks, the bee-keeper will see the means on which he must rely, to protect his hives from the moth. Knowing that strong colonies which have a fertile queen, can take care of themselves in almost any kind of hive, he should do all he can to keep them in this condition. They will thus do more to defend themselves than if he devoted the whole of his time to fighting the moth.*

---

* Inexperienced bee-keepers, who imagine that a colony is nearly ruined when they find a few worms, should remember that almost every colony

It is hardly necessary, after the preceding remarks, to say much upon the various contrivances to which some resorted as a safeguard against the bee-moth. The idea that gauze-wire doors, to be shut at dusk and opened again at morning, can exclude the moth, will not weigh much with those who have seen them on the wing, in dull weather, long before the bees have ceased their work. Even if they could be excluded by such a contrivance, it would require, on the part of those using it, a regularity almost akin to that of the heavenly bodies.

An ingenious device has been invented for dispensing with such close supervision, by governing the entrances of all the hives by a long lever-like hen-roost, so that they might be regularly closed by the crowing and cackling tribe when they go to rest at night, and opened again when they fly from their perch to greet the merry morn. Alas! that so much skill should have been all in vain! Some chickens are sleepy, and wish to retire before the bees have completed their work, while others, from ill-health or laziness, have no taste for early rising, and sit moping on their roost, long after the cheerful sun has purpled the glowing East. Even if this device could entirely exclude the moth, it could not save a colony which has lost its queen. The truth is, that such contrivances are equivalent to the lock put upon the stable door after the horse has been stolen ; or, to attempts to banish the chill of death by warm covering, or artificial heat.

The prudent bee-keeper, remembering that "prevention is better than cure," will take pains to destroy the larvæ of

(especially black bees) however strong or healthy, has some of these enemies lurking about its premises.

The late Mr. M. Quinby, of New York, whose common-sense treatise on Bee-keeping, lately revised by his son-in-law, L. C. Root, will richly repay perusal, is of opinion that some of the imperfect bees carried out of the hive in the Spring, have been destroyed by the worms, which have made their way through the comb.

the moth as *early* in the season as he can, while swarming
his bees. The destruction of a single female worm may
thus be more effectual than the slaughter of hundreds at a
later period.

816. MICE. It seems almost incredible that such puny
animals as mice should venture to invade a hive of i.ees ;
and they often slip in when cold compels the bees to retreat
from the entrance. Having once gained admission, they
build a warm nest in their comfortable abode, eat up the
honey and such bees as are too much chilled to offer resistance,* and fill the premises with such a stench, that the bees,
on the arrival of warm weather, often abandon their polluted
home. The entrance should never be made deep enough to
allow mice to pass (**348**).

817. BIRDS. Very few birds are fond of bees. The
King-bird (*Tyrannus musicapa*), which devours them by
scores, is said—when he can have his choice—to eat only the
drones ; but as he catches bees on the blossoms—which are
never frequented by these fat and lazy gentlemen — the
industrious workers must often fall a prey to his fatal snap.
There is good reason to suspect that this gourmand can
distinguish between an empty bee in search of food, and
one which, returning laden to its fragrant home, is in excellent condition to glide — already sweetened — down his
voracious maw.

818. The bee-keepers of England complain of the sparrows, which they accuse of eating bees. If these birds
add this mischief to so many others of which they are guilty,
the bee-keepers should find some means of getting rid of
them. In the Vosges (France) most of the farmers suspend
earthen pots to the walls of their barns in which the sparrows make their nests. These jug-shape pots are examined

---

* In eating bees, the mice eat the head and corselet, but not the abdomen,
probably because of the smell of the poison sack.

every week and the young birds are killed as soon as they are ready to fly out, and are put into the frying-pan. We have seen as many as five or six dozen pots on the same wall, nearly all filled with nests, for sparrows raise many broods every year.

In Italy the consumption of these birds is carried on, on a large scale. Not only are the churches riddled with thousands of holes, in which the sparrows make their nests, but there are, at the road crossings, high square towers, which are built for this purpose. An overseer has them locked; He climbs inside, and clips the wings of the young, to compel them to stay till they are full grown.

During the Franco-Italian war against Austria, the French soldiers bought the young sparrows, which they found delicious eating. If the sparrows destroy our bees, can we not destroy them? It is better to eat than to be eaten!

If—as in the olden time of fables— birds could be moved by human language, it would be worth while to post up, in the vicinity of our Apiaries, the old Greek poet's address to the swallow:

> " Attic maiden, honey fed,
>   Chirping warbler, bears't away
> Thou the busy buzzing bee,
>   To thy callow brood a prey ?
> Warbler, thou a warbler seize ?
>   Winged, one with lovely wings ?
> Guest thyself, by Summer brought,
>   Yellow guests whom Summer brings?
> Wilt not quickly let it drop ?
>   'Tis not fair; indeed, 'tis wrong,
> That the ceaseless warbler should
>   Die by mouth of ceaseless song. "

819. No Apiarist ought ever to encourage the destruction of any birds, except the too-plentiful sparrows, because of their fondness for bees. Unless we can check the custom of destroying, on any pretense, our insectivorous birds, we shall soon, not only be deprived of their ærial melody among the leafy branches, but shall lament, more and more, the increase of insects from whose ravages nothing but these birds can protect us. Let those who can enjoy no music made by these winged choristers of the skies, except that of their agonizing screams as they fall before their well aimed weapons, and flutter out their innocent lives before their heartless gaze, drive away, as far as they please from their cruel premises, all the little birds that they cannot destroy, and they will, eventually, reap the fruits of their folly, when the caterpillars weave their destroying webs over their leafless trees, and insects of all kinds riot in glee on their blasted harvests.

820. Tame chickens eat drones, but not workers. Once we noticed a rooster seemingly eating bees at the entrance of a hive. The bees were then killing their drones (**192**). On approaching the hive, we saw him carefully pick out a drone from among the bees, shake off a worker-bee which had clung to him and swallow the drone. Young drones can be fed to chickens, who soon learn to eat them greedily, but if a worker bee is found among them they will shake their heads at her, with a knowing look of disgust. Young ducks, if insufficiently fed, will eat bees and are often killed by being stung while swallowing them.

821. OTHER ENEMIES.—The toad is a well-known devourer of bees. Sitting, towards evening, under a hive, he will sweep into his mouth, with his swiftly-darting tongue, many a late returning bee, as it falls, heavily laden, to the ground; but as he is also a diligent consumer of various injurious insects, he can plead equal immunity with the insectivorous birds.

It may seem amazing that birds and toads can swallow bees without being stung to death. They seldom, however, meddle with any, except those returning fully laden to their hives, or such as, being away from home, are indisposed to resent an injury. As they are usually swallowed without being crushed, they do not instinctively thrust out their stings, and before they can recover from their surprise, they are safely entombed.

822. Bears are exceedingly fond of honey; and in countries where they abound, great precautions are needed to prevent them from destroying the hives.

In that quaint but admirably common-sense work, entitled, " *The Feminine Monarchie, written out of Experience, by Charles Butler; printed in the year 1609,*" we have an amusing adventure, related by a Muscovite ambassador to Rome:

" A neighbor of mine," saith he, " in searching in the woods for honey, slipped down into a great hollow tree, and there sunk into a lake of honey up to the breast; where—when he had stuck fast two days calling and crying out in vain for help, because nobody in the meanwhile came nigh that solitary place—at length, when he was out of all hope of life, he was strangely delivered by the means of a great bear, which, coming hither about the same business that he did, and smelling the honey, stirred with his striving, clambered up to the top of the tree, and then began to lower himself down, backwards, into it. The man bethinking himself, and knowing that the worst was but death, which in that place he was sure of, beclipt the bear fast with both hands about the loins, and, withal, made an outcry as loud as he could. The bear being thus suddenly affrighted, what with the handling and what with the noise, made up again with all speed possible. The man held, and the bear pulled, until, with main force, he had drawn him out of the mire; and then, being let go, away he trots, more afeard than hurt, leaving the smeared swain in a joyful fear."

823. The *braula cœca* or bee-louse, exists in Italy and other warm countries. Dr. Dubini has seen queens so completely covered with them, that only their legs could be seen. These lice, whose second name, *cœca*, means *blind*, have been often found by us on imported queens on their arrival. They are **so** large that they can easily be taken off the queen and killed. It appears that they can only propagate in warm countries, for they exist in the South of Europe and are unknown either in Russia, or in North America.

824. Small ants often make their nests about hives, to have the benefit of their warmth. They are annoying to the Apiarist, but neither molest the bees nor are molested by them.

Our limits forbid us to to speak of wasps, hornets, millepedes (or wood-lice), spiders, libellulas and other enemies of bees. These lesser enemies are detailed at length and in a scientific manner, with engravings, in the work of Prof. Cook, "The Bee-Keeper's Guide," to which we refer the lovers of entomological study. If the Apiarist keeps his colonies strong, they will usually be their own best protectors, for, unless they are guarded by thousands ready to die in their defense, they are ever liable to fall a prey to some of their many enemies, who are all agreed on this one point, at least—that stolen honey is much sweeter than the slow accumulations of patient industry.

CHAPTER XX.

HONEY HANDLING.

Marketing Honey.

825. The quality of honey depends very little, if at all, upon the secretions of the bees; and hence, apple blossom, white clover, buckwheat, and other varieties of honey, have each their peculiar flavor, and color. The difference between the honey of one blossom, and that of another, is so great, that persons unacquainted with this diversity, when tasting honey different from that to which they are accustomed, imagine that either the one or the other is adulterated.

The whitest * and best flavored honey produced in this country, is that from white clover blossoms (**712**). Basswood honey, if unmixed with any other grade, is too strong in taste, but a slight quantity of it in clover honey makes a de'icious dish. Both these grades, being very white, sell more readily than any other, in the comb (**719**).

Smart-weed honey, — wh'ch should properly be called knot-weed or Persicaria honey, — is of a pale yellow color and very fine in flavor. Asters produce honey nearly as white as clover. Different grades of fall-honey, from Spanish needles, golden-rod, iron-weed, etc., are of a yellow color, and strong in taste. Buckwheat honey is dark; boneset honey and honey dew are the ugliest and poorest in quality, looking almost like molasses.

Some kinds of honey are bitter, and others very unwhole-

---

* The honey of Hymettus, which has been so celebrated from the most ancient times, is of a fair golden color. The lightest-colored honey is by no means always the best.

some, being gathered from poisonous flowers. A Mandin-
go African informed a lady of our acquaintance that his
countrymen eat none that is *unsealed* until it has been *boiled*.
The noxious properties of honey gathered from poisonous
flowers would seem to be mostly evaporated before it is
sealed over by the bees. Heating, however, expels them
still more effectually, for some persons cannot eat even the
best, when raw, with impunity. Well ripened honey is
more wholesome than that freshly gathered by the bees.
When it is taken from the bees, it should be put where it
will be safe from all intruders. The little red and the large
black ant are extravagantly fond of it, and will not only
carry off large quantities if within their reach, but many of
them will drown in it, spoil its appearance, and render it
unfit for use.

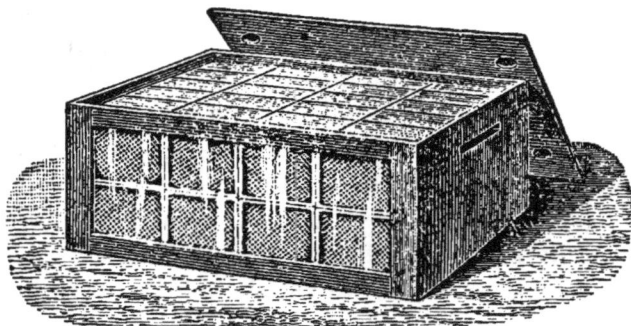

Fig. 195.
TWO-TIER HONEY CRATE.

826. *Comb honey,* in sections (**721**), put up in crates
of 12, 16, 24, or 40 sections, with glass on the side, sells
most readily ; and were it not for the greater cost of production, and the difficulty of safe transportation, this kind
would be raised exclusively. One objection to it, by large
producers, is that it cannot always be kept in good shape,
from one year to another, owing to its tendency to "sweat."
Sweating takes place in comb-honey which has been sealed

by the bees before it was fully ripened or evaporated (**744**), during a plentiful honey harvest. The changes of temperature in Spring and Summer cause a certain amount of fermentation in it, exactly as in the housekeepers' sealed preserves, when not sufficiently heated or sweetened. The result is a bursting of the cappings, by the pressure of the expanding honey, which runs out and over the comb and renders it unsalable. The same expansion sometimes takes place in granulated extracted honey accompanied by a slight fermentation.

827. It is also held, by some leading Apiarists, that the cells, although sealed are not moisture-proof, and that comb-honey gathers water from the air, till it overfills the cell and escapes through its pores. For this reason they keep their comb-honey in *a warm dry room. This is a good thing to do in every case.* Honey is hygrometric, and whenever exposed, gathers moisture rapidly, so that when kept in a damp place, a few unsealed or damaged cells very readily overflow, with watery honey, that daubs everything. Therefore, whether we believe that the sealed cells are air-tight or not (**262**), we should keep our honey in a dry place at all times.

To prevent the leaking honey in sections from running out of a crate and daubing other boxes, a sheet of strong manila paper should be placed at the bottom of each case, with the edges folded up slightly, say half an inch.

" The cases for shipping and retailing honey, should be light, and glazed on one or both sides. Those holding but one tier are best. The sections should rest on narrow strips of wood ¼ inch thick, tacked to the bottom of the case over a sheet of manila paper. This is to preserve the boxes from being daubed, in case the honey drips.

" These cases should be in readiness before the honey is ready to be taken off. " — (OLIVER FOSTER).

828. " Glazed sections " — one glass on each side of each section — have been largely sold in the East; but this mode of putting up honey, being very expensive, will only do for fancy trade. The producer can best tell what his trade requires.

When shipping comb-honey to the large cities, Mr. Hutchinson, who is one of the successful comb-honey producers, wraps each crate separately in paper, to protect it against dirt, dust, or coal-smoke, along the way. By this method his crates arrive on the market, as fresh and neat-looking on the outside, as when first put up.

As the careful handling of comb-honey during shipment is very important, it is best to mark each case with a large label or a stencil, bearing the words: ·

HONEY IN GLASS. HANDLE WITH CARE.

Very small lots ought never to be sent by rail, at least until we get better railroad regulations, concerning the handling of goods in transit, than we have at present.

829. The barrels that we use for extracted honey are oak barrels, which have contained alcohol. They are gummed inside, with some composition, to prevent the alcohol from soaking through the wood, and this gum, or glue, prevents the leakage of honey. Whisky barrels are often unfit to contain honey, for they are usually charred on the inside, and motes of charcoal fall into the honey and spoil its appearance. We keep our empty barrels in a dry place. As soon as filled, they are bunged and rolled into a cool and dry cellar, where they remain until the honey selling season, which begins in September, or October. Any dry room will do, when a dry cellar is not at hand, but a cellar has a more even temperature when cold weather comes.

Some Apiarists use cheap syrup barrels, made of soft wood, which are said to leak less than oak barrels. Messrs. New-man of Chicago have, for years, manufactured soft wood honey kegs, which have proved satisfactory to many of our friends, as they are more easily handled than larger barrels.

PLATE 15.

A. J. COOK,

Professor of Entomology at the Michigan Agricultural College;
Author of " *The Bee-keepers' Guide.*"

This writer is mentioned, pages 6, 10, 37, 51, 91, 140, 254, 459, 476.

They will do very well when the honey is to be sold at wholesale, as the barrel is usually lost by the shipper; but we have an objection to them for our own use. We generally have to take the honey out of them after it is granulated, to put it up for retail trade; and the cheap barrels are so easily damaged, by taking the head out, that they cannot be used more than one or two seasons, while good ironbound oak barrels will last for years, and will never leak, if managed properly. To take the head out, it should be marked, with a chisel, so as to replace it afterwards exactly in the same position. A strong gimlet is screwed into the middle of it, for a handle. After the hoops have been chased off, the head can be pulled out readily, and it is replaced in the same manner, when the barrel is empty.

If the barrels are damp, when the honey is put in, and are removed to a dry place afterwards, they will soon leak; for honey does not keep the wood from drying and shrinking. Honey barrels, then, should not be treated in the same way as wine or cider barrels; and swelling them, with steam, or hot water, previous to filling them with honey, will not be of any benefit, unless they are kept damp afterwards.

830. In October, the honey of the July crop is all granulated, and that of the September crop is beginning to granulate. There are many different opinions in regard to the causes of granulation. Some think that it is effected by the action of light, but this is certainly a mistake, for our honey only sees the light when extracted, and is then kept in the dark until sold. We are more inclined to think that it is the action of cold air which causes granulation; for sealed comb-honey generally remains liquid. The extracted honey, which we harvest, *always* granulates. We have handled liquid honey, however, several times, but we have *always* found it to be unripe; and have laid it down as a rule for ourselves, that good honey should be granulated after Nov-

31

ember. We speak of honey harvested in the Mississippi valley; such as clover, basswood, knot-weed, golden rod, buckwheat, Spanish-needle, etc.

831. Of California honey, we can say nothing, having never handled it. But we have handled Louisiana honey, which, we were told, would not granulate before a year, and we had scarcely had it three weeks in our cold climate, before it began to granulate. The only ripe honey which did not granulate, was a lot of Spanish–needle honey, which had been extracted late in November. It remained liquid until sold, a month or two later, and we ascribed its not granulating to the late harvesting of it.

832. Every bee-keeper has noticed that, at times, honey hardens in very coarse and irregular granules, that look like lumps of sugar, and have no adherence with one another, with a small amount of liquid honey interposed between them; and that at other times, the candying is compact, and can be compared to the hardening of lard.

The first kind of granulation is always produced in honey harvested, like clover or basswood, during the warm months of the year; while the soft candying is prevalent in the honey extracted in the Fall. In France, coarsely granulated honey is held as less valuable than the fine grained honey, and there is a good reason for this preference, for the coarsely granulated honey cannot be kept as well as the fine grained.

In this country also, coarsely granulated honey sells with less facility — especially because many ignorant persons imagine that it has been adulterated with sugar, and that the coarse grains are lumps of sugar.

We think that this coarse granulation is the result of an aggregation of particles, which, having an affinity for each other, unite, while the honey remains liquid in Summer.

In such honey, the liquid parts come to the surface, and

absorbing moisture from the air, are very apt to become acid by fermenting. But, even after granulation, it can easily be brought to a fine grain by melting it, and exposing it to the cold of our Northern Winters. Basswood honey would even be benefited by this, as it would lose a little of its too strong flavor.

Basswood and clover honey are more apt to ferment than any other class of honey, even when thoroughly granulated, if they remain exposed to the heat of the following Summer, and it is advisable to keep these two kinds in a cool, dry place during the hot weather. A damp cellar would be objectionable, since honey readily absorbs moisture from the air.

833. Those bee-keepers who will follow our methods, of extracting (**751**) after the honey crop, will have but little trouble with honey fermenting, even if they have to keep it through the following Summer. If any honey should ferment, however, let them not think that it is spoiled, unless it was really unripe and has turned quite sour. A slight amount of alcoholic ferment can be evaporated readily by melting the honey over water, when the ferment escapes in the shape of *foam*. As this fermentation is caused by the presence of unripe honey, some of our friends succeed in entirely preventing it by melting all their honey *immediately after granulation*. The melting evaporates all excess of moisture contained in it, and we highly *commend this method*.

Mr. C. F. Muth of Cincinnati, whose large experience in handling honey makes him a high authority, ripens all his honey by keeping it in open vessels in a dry and ventilated room, for a month or two after extracting.

834. *Melting Honey.* Honey should never be placed directly over a fire to melt it. The least over-heating will evaporate its essential oils, and give it the burnt taste of dark molasses instead. It should be put in a tin or copper

vessel, and this in another large vessel containing water.
This heating *au bain-marie*, as the French call it, is resorted
to by cooks, confectioners and others, whenever there is
any danger of scorching the substance heated.

835. The increase of honey production has been so great,
in a few years, that the consumption has barely kept pace
with it. But it will soon take its rank among necessities,
like butter or syrups; and change from a luxury to a staple.

836. Our first crops of extracted honey, were sold readily at wholesale, and at good prices; for it was then that
the wholesale dealers and manufacturers were making the
largest profits, by mixing the honey, which they bought
from bee-keepers, with cheap substances, like glucose,
which kept the honey from granulating, and by putting it
up in tumblers, with a small piece of comb honey in the
center. This honey, or rather mixture of honey, was sold
by them usually at lower prices than they had paid for the
pure honey. But ready sales in this way did not last long;
for, after a year or two, the markets were crowded with this
drug; and we were left to market our honey alone; if we
did not want to sell it for little above nothing.

Should our readers ever come across suspicious-looking
honey, they will find the following a cheap recipe to recognize adulteration:

"Put in a small vial about one ounce of the honey to be
tested, fill the vial with pure cistern water, shake thoroughly,
to dissolve the honey; then add to the mixture about a thimble-
ful of pure alcohol. If the honey is pure the solution will remain
unchanged, but if adulterated with glucose, it will be turbid and
whitish.

"This is the means used by the honey dealers of Paris, to detect
adulterated honey." — (*Annales de la Société d'Apiculture de l'Aube.*)

The present low prices have put an end to adulteration,
for, *a fair grade of Southern or California honey can now
be bought as cheaply, at wholesale, as the vile, unhealthy*

compound, adorned with the name of golden syrup, golden drip, etc.

837. But a slight prejudice remains in the minds of some buyers, against honey, unless they are acquainted with the producer. This prejudice has been helped by idle writers whose sensational stories found their way in the newspapers, concerning the supposed manufacture of artificial comb-honey.

Alas! that so many sensible people should give credit to such ridiculous *canards!* A minute's examination of a sealed honey comb, will convince any sensible person of the utter impossibility of its artificial manufacture. Nevertheless, we knew of grocers who bought and sold beautiful comb-honey believing it to be artificial, on the strength of those newspaper stories. These willful and silly lies were finally put an end to by an authoritative article in the "*American Grocer*" of November 10th, 1886, concerning manufactured honey and manufactured eggs. We quote a few passages of this lengthy article:

"Glucose at all fit for adulteration is worth from 4½ to 5 cents per pound. In California, excellent honey is now sold for 3 cents (*) per pound. This state of affairs makes it more feasible and more likely that glucose should be adulterated with honey, than that honey should be adulterated with glucose. We now come to artificial comb-honey. The only way in which it is possible to put a spurious article of comb-honey on the market would be by feeding the bees glucose or some other substitute; and there would be a greater probability of this being done were it not for the fact that the bees must consume a very large quantity of honey or other sweets to enable them to secrete a very small quantity of white wax from which the comb is made. . . .

"Our last point is in reply to the newspaper statements that were so widespread a year or two ago, to the effect that our comb

---

* We have before our eyes the price-list of a San Diego, Cal. firm, who offered extracted honey (October 1st, 1886), as low as 3¼ cents per pound; with a discount of 3 per cent. on car load lots.

honey on the market was made by machinery, and that neither
comb nor contents ever came from a bee-hive. So widespread
was this falsehood, that in our journal of November 1, 1885,
page 738, I offered $1,000 to anybody who would tell me where
such spurious comb-honey was made. No one has ever given
the information, neither has one ounce of manufactured comb-
honey ever been forthcoming. It is a mechanical impossibility,
and will, in my opinion, always remain so.... I hardly need
add, that the above slanderous report in regard to bogus comb-
honey was very damaging to the bee-keeping industry. It probably obtained wider credence because one Prof. Wiley, some
years ago, started it by what he termed a ' scientific pleasantry '.
  "In regard to the artificial eggs, I believe this will be a feat
still more difficult to accomplish than making artificial honey-
comb, especially if these artificial eggs are expected to hatch.
Some of the newspapers have jocosely declared that such eggs
would hatch, but that the chickens did not have any feathers on
them, the invention not yet being sufficiently ' perfected ', etc."
— A. I. Root.

838. The *granulation* of *honey* was objected to by many
consumers, at first, from the prejudiced idea that granulated honey had been mixed with sugar. It has ceased to be
an objection, for, in our neighborhood, nearly all honey
consumers now know that good ripe honey generally granulates in cold weather. But, now and then, a person is
found who wants liquid honey, or comb honey, thinking
that no other is pure.

We were told that the judges at an agricultural exposi-
tion refused to give a premium to a bee-keeper for his honey,
because it was spoiled by granulating. These competent
judges probably think that water is spoiled by freezing, for
granulated honey if carefully melted (**834**), is as good as
before hardening.

839. We have always found an easy sale for extracted
honey among foreigners — especially German or French;
as they have been used to granulated strained honey,
which has been produced for centuries in almost all parts of
Europe. Some of them are so well acquainted with it, that

they prefer it to the finest comb-honey, saying that comb is not made to be eaten.

Once, having received a favor from a French farmer, living a short distance from us, we selected a beautiful large comb of nicely sealed clover honey, while extracting, and sent it to this family after having carefully laid it on a dish. Much to our astonishment, we learnt, a few days after, that the good French housewife had put our nice comb in a clean towel, carefully pressed the honey out, and melted the wax; and besides, that she was very much astonished at our having sent comb honey to her, when we had such nice extracted honey on hand. The reader may readily imagine that thenceforth we never sent to them anything but extracted honey, much to their satisfaction and ours.

Every bee-keeper who understands his business, should try to sell his honey when granulated, explaining to his customers that adulterated honey does not granulate, and that granulation is the best proof of purity. We have these words printed on all our labels.

840. To improve the present prices of honey, which are in some cases lower than the prices of second class sweets, it is necessary that the masses should be induced to buy it. Thus far it is an article which few persons will buy regularly. Consumers will go to the grocery for tea, coffee, sugar, flour, meal, butter, etc., but very few make it a custom to buy honey — not that they dislike it, for " what is sweeter than honey? " but *because they are not used to it.*

All children, even in the heart of our manufacturing centers, have heard of " honey," but how many have never tasted it! Why? Twenty-five years ago honey was thirty cents per pound. Ten years ago the very cheapest grades retailed higher than the best sugars. To-day, in many places, honey is still retailed at from fifteen to twenty cents, while fourteen pounds of the best sugar are sold for a dollar. Yet

the Apiarists crowd it to the markets at prices ranging as
low as three cents. What is lacking?   Proper distribution.
Instead of shipping our honey to the cities, whence it will
be partly shipped back to our village retailers after having
passed through the hands of commission men, and wholesale
merchants, we must cultivate home consumption. We must
show our neighbors, our farmers, our mechanics, *at home*,
that our progressive methods enable us to furnish to them
the *sweetest of all sweets*, at nearly as low a price as syrups.
The occasional depression of the honey markets is but temporary and its termination is only a question of time.

841. It is important, in offering honey, whether to grocers or to consumers, to have it put up in neat and attractive shape.   Comb-honey in
sections weighing only a pound
sells best, because it is, and always
will be, a fancy article.

But in putting up extracted
honey, a one-pound package is
now too small. We must encourage a consumption in which the
expense of packing will not materially advance the cost, and we
find that, owing to this advance
of cost, the one or one and-a-
quarter-pound package is less in
demand than it was a few years
ago.

842. Tin is the cheapest package for honey, in small quantities.
Our favorite sizes are two and-a-
half-pound, five-pound, and ten-
pound pails. The two and-a-half-

Fig. 196.
HONEY PAILS.

pound pail is in great demand, and in the Winter of 1886-7,
the bulk of our crop of that year, about 24,000 lbs., was sold

in this package, at twenty-three cents per pail, or about nine cents per pound.

Some of our readers will ask why we do not put up our honey in these pails from the first, instead of putting it up in barrels. We never do so, because we do not know what proportion of each size will be required by the trade; because honey in cans occupies too much room, and is not so easily moved out of the way; and especially because we keep honey from the best seasons for the years of poorer crop, and it keeps best in barrels. We have kept honey in pails for two years or more, but the pail often rusts on the outside, and becomes unsalable. The objections above given are very weighty, in extensive production, when tens of thousands of pounds have to be cared for, but the small producer may, if he chooses, put up his honey, at once, in retail packages.

843. To stop the accidental leakage of honey in pails — for, owing to its weight, it will leak through seams that are water-tight — we simply rub over the leaky spot a little tallow-wax, prepared by melting beeswax with tallow or lard, in varied quantities. We also prevent the running over of pails of liquid honey, when transported in hot weather, by dipping the top edge of the pail in melted tallow-wax, before filling it. This puts a small rim of the ingredient around the outer edge of the pail, and the cover fits over it, air-tight.

A great deal of honey is sold in glass jars, but our objection to them is that granulated honey does not look well in them, and they are more costly than tin. Honey, in tin, can be put up gross weight, and although no one objects to the weight of the pail, this weight helps to pay for its cost. Those who use glass as a honey package, melt the honey before bottling it.

For shipping honey in small packages, Mr. Aug. Christie, a large producer of Iowa, puts it up in soldered cans. But the honey must be very ripe, or else must be previously heated, for the least fermentation would burst the can.

844. In every case when honey is sold, it should be neatly labeled with the name and address of the producer, which is, in itself, a guarantee of its quality.

When you go into a strange grocery, where you are unknown, the immediate answer of the grocer, to your mention of honey is: "I don't want any honey; I have no sale for it, and I don't like to handle it." Should you then take your leave and go, there would be but little hope of increasing your sales. You have to study, and learn to imitate the cunning and the perseverance of the traveling agent, and quietly talk it out. You first have to assure the grocer that you only wish to show him your goods and your prices at his leisure, and that he can then refuse to buy, if he chooses. You must show him why he has no sale for honey. You tell him that pure honey is one of the best sweets in the world, to which he readily agrees. You then explain that honey, not being a staple, his customers never come on purpose to buy it, but that when they see it, they are tempted to buy; that, for this reason, it should be put up with large and showy labels, and placed in a conspicuous position, so that it will readily catch the eye.

845. White honey in nice sections (**721**) will generally sell at sight, unless the grocer has had some leaky packages, which dripped honey on the counter, left a sticky reminiscence of their presence, and attracted flies and bees. But if your honey is put up carefully, according to directions given, the first sale alone will be difficult. In selling extracted honey it may be necessary for you to explain the difference between *extracted* honey, and the *strained* (**276**) honey of old; for now and then some persons are found who do not know any thing about this, or about the facility with which granulated honey may be liquefied.

With grocers that were unacquainted with us, we usually

began by supplying them with yellow honey, such as buckwheat, or heartsease, or golden rod. This honey, strong in flavor, sells better to the inexperienced, who are afraid of getting sugar, or glucose. It is only after one or two years that we venture to offer to such grocers our whitest clover and bass-wood, which, though of superior flavor, are objected to, on account of their very beauty and quality. In every case we try to furnish some good reference to the grocer, and we give him a full guarantee of satisfaction, with an agreement to take the honey back, if it does not prove altogether as we represent it. When a dealer is well satisfied that the merchandise which he sells is pure, his customers are quite likely to have confidence in it themselves; but, on the other hand, if he is in doubt as to the quality and purity of it, he will have but little chance of selling it, unless he does not care for the satisfaction of his patrons.

846. We must therefore spare no pains to fully convince our grocers of the quality of our goods.

After the first sales have been made, the sales always become larger and easier. Of course, occasional objections are made, by persons who are unacquainted with the properties and qualities of good honey; but these are easily overcome, when you have once gained the confidence of the dealers.

Extracted honey is usually sold at between half and two-thirds of the price of comb-honey. It ships better, leaks less, and keeps more easily than comb-honey; and its lower cost of production will sooner or later make it the *honey for the masses.*

Uses of Honey.

847. The traditions of the remotest antiquity show that honey has always been considered a pleasant and healthy food. For several thousand years, it was the only sweet known.

Now that the sap of the cane, or the beet, converted into sugar, has become a necessity in every family, let us see what place honey may occupy in our diet, not only as a condiment like sugar, but as food, drink, and medicine.

As Food.

Honey as food is very healthy. It is admitted that those who use honey freely at meal time, find in it health and long life.[*]

" It is Nature's offering to man — ready for use, distilled drop by drop in myriads of flowers, by a more delicate process than any human laboratory ever produced."— (T. G. Newman, " Honey as Food and Medicine.")

[*] The following extract from the work of Sir J. More, *London*, 1707, will show the estimate which the old writers set upon bee-products :

'' Natural wax is altered by distillation into an oyl of marvellous vertue; it is rather a Divine medicine than humane, because, in wounds or inward diseases it worketh miracles. The bee helpeth to cure all your diseases, and is the best little friend a man has in the world...... Honey is of subtil parts, and therefore doth pierce as oyl, and easily passeth the parts of the body; it openeth obstructions, and cleareth the heart and lights of those humors which fall from the head; it purgeth the foulness of the body, cureth phlegmatick matter. and sharpeneth the stomach; it purgeth those things which hurt the clearness of the eyes, ı reedeth good blood, stirreth up natural heat, and prolongeth life; it keepeth all things uncorrupt which are put into it, and is a sovereign medicament, both for outward and inward maladies; it helpeth the greif of the jaws, the kernels growing within the mouth, and the squinancy; it is drank against the biting of a serpent or a mad dog; it is good for such as have eaten mushrooms, for the falling sickness, and against the surfeit. Being boiled, it is lighter of digestion, and more nourishing. ''

848. When Augustus-Julius-Cæsar, dining with Pollio-Rumilius on his hundredth birthday, inquired of him how he had preserved both vigor of body and mind, Pollio replied: "*Interius melle, exterius oleo.*"—Internally by honey, externally by oil.

Honey is in daily use on our table, and we find that children prefer it to sugar. The only cause of its not being in general use in place of "vile syrups" is the high price at which it was formerly sold.

Mr. Newman in his little pamphlet above quoted, says:—

"It is a common expression that honey is a luxury, having nothing to do with the life-giving principle. This is an error— honey is food in one of its most concentrated forms. True, it does not add so much to the growth of the muscle as does beef-steak, but it does impart other properties no less necessary to *health* and vigorous physical and intellectual action! It gives warmth to the system, arouses nervous energy, and gives vigor to *all* the vital functions. To the laborer it gives strength— to the business man, mental force. Its effects are not like ordinary stimulants, such as spirits, &c., but it produces a healthy action, the results of which are pleasing and permanent—a sweet disposition and a bright intellect."

These words are so true that we have found them translated, in European books, by noted Apiarists.

849. As a condiment it can be used in many ways. In candies it will finally replace the unhealthful glucose of commerce. The confectioners who now use it, increase their trade every year.

In France, "*pain-d'épice,*" "ginger bread," is sold in immense quantities at the fairs. The best makes are sold at the most important fairs through the country. It keeps an indefinite length of time, and farmers' wives are wont to buy enough to last for months. The following is the recipe:

850. Dissolve 4 ounces of soda, in a glass of warm skimmed milk. Take 4 pounds of flour and pour in the milk and enough warm honey to make a thick dough, flavor with anise and coriander seeds, cloves, and cinnamon, all powdered fine. Knead carefully, as you would bread. Let it rise two hours in a warm place, spread in pans and bake in a moderately warm oven. Ten or twelve minutes will do, if the cakes are thin. As soon as the cake resists to the touch of the finger it is done. Before baking, it may be decorated with almonds, preserved lemon peel, etc. Wheat flour makes good "*pain-d'épice*," but some prefer rye flour. Fall honey is preferable for it, on account of its stronger taste."—*L'Apiculteur.*

The spices may be varied according to taste. Some add powdered ginger, or grated lemon or orange peel.

851. Crisp ginger bread can be made by mixing in it a quantity of broken almonds, blanched by dipping in boiling water, hazel-nuts, English walnuts, etc. The same dough, in skilled hands, with different seasonings, will make a variety of dainties, all with honey.

Instead of lard or butter, artistic cooks use olive oil to grease the pans; in America, cotton seed oil takes its place, and is good. The Italians sometimes use *beeswax.*

852. *Alsatian Ginger Bread:* " Take, yellow honey 1 pound, flour 1 pound, baking soda ½ ounce. Dissolve the soda in a tablespoonful of brandy, heat the honey and put in the flour and the soda. Knead the whole carefully, and cut in lumps before putting in the oven.

" This mixture can be kept in the cellar for months and can be used to make the

"*Leckerli*: Add to the dough, chopped almonds ½ .lb., preserved orange peel 2 drams, ditto lemon 1 dram, cinnamon ½ dram, and 20 cloves, all finely powdered. Mix well and bake." (DENNLER, " Honey and its Uses.")

853. *Honey Cake.* Warm half a glass of milk with ¼ pound of sugar in a stew pan. Put in ¾ of a pound of honey and boil slowly. Then add 1 pound of flour, ½ dram of soda, and knead, spread on a pan and bake for an hour.

854. *Italian* "*Croccante Di Mandòrle*": " Blanch two pounds of almonds, by dipping in boiling water. Slice them with a knife. Add the yellow peel of a lemon cut fine, some powdered vanilla, and a few lumps of sugar flavored by rubbing them on orange peel. Boil 2 pounds of good honey with an ounce of olive

oil or good unsalted butter, till it is reduced to thick syrup
Then add the almonds, lemon, etc., a little at a time, mix well,
pour in a buttered tin pan and press the mixture against the
sides with a lemon peel. It should not be more than half an
inch thick. When cool take the crisp cake out of the vessel by
warming it a little." (SARTORI & RAUSCHENFELS, *L'Apicoltura in
Italia.*)

855. *Muth's Honey Cake:* 4 quarts of hot honey and 10 pounds
of flour, with ground anise seed, cloves and cinnamon to suit the
taste. This is made into a dough and left to *rest* for a week or
two, when it is rolled out in cakes and baked. The longer the
rest, the better the cakes.

Fruit jellies with honey. Take the juice of currants or other
fruits, and after adding a like quantity of honey, boil to a jelly.
Put in small tumblers, well sealed, in a dry room.

856. *Honey-vinegar* is superior in quality to all other
kinds, wine vinegar included.

It takes from one to one and a half pounds of honey to
make one gallon of vinegar. Two good authorities on
honey vinegar, Messrs. Muth and Bingham, advise the
use of only one pound of honey with enough water, to make
each gallon of vinegar. We prefer to use a little more
honey, as it makes stronger vinegar, but the weaker grade
is more quickly made. If the honey water was too sweet,
the fermentation would be much slower, and with difficulty
change from the alcoholic, which is the first stage, into the
acetic. This change of fermentation may be hurried by the
addition of a little vinegar, or of what is commonly called
vinegar mother.

If honey water, from cappings, is used, a good test of
its strength is to put an egg in it. The egg should float,
coming up to the surface at once. If it does not rise
easily, there is too little honey. As vinegar is made by the
combined action of air and warmth, the barrel in which it
is contained must be only partly filled, and should be kept
as warm as convenient. It is best to make a hole in each
head of the barrel, about four or five inches below the up-

per stave, to secure a current of air above the liquid.
These, as well as the bung hole, should be covered with
very fine wire screen, or with cloth, to stop insects.

A very prompt method consists in allowing the liquid to
drip slowly from one barrel into another, as often as possible during warm weather.

As we make vinegar not only for our own use, but also to
sell to our neighbors, we keep two barrels, one of vinegar
already made, the other fermenting. When we draw a gallon of vinegar, we replace it with a gallon from the other
barrel. This keeps up the supply.

Vinegar should not be kept in the same cellar with wines,
as its ferment would spoil the wines sooner or later.

Honey as Medicine.

857. In Denmark and Hanover, the treatment of Chlorosis, by honey, is popular. The pale girls of the cities are sent
to the country, to take exercise and eat honey. The good
results of this treatment have suggested to Lehman the
theory that the insufficiency of hepathic sugar is the cause
of Chlorosis, which thus explains the curing effect of honey.
(JACCOUD, as quoted by the *Revue Internationale d'Apiculture.*)

Honey, mixed with flour, is used to cover boils, bruises,
burns, etc. ; it keeps them from contact with the air, and
helps the healing. Beverages, sweetened with honey, will
cure sore throat, coughs, and will stop the development of
diphtheria, especially if taken on an empty stomach, at bed
time. A glass of wine or cider, strongly sweetened with
honey, is advised in *L'Apiculteur*, as a cure for colds.
(1886.)

Suckling babies are cured of constipation, by a mixture
of bread and honey given them, tied in a '' sugar teat.''

PLATE 16.

THOS. G. NEWMAN,

Editor of " *The American Bee Journal* " ; Author of " *Bees and Honey.*"

This writer is mentioned pages 374, 375, 383, 404, 440, 479, 492, 493, 497.

A constant use of honey, at meal time, cures some of the worst cases of piles.

"According to Mr. Woiblet, washing the hands with sweetened water will kill warts. Having heard of this healing he put honey plasters on the hands of a child who had a large wart in the palm of the hand, and after a few days of treatment the wart disappeared." — BERTRAND, (*Revue Internationale d'Apiculture.*)

To these many uses of honey, we might add other recipes, but our limits forbid. For all sorts of honey-cakes, wines, metheglin, mead, etc., we will refer our readers to the already mentioned pamphlet of Mr. T. G. Newman, of Chicago, "Honey as Food and Medicine." The price is a trifle. It contains many good things.

32

CHAPTER XXI.

BEESWAX, AND ITS USES.

Melting Wax.

858. We will now describe the different processes used by bee-keepers to render the combs into wax. To melt every comb, or piece of comb, as it is taken from the hive, would increase the work, and, as it is preferable to choose our time for this operation, we have to preserve them from the ravages of the moths (**802**) by some of the methods that we have given (**812**).

859. The cappings (**772**) after extracting (**775**), are allowed to drain in a warm place for several weeks; very nice honey being obtained from them. They are then washed in hot water, and the sweet water obtained can be used for cider, or wine, or vinegar (**856**). These cappings, as well as the broken pieces of white comb in which brood was never raised, should be melted apart from the darker combs, for, not only are they easier to melt, but, the wax obtained being very bright in color, is unsurpassed for making comb-foundation (**674**) for surplus boxes (**688**).

860. When the combs are blackened by the dejections of the worker bees (**784**), or of the drones (**40**), and by the skins and cocoons of the larvæ (**167**), it is so difficult to render the wax, that many bee-keepers think it is not worth the trouble. We advise washing these combs and keeping them under water for about twenty-four hours. Then the cocoons and other refuse being thoroughly wet and partly dissolved, will not adhere to the wax. This will be lighter colored, if the combs are melted with clear water and not with the water already darkened by the washing.

But as this method always leaves some wax in the residues, for some of it goes into the cells during the melting, and it is impossible to dislodge it, a better result is obtained by crushing the combs before washing them. But this pulverizing can be done only in Winter, when the wax is brittle.

861. The combs should be melted with *soft or rain water*, the boiler kept about two-thirds full, and heated slowly, to prevent boiling over. If the floor, around the stove, is kept wet, any wax that may drop will be easily peeled off.

During the melting carefully stir till all is well dissolved. Then lower into the boiler a sieve made of a piece of wire cloth, bent in the shape of a box, from which the wax can be dipped as it strains into it. If the whole is thoroughly stirred for some time, very little wax will be left in the residues. This is the cheapest and best method of rendering wax, without the help of a specially made wax-extractor.

862. To obtain as much wax as possible from the combs, the large wax manufacturers of Europe empty the contents of the boiler into a bag, made of horse-hair or strong twine, and place the bag under a press while boiling hot. All the implements used, as well as the bag, are previously wetted, to prevent their sticking.

863. Some bee-keepers use a wax-boiler in which the wax is melted by steam.

But the best wax can be rendered by a solar extractor (fig. 197), yet, by its use, some wax is always left in the refuse, for the cocoons, skins of larvæ, etc., being dry, always absorb more or less of it. This implement however is destined to overthrow all others for the rendering of wax in all countries where the heat of the sun is sufficiently powerful. At this latitude, the 42°, sun-extractors can be efficiently used during the months of May, June, July, and

August. The sun-extractor requires no labor from the Apiarist, other than filling it with combs and removing the melted wax.

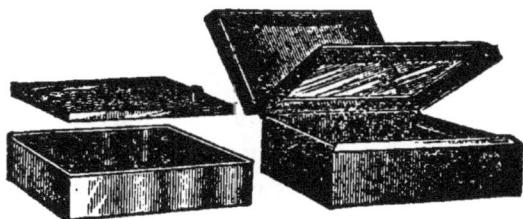

Fig. 197. (From 'Gleanings.'')
SUN-EXTRACTOR.

864. The dealers in France buy, from the bee-keepers, for little or nothing, the residues of their melted combs. They dissolve them in turpentine, press the pulp dry, and distill the liquid, to separate the turpentine. As the wax is not volatile, it remains in the still. It is said that, when wax was dearer than it is now, large profits were realized by this operation.

865. To cleanse beeswax from its impurities, we melt it carefully with cistern water and pour it into flaring cans (wider at the top than at the bottom) containing a little boiling water. This wax is kept in the liquid state, at a high temperature, for twenty-four hours. During this time, the impurities drop to the bottom and can be scraped from the cake when cold. Some wax can be obtained from this refuse, but some of it is always left in the dregs, as is proven by the impossibility of dissolving them by exposure. We have lumps of this refuse, as dark as ink, which were scattered on our farm, with manure, ten years ago, and are just as they were when put in the fields. Nothing can destroy beeswax, except fire, or the ravages of the bee-moth. Exposure to the weather does not affect it, but only bleaches it.

To prevent the cakes of wax from cracking, it should be

poured into the molds or cans when only 165° Fahr. and should be kept in a warm place to cool slowly.

866. The utmost care is necessary not to spoil wax in melting it. If heated too fast, the steam may disaggregate it. Then its color is lighter, but very dim ; the wax having lost its transparency, resembles a cake of corn meal. When it is in this condition, water will run out of it if a small lump is pressed between the fingers. The best way to restore it is to melt it slowly in a solar wax extractor (fig. 197). We have succeeded also by melting it with water, and keeping the water boiling slowly till all the water contained between the particles of wax had evaporated. But this work is tedious and cannot be accomplished without the greatest care and a skillful hand. Whatever the means used, you may rely on more or less waste.*

Wax-bleachers draw wax into small ribbons which are exposed to the rays of the sun for several weeks, or melted with chemical acids ; but wax-bleaching is beyond the purpose of this book.

Uses of Wax.

867. Before the invention of parchment, prepared as a material for writing, from the skins of goats, sheep, calves, etc., tablets covered with a light coat of wax were used. A style—an instrument sharp at one end to engrave characters in the wax, and broad and smooth at the other end to erase them—was used in place of a pen. The Latin poet Horatius, born sixty-five years before Christ, probably used these tablets, for, in his admonition to poets, he writes : *"Sœpè stylum vertas.*—" turn often your style ;" thereby meaning : "Carefully correct your writings."

* Whenever beeswax is melted in water, even with the utmost care, some small portions of it are water-damaged and settle to the bottom of the cake with the dregs. This water-damaged beeswax has often been mistaken for *pollen residues.*

Several nations of old, having noticed that beeswax does not rot, used it to embalm their dead. Alexander the Great was embalmed with wax and honey.

868. Beeswax is largely used by the Catholic churches, for lights, during the ceremonies, for it is prescribed to priests to use exclusively *wax produced by bees.*

869. In several countries of Europe the floors and stairs, instead of being covered with carpets, are rubbed with wax and carefully scrubbed with a dry brush every day till they shine. In Paris, floor scrubbing is a business which supports many working families.

Beeswax is used also by the sculptors and painters to varnish their work, to model wax figures; by dentists to take imprints of jaw-bones. It is retailed in small lumps and used to give smoothness and stiffness to thread for sewing.

870. The casting of bronze statues and works of art, *à cire perdue*, has been largely practiced in France since the Renaissance. This process is mentioned in *Harpers' Monthly* for September, 1886.

871. Beeswax forms part of a great many medicines, and pomades for the toilet. Here are a few recipes selected among hundreds of others:

1. Salve or Cerate for Inflamed Wounds.

Beeswax...................... 1 part,
Sweet almond oil.............. 4 parts.

Dissolve the wax in the oil and stir well till cold. Sweet almond oil can be replaced by olive, or cotton seed, or linseed oil, or even by fresh unsalted butter.

This cerate, may be used as a vehicle by the endermic method — we mean by frictions on the thin parts of the skin — to introduce into the blood several substances, such as quinine, against fever; sulphur, for itches; camphor, henbane, opium, as sedatives; iodine, as depurative; and so on, the only care being to have the drugs carefully mixed.

2d. Turpentine Balm for Atonic Wounds, (without inflammation):

Yellow Beeswax.........
Turpentine............. } Equal parts.
Essence of Turpentine...

Melt the wax, add the turpentine, then the essence.

3d. Salve for the Lips:

Wax........................one part,
Sweet Almond Oil............two parts.

·Add a small quantity of Carmine to color it, strain and add, when melted again and half cold, some volatile Oil of Rose.

4th. Adhesive Plaster for Cuts (sweet-scented):

Colophony..... ....40 parts,
Wax ....45 "
Elemi rosin.................25 "

Melt and add:

Oil of Bergamot..............5 parts
 " Cloves.................2 "
 " Lemon.................2 "

5th. Green Wax for Corns:

Yellow wax.................4 parts,
White pitch.................2 "
Venice Turpentine............1 "
Sub-acetate Copper (finely powd.) 1 "

Melt the wax and the white pitch, add the acetate of copper well mixed with the turpentine, and stir till cold. If too hard to be spread on small pieces of cloth, add a little olive, or cotton seed, oil.

6. Balm of Lausanne, for Ulcerated Chilblains and Chaps of the Mammæ or Teats:

Olive or Cotton seed oil	500
Rosin of Swiss Turpentine	100
Yellow Wax	133
Powdered Root of Alkanet	25

Keep it melted *au bain-marie* (**834**) for half an hour and add:

Balsamum Peruvianum	16
Gum Camphor	1

7. Mixture to Remove the Cracks in Horses' Hoofs:

Melt equal parts of wax and honey on a slow fire, and mix thoroughly.

Clean carefully the hoof with tepid water and rub the mixture in it with a brush. The cracks will disappear after several applications and the hoof will be softened.

8. To Keep the Luster of Polished Steel Tools:

Oil of Turpentine	8
Wax	1
Boiled Linseed Oil	½

JAN SWAMMERDAM

CHAPTER XXII

BEES AND FRUITS AND FLOWERS.

871. We have shown, in the chapter on Physiology (**43**), that bees cannot injure sound fruits, and in the chapter on Food (**268**), that they help the fecundation of flowers; but this accusation of bees injuring fruits has become of so much importance in the past few years, especially in the best fruit and bee country of the world, California, that we deem it necessary to give it a whole chapter.

While the honey-bee is regarded by the best informed horticulturists as a friend, a strong prejudice has been excited against it by many fruit-growers; and in some communities, a man who keeps bees, is considered as bad a neighbor, as one who allows his poultry to despoil the gardens of others. Even some warm friends of the "busy bee," may be heard lamenting its propensity to banquet on their beautiful peaches and pears, and choicest grapes and plums.

That bees do gather the sweet juice of fruits when nothing else is to be found, is certain; but it is also evident that their jaws being adapted chiefly to the manipulation of wax, are too feeble to enable them to puncture the skin of the most delicate grapes.

872. We made experiments in our Apiary on bees and grapes, during the season of 1879, — one of the worst seasons we ever knew for bees. The Summer having been exceedingly dry, the grape crop was large and the honey crop small. In every vineyard a number of ripe grapes were eaten by bees, and the grape-growers in our vicinity were so positively certain that the bees were guilty, that they held a meeting, to petition the State Legislature, for a law preventing any one from owning more than ten hives of bees.

This serious charge called our attention to the matter, and we decided to make a thorough investigation, in our own vineyard. But although many bees were seen banqueting on grapes, not one was doing any mischief to the *sound* fruit. Grapes which were bruised on the vines, or lying on the ground, and the moist stems, from which grapes had recently been plucked, were covered with bees; while other bees were observed to alight upon bunches, which, when found by careful inspection to be sound, they left with evident disappointment.

Wasps and hornets, which secrete no wax, being furnished with strong, saw-like jaws, for cutting the woody fibre with which they build their combs, can easily penetrate the skin of the toughest fruits. While the bees, therefore, appeared to be comparatively innocent, multitudes of these depredators were seen helping themselves to the best of the grapes. Occasionally, a bee would presume to alight on a bunch where one of these pests was operating for his own benefit, when the latter would turn and " show fight," much after the fashion of a snarling dog, molested by another of his species, while daintily discussing his own private bone.

During grape picking, the barrels in which our grapes were hauled to the wine cellar, were covered with a cloud of bees feeding on the damaged clusters, and they followed the wagon to the cellar. After removing the barrels to a place of safety, we left *one bunch of sound grapes*, on the wagon, puncturing *one of the grapes* with a pin. This bunch, being the only one remaining exposed, was at once covered with such a swarm of bees that it was entirely hidden from sight. It was three o'clock in the afternoon. At sunset the bees were all gone, except three, who were too exhausted to fly off. The bunch had lost its bloom, the grapes were shiny, but entirely sound. The one punctured grape had a slight depression at the pin hole, showing that the bees had sucked all the juice they could reach, but they had not even enlarged the hole.

We also placed bunches of sound grapes inside of some four or five hives of bees, directly over the frames, and three weeks after we found that the bees had glued them fast to the combs, as they glue up anything they cannot get rid of, but the grapes were perfectly intact. This test can be made by every Apiarist.

Mr. McLain, in charge of the U. S. Apicultural Station, was instructed to test this matter thoroughly by shutting up bees with sound fruit, and the results were the same as in our case. (See the Agricultural Reports for 1885.)

873. The main damage to grapes is done by birds. Hence, the borders of a large vineyard are first to suffer, especially when in proximity to hedges, orchards or timber.

Even in small cities, the number of birds that feed on fruit is extraordinary, and one can have no idea of their depredations until he has watched for them at day-break, which is the time best suited to their pilfering.

After the mischief has been *begun* by them or by insects, or wherever a *crack*, or a spot of *decay* is seen, the honey-bee hastens to help itself, on the principle of " gathering up the fragments, that nothing may be lost." In this way, they undoubtedly do some mischief, but they are, on the whole, far more useful than injurious.

875. Among thousands of testimonials, we translate the following from *L'Apicoltore*, of Milan, Italy, May 1874, page 181:

" Being a lover of good wine, I manufacture mine from wilted grapes; my crop amounts annually to from thirty to forty hectolitres* of wine, worth on average, one franc seventy-five

* One hectolitre is twenty-five gallons.

centimes per litre.* When my grapes are gathered, I spread them on mats of reed or straw in a sunny place in front of my Apiary, where they remain about two weeks. For the first two or three days the mats are covered with bees, but I pay no attention to this, for I have ascertained that they gather only the juice of the berries that are damaged. As soon as the injured berries are sucked dry, the bees cease visiting the mats, for they cannot open sound berries. Instead of doing me any damage, they help me greatly, as they take away from my grapes the otherwise souring juices, which would give a bad taste to my wine.— GAETANO TAXINI, Coriano, Italy, February 1874.

876. Those who handle grapes, apples, etc., in times of honey-dearth, should avoid attracting the bees, by unnecessarily exposing the crushed fruit, in warm weather, as the presence of bees in press-houses and sheds, where fruit is either made into wine, or otherwise prepared for use, is the greatest annoyance that they can cause the horticulturist.

With a little care, a wine-grower may escape all trouble, even if his press-house is in reach of a large Apiary. But let him not imitate the grocer who had an open box of comb-honey at his door "for show," and tried to "shoo" the bees off, when they, in turn, deputized a few of their number to "shoo" him off, with great success.

877. In these depredations, the wine-growers who do not own bees are often very much incensed, because they believe that the Apiarist is making a profit out of their loss. But such is not the case. The Apiarist loses more than the wine-grower, for many of the bees are destroyed, and the juice which the others bring home is worse than useless, as it is bad Winter food (**627**).

It is therefore, to the interest of the Apiarist, as well as of the fruit-grower, to prevent the bees, in all possible ways, from getting a taste of the forbidden juices, in seasons, — luckily scarce, — when there is a dearth of honey during wine-making time.

---

* This is about one dollar and forty cents per gallon, a high price for Italy.

878. Some ignorant people have also contended that the numerous visits of bees to flowers, injure the latter and cause them to abort. This is the greatest of all delusions. White-clover, knot-weed, and Spanish-needles, which are among the plants most visited by bees, are also the most abundant, and if they were damaged, by being deprived of the honey which they yield, they would sooner or later disappear. All the observations that have been made, whether scientific or practical, show that the contrary is the truth **(269).**

In 1885, at the earnest request of our enthusiastic friend, Jas. Heddon, a Bee-keepers' Union was formed to defend the interests of Apiarists in North America. Some such association is as necessary to Bee-keepers as Trades Unions to any group of laborers. "United we stand, divided we fall."

CHAPTER XXIII.

BEE-KEEPER'S CALENDAR.

This chapter gives to the inexperienced bee-keeper brief directions for each month in the year,* and, by means of the full Alphabetical index, all that is said on any topic can easily be referred to.

879. JANUARY.— In cold climates, bees are now usually in a state of repose. If the colonies have had proper attention in the Fall, nothing will ordinarily need to be done that will excite them to an injurious activity.

In January there are occasionally, even in very cold latitudes, days so pleasant that bees can fly out to discharge their fæces ; do not confine them, even if some are lost in the snow.

It is advisable to arouse them early so as to cause them to fly (**639**) if the day is sufficiently warm. Otherwise, disturb them as little as possible. In very cold climates, where cellar wintering (**646**) is resorted to, all that is required is to keep the temperature as even and as near 42° to 45° as possible (**648**), with quietude and darkness (**650**). The Winter months are those, in which the bee-keeper should prepare his hives, sections, foundation, &c. for the coming busy season.

880. FEBRUARY.—This month is sometimes colder than January, and then the directions given for the previous month must be followed. In mild seasons, however, and in warm regions, bees begin to fly quite lively in February,

---

Palladius, who wrote on bees nearly 2,000 years ago, arranges his remarks in the form of a monthly calendar.

and in some locations they gather pollen (**263**). The bottom-board should be cleaned of the dead bees and other rubbish (**663**) that sometimes obstruct the entrance, and prevent the bees from flying out; as their worry in finding themselves imprisoned does them much harm. If any hives are suspiciously light, food (**607**) should be given them; this only in mild climates.

Strong colonies will now begin to breed slightly, but nothing should be done to excite them to premature activity.

884. MARCH.—In our Northern States, the inhospitable reign of Winter still continues, and the directions given for the two previous months are applicable to this. If there should be a pleasant day, when bees are able to fly briskly, seize the opportunity to remove the covers (**636**); carefully clean out the hives (**663**), and learn the exact condition of every colony. See that your bees have water (**271**), and are well supplied with rye-flour (**265**). In this month, weak colonies commonly begin to breed, while strong ones increase quite rapidly.

If the Winter has been very severe, this month is the most destructive to unhealthy bees. The hives of dead colonies should be throroughly cleaned, and closed tightly to keep robbers (**664**) out, or they would carry off what honey may remain in them. Spring dwindling (**659**) should be guarded against by shutting off all upward ventilation (**352**), and reducing the space in the brood-chamber (**349**) to the number of combs actually occupied by the bees. The entrance of the hives, especially of the weak colonies, should also be narrowed (**348**).

If the weather is favorable, colonies which have been kept in a special Winter depository, may now be put upon their proper stands.

The time of removal from cellars (**646**) must depend altogether on the locality. Dr. C. C. Miller removes his bees *when the first maple tree blooms*. In Canada, they are sometimes left in the cellar till May. As a rule, bees are not, and should not be, wintered in cellars, south of the 39th degree of latitude.

882. APRIL. — Bees will ordinarily begin to gather much pollen (**263**), in this month, and sometimes considerable honey. As brood is now very rapidly maturing, there is a largely increased demand for honey, and great care should be taken to prevent the bees from suffering for want of food (**607**). If the supplies are at all deficient, breeding will be checked, even if much of the brood does not perish, or the whole colony die of starvation. If the weather is propitious, and the bees do not have a liberal supply of stores on hand, feeding to promote a more rapid increase of young may now be commenced (**605**). Feeble colonies must now be reinforced (**480**), and should the weather continue cold for several days at a time, the bees ought to be supplied with water (**271**) in their hives.

This point is much neglected, by ourselves, as well as by others, in practice, but we are convinced that much of our April loss is due to the bees going in search of water in inclement weather (**662**). At this time, if not before, the larvæ of the bee-moth will begin to make their appearance, and should be carefully destroyed, not that they are very damaging to bees in a carefully-conducted Apiary, but only that they give annoyance by their presence on the combs or comb-honey, removed from the bees, in the latter part of the season (**812**). "One stitch in time saves nine." One moth killed in April, prevents several thousand in October.

It is at this time, that the hives should be inspected, to remove all drone comb (**228**) that can be found, as well as crooked combs and broken pieces,—to be replaced by straight worker comb (**676**), or strips of foundation (**674**). At this time, also, the hives that are intended for drone raising (**511**), should be supplied with sufficient drone comb for the purpose. Queenless colonies should be given young brood to raise a new queen (**489**).

883. MAY. — As the weather becomes more genial, the increase of bees in the colonies is exceedingly rapid, and drones, if they have not previously made their appearance, begin to issue from the hives that have been allowed to retain a notable amount of drone comb, and this is the time to raise queens for increase, or for improvement (**489**).

The breeding space of weak colonies, which has been previously reduced, should again be enlarged as their needs may demand (**349**). If their combs are judiciously increased with a proper amount of stimulative food (**606**), and a little help from the stronger colonies (**480**), they may become as strong as any for the June harvest. In some localities, the strongest colonies may already gather much honey, and it will often be advisable to give them the spare honey receptacles (**724**); but in some seasons and localities, either from long and cold storms, or a deficiency of forage, hives not well supplied with honey will exhaust their stores, and perish, unless they are fed. In favorable seasons, swarms (**406**) may be expected in this month, even in the Northern States. These May swarms often issue near the close of the blossoming of fruit-trees, and just before the later supplies of forage, and if the weather becomes suddenly unfavorable, may starve, unless they are fed, even when there is an abundant supply of blossoms in the field.

884. JUNE. — This is the great swarming month in all our Northern and Middle States. As bees keep up a high temperature in their hives, they are by no means so dependent upon the weather for forwardness, as plants, and as most other insects necessarily are. We have had as early swarms in Northern Massachusetts, as in the vicinity of Philadelphia.

33

If the surplus cases (**724**) have been put on before the honey crop, there will be a less number of swarms, especially if the boxes have been furnished with combs, as baits, and the entrance enlarged to help ventilation (**344**).

If the Apiary is not carefully watched, the bee-keeper, after a short absence, should examine the neighboring bushes and trees, on some of which he will often find a swarm clustered, preparatory to their departure for a new home (**419**).

"As it may often be important to know from which hive the swarm has issued, after it has been hived and removed to its new stand, let a cup-full of bees be taken from it and thrown into the air, near the Apiary, after having sprinkled them with flour; they will soon return to the parent colony, and may easily be recognized, by their standing at the entrance, fanning, like ventilating bees." — DZIERZON.

This is the quickest method to discover the home of a swarm.

As fast as the surplus honey receptacles are filled, more room should be given (**763**). Careless bee-keepers often lose much, by neglecting to do this in season, thereby condemning their colonies to a very unwilling idleness. The Apiarist will bear in mind, that all small swarms which come off late in this month, should be either aided, doubled or returned to the mother-colony. With movable-frame hives, the issue of such swarms may be prevented, by removing, in season, the supernumerary queen-cells. During all the swarming season, and, indeed, at all other times when young queens are being bred, the bee-keeper must ascertain seasonably. that the hives which contain them, succeed in securing a fertile mother (**152**).

885. JULY.—In some seasons and districts, this is the great swarming month; while in others, bees issuing so late, are of small account. In Northern Massachusetts, we have known swarms coming after the Fourth of July, to fill their hives, and make large quantities of surplus honey besides. In this month, or as soon as the first crop is over, all the spare honey should be removed from the hives, before the delicate whiteness of the combs becomes soiled by the travel of the bees, or the purity of the honey is impaired by an inferior article gathered later in the season (**782**). For the same reason, the honey extracted after this crop should not be mixed with that harvested later. In all the localities where a second crop is expected, the bees should again be incited to breed (**606**) to be ready for this second crop.

The bees should have a liberal allowance of air during all extremely hot weather, especially if they are in unpainted hives, or stand in the sun (**344**).

The larger the amount of honey they contain, the greater the danger of combs breaking down from the intense heat (**369**). The end of the honey crop can be told by the presence of a few robbers who immediately begin lurking about the hives (**664**).

886. August.—In most regions, there is but little forage for bees during the latter part of July, and the first of August, and they being, on this account, tempted to rob each other, the greatest precautions should be used in opening hives (**666**). In districts where buckwheat is extensively cultivated, on flat prairies, or in the low land surrounding our rivers, in which Fall-blossoms grow, the main harvest is sometimes gathered, during this month and the next, and swarming (**406**) may be resumed. In 1856, we had a buckwheat swarm as late as the 16th of September!

The bee-keeper who has queenless hives (**499**) on hand as late as August, must expect, as the result of his ignorance or neglect, either to have them robbed (**664**) by other colonies, or destroyed by the moth (**802**).

887. September.—This is often a very busy month with bees. The Fall flowers are in full blossom, and in some seasons, colonies which have hitherto amassed but little honey, become heavy, and even yield a surplus to their owner. Bees are quite reluctant to build comb so late in the season, even if supplies are very abundant; but if empty combs are provided, they will fill them with astonishing celerity (763).

As soon as the first frost takes place, or whenever the crop is at end, the entire surplus must be removed, whether it be comb or extracted honey. If our method of extracting (781) is resorted to, the supers that have been returned to the bees, for cleaning, after the honey is extracted, may be left on the hives till October, as they are safer from the moths, when in care of the bees.

If no Fall supplies abound, and any colonies are too light to winter with safety, then, in the Northern States, the latter part of this month is the proper time for feeding (608) them. We have already stated, that it is impossible to tell how much food a colony will require (623), to carry it safely through the Winter; it will be found, however, very unsafe to trust to a bare supply, for, even if there is food enough, it may not always be readily accessible (631) to the bees. Great caution will still be necessary to guard against robbing; but if there are no feeble, queenless or impoverished colonies, the bees, unless tempted by improper management, will not rob each other (664).

888. OCTOBER. — Forage is now almost entirely exhausted in most localities, and colonies which are too light should either be fed, or have surplus honey from other hives given to them, early this month.

The extracting cases (781) should be removed previous to cold weather, as some bees may cluster in them and starve. These cases must be piled up carefully in the coldest room (810) of the honey house, safe from mice (816). The exact condition of every hive should be known now, at the latest, and, if any are queenless, they should be broken up. Small colonies ought to be promptly united.

The honey-selling season is now at hand, and from this time till the end of the holidays, the producer must look for a honey market. He should not rely on sale in large cities, for they are always crowded, but a home market must be cultivated (**840**).

889. NOVEMBER. — The hives should now be put in Winter quarters, the quilt removed, and absorbents placed in the upper story (**636**).

All possible shelter should be given (**635**). For cellar-wintering (**646**), the time of removing the bees should be at the opening of cold weather. The later in the season that the bees are able to fly out and discharge their fæces, the better. The bee-keeper must regulate the time of housing his bees by the season and climate, being careful neither to take them in until cold weather appears to be fairly established, nor to leave them out too late. A cold day, immediately after a warm spell is the best time (**647**).

890. DECEMBER.—In regions where it is advisable to house bees, the dreary reign of Winter is now fairly established, and the directions given for January are for the most part equally applicable to this month. It may be well, in hives out of doors, to remove the dead bees and other refuse from the bottom boards if the weather is warm enough for them to fly; but, neither in this month nor at any other time should this be attempted with those removed to a dark and protected place. Such colonies must not, except under the pressure of some urgent necessity, be disturbed in the very least.

We recommend to the inexperienced bee-keeper to read this synopsis of monthly management, again and again, and to be sure that he fully understands, and punctually discharges, the appropriate duties of each month, neglecting

nothing, and procrastinating nothing to a more convenient
season; for, while bees do not require a large amount of
attention, in proportion to the profits yielded by them, they
must have it at the *proper time* and in the *right way.* Those
who complain of their unprofitableness, are often as much
to blame as a farmer who neglects to take care of his stock,
or to gather his crops, and then denounces his employment
as yielding only a scanty return on a large investment of
capital and labor.

In Short.

891. SPRING.—Keep hives warm, give plenty of food,
help weak colonies, look out for robbers, remove drone-
comb, prepare for queen-breeding, and for the honey crop.

892. SUMMER.—Watch for swarms; and make divisions,
if increase is wanted. Give sufficient storage-room. Give
additional ventilation if needed. Whenever the crop is
over, remove the surplus.

893. FALL.—Look out for robbers, and for moths on
unoccupied combs. See that all hives have sufficient stores
for Winter, and unite worthless colonies to others.

894. WINTER.—For out of doors, pack absorbents in
upper story, removing air-tight quilts. Shelter as much as
possible from winds. Leave the bees quiet in cold weather,
and see that they have a flight in warm weather. Do not
be confident of safe wintering till *March* is over. Then
proportionate the room to the strength of the colony. For
cellar wintering, take the bees in, after a warm day, leave
them quiet, in the dark, with an even temperature; take
them out on a warm day, and decrease the brood-chamber
to suit the strength of the colonies.

Mistakes that Beginners Are Liable to Make.

895. *1.*—They are apt to think themselves *posted* after they have read the *theory*, and before they get the *practice*.

2.—Hence they are apt to *invent or adopt new hives*, that are lacking in the most important features (**358**).

3.—They are apt to think that bees are harvesting honey, at times when they are starving. They should remember that each honey crop usually lasts only a few days,—a few weeks at most.

4.—They are apt to mistake young bees on their first trip for robbers and *vice versa*. Young bees fly out *in the afternoon only*, and do not hunt around corners. Robbers are gorged with honey when coming out of the plundered hive, and a number of them are slick, hairless and shiny. Bees that have been fed in the hive or whose combs have been damaged, or extracted, and returned to the hive, act like robbers, and incite robbing (**664**).

5.—They are apt to overdo artificial swarming (**481**).

6.—They are apt to extract too much honey from the brood-combs (**771**).

7.—They underestimate the value of good worker comb. (**676**).

8.—They do not pay sufficient attention to the removal of the excess of drone-comb (**675**).

9.—They become early discouraged by Winter losses and Spring dwindling. Some of our most successful Apiarists periodically lose a large portion of their colonies, and promptly recruit again, by the help of their empty worker-combs (**676**).

10.—When they find bee-keeping successful, they are liable to rush into it on too large a scale before being sufficiently acquainted with it. "If there is any business in this world that demands industry, skill and tact, to insure success, it is this of ours."—(HEDDON.)

11.—They are apt to try two or three different styles of hives, before they find out that it is important to have all the hives, frames, caps, crates, etc., in an Apiary, alike, and interchangeable, except for purposes of experiment.

12.—They are liable to attempt to winter their bees in a cold room, or in some repository in which the temperature goes below the freezing point (**648**). Many a colony has been thus innocently murdered, by misguided solicitude.

BEE-KEEPERS' AXIOMS.

896. There are a few *first principles* in bee-keeping which ought to be as familiar to the Apiarist as the letters of his alphabet:

1st. Bees gorged with honey never volunteer an attack. Thus, bees that come back loaded from the field, or bees that have gorged themselves for swarming, are not dangerous.

2d. The bees that are to be feared are those that have joined a swarm without fully gorging themselves. In the hive, the guardians, and the old bees that are ready to depart for the field, are the most dangerous.

3d. During a good honey harvest, the bees are nearly all filled with honey and there is but little danger from stinging.

4th. Those races of bees that cannot be compelled, by smoke, to fill themselves with honey, are the most dangerous, to handle.

5th. Bees dislike any *quick* movements about their hives, especially any motion that *jars* their combs.

6th. The bee-keeper will ordinarily derive all his profits from colonies, strong and healthy in early Spring.

7th. In districts where forage is abundant only for a short period, the largest yield of honey will be secured by a *very moderate* increase of colonies.

8th. A moderate increase of colonies in any one season, will, in the long run, prove to be the easiest, safest, and cheapest mode of managing bees.

9th. Queenless colonies, unless supplied with a queen, will inevitably dwindle away, or be destroyed by the bee-moth, or by robber-bees.

10th. It must be obvious, to every intelligent bee-keeper, that the perfect control of the combs of the hive is the soul of a system of practical management, which may be modified to suit the wants of all who cultivate bees.

11th. A man, who *knows* " *all about bees,*" and does not believe that anything more can be gained by reading Bee-Journals, new bee-books, etc., will soon be far behind the age. Yet, as what is written in the journals and books, ours included, is not always perfectly correct, every bee-keeper should try to sift the grain from the chaff.

12th. The formation of new colonies should ordinarily be confined to the season when bees are *accumulating* honey; and if this, or any other operation must be performed, when forage is scarce, the greatest precautions should be used to prevent robbing.

The essence of all profitable bee-keeping is contained in Oettl's Golden Rule: KEEP YOUR COLONIES STRONG. If you cannot succeed in doing this, the more money you invest in bees, the heavier will be your losses; while, if your colonies are strong you will show that you are a *bee-master*, as well as a bee-keeper, and may safely calculate on generous returns from your industrious subjects.

INDEX OF ENGRAVINGS.

ɪ—PLATES.

II—TEXT ILLUSTRATIONS.

[The engravings of plants whose original is not indicated, came from Messrs. **J. B.** Baillière et Fils, and Vilmorin Andrieux et Cie of Paris.]

INDEX.

www.ingramcontent.com/pod-product-compliance
Lightning Source LLC
Chambersburg PA
CBHW020853210326

41598CB00018B/1654